DEVELOPMENTS IN THE STRUCTURAL CHEMISTRY OF ALLOY PHASES

The Metallurgical Society of AIME Proceedings
published by Plenum Press

1968–Refractory Metal Alloys: Metallurgy and Technology
 Edited by I. Machlin, R. T. Begley, and E. D. Weisert

1969–Research in Dental and Medical Materials
 Edited by Edward Korostoff

1969–Developments in the Structural Chemistry of Alloy Phases
 Edited by B. C. Giessen

A Publication of The Metallurgical Society of AIME

DEVELOPMENTS IN THE STRUCTURAL CHEMISTRY OF ALLOY PHASES

Based on a symposium sponsored by the Committee on Alloy Phases of the Institute of Metals Division, The Metallurgical Society, American Institute of Mining, Metallurgical and Petroleum Engineers, Cleveland, Ohio, October, 1967.

Edited by

B. C. GIESSEN

Department of Chemistry
Northeastern University
Boston, Massachusetts
and
Department of Metallurgy and Materials Science
Massachusetts Institute of Technology
Cambridge, Massachusetts

 Springer Science+Business Media, LLC 1969

ISBN 978-1-4899-5566-1   ISBN 978-1-4899-5564-7 (eBook)
DOI 10.1007/978-1-4899-5564-7

Library of Congress Catalog Card Number 77-94080

FOREWORD

A basic problem in the structural chemistry of alloy phases concerns the relative stability of observed and hypothetical phases with different compositions (which determines the constitution diagram) and crystal structures. The classical approach to this question has been the application (mostly qualitative) of several concepts, such as geometrical factors, electron concentrations (using a number of counting schemes), electronegativities, etc. Fundamentally, these concepts involve simplified aspects of a complete quantum mechanical description of the behavior of all electrons in the crystal; to be valid, they must be compatible with the quantum mechanical foundations as well as with the experimental experience. At present, the complete description cannot be carried out exactly; however, it lies hidden behind the various concepts used. The latter are somewhat reminiscent of the parts of the legendary elephant which is only insufficiently described by its trunk, legs or tail. The majority of current treatments use only the aforementioned criteria, although encouraging attempts have been made to treat the structural problem in its entirety.

There has been continued growth of the literature; this includes crystal structure determinations, which have been extended to metastable or highly complex alloy phases, electronic measurements, and frequently successful correlations of these properties with empirical concepts. The need persists to collect and review this information. The present volume offers comprehensive treatments of selected areas of structural alloy chemistry in which considerable progress has been made in recent years. The reviews are written from the varied standpoints of the physicist, chemist or materials scientist; for the reasons given above, the treatment is, of necessity, largely descriptive.

The content of the book is loosely composed of three sections of related content. In the first section, which deals essentially with physical concepts underlying structure formation, there is an introductory paper representing the physicist's viewpoint. Next, an exposition of the controversial Engel-Brewer theory is given, with two written discussions on the problem of the NaTl structure, followed by a paper on the concept of partial valence electron concentration.

In the second section, geometrical and chemical considerations are emphasized. It is composed of two papers in which extensive crystallographic work on complex alloy phases is reviewed.

In the third section, there are surveys of alloy phase families related by common structure types, common constituent elements, or the common criterion of metastability. First, interstitial compound phases are treated; next, actinide compounds are discussed systematically. Finally, a listing of terminal and intermediate nonequilibrium alloy phases is given.

The contributions in this volume are based on a symposium organized by the Committee on Alloy Phases, Institute of Metals Division, the Metallurgical Society, American Institute of Mining, Metallurgical, and Petroleum Engineers. The Symposium was held on October 16, 1967, in Cleveland, Ohio. In addition to the authors listed, Dr. A. U. MacRae of the Bell Telephone Laboratories, Murray Hill, New Jersey, presented a paper on "Surface Structures by Low Energy Electron Diffraction (LEED)" which was withdrawn from the proceedings at the author's request. The editor is grateful for the advice and cooperation of members of the committee, especially Messrs. Karl A. Gschneidner, Jr., whose active help is sincerely appreciated, Paul A. Beck, and W. B. Pearson, as well as J. F. Smith and Austin Dwight who assisted in the planning of this symposium; Mr. Dwight also chaired one of its sessions. The editor is pleased to acknowledge partial financial support by the Office of Naval Research and editorial assistance by Mr. Dan Szymanski; he thanks Mr. J. V. Richards, Secretary of the Metallurgical Society and Mr. H. Hutchins, editor of the Journal of Metals for their efforts. Finally, he thanks the authors whose contributions made this symposium possible.

CONTRIBUTORS

Lawrence H. Bennett
   Metallurgy Division, Institute for Materials Research,
   National Bureau of Standards, Gaithersburg, Maryland

A. E. Dwight
   Metallurgy Division, Argonne National Laboratory, Argonne,
   Illinois

B. C. Giessen
   Department of Chemistry, Northeastern University,
   Boston, Massachusetts, and Department of Metallurgy and
   Materials Science, Massachusetts Institute of Technology,
   Cambridge, Massachusetts

Niels N. Engel
   Division of Metallurgy, School of Chemical Engineering,
   Georgia Institute of Technology, Atlanta, Georgia

E. Parthé
   School of Metallurgy and Materials Science, and Laboratory
   for Research on the Structure of Matter, University of
   Pennsylvania, Philadelphia, Pennsylvania

W. B. Pearson
   Division of Pure Physics, National Research Council of Canada,
   Ottawa, Canada

Sten Samson
   Gates and Crellin Laboratories of Chemistry, California
   Institute of Technology, Pasadena, California

Clara B. Shoemaker
   Department of Chemistry, Massachusetts Institute of Technology,
   Cambridge, Massachusetts

David P. Shoemaker
   Department of Chemistry, Massachusetts Institute of Technology,
   Cambridge, Massachusetts

H. H. Stadelmaier
   Department of Engineering Research, North Carolina State
   University, Raleigh, North Carolina

OFFICERS

Committee on Alloy Phases*

    K. A. Gschneidner, Jr., Chairman

Institute of Metals Division*

    M. E. Fine, Chairman

    J. A. Fellows, Past Chairman

    D. J. McPherson, Senior Vice Chairman

    J. H. Rizley, Vice Chairman

    V. H. Branneky, Secretary-Treasurer

The Metallurgical Society of AIME*

    J. H. Jackson, President

    A. E. Lee, Jr., Past President

    M. Tenenbaum, Vice President

    C. C. Long, Treasurer

    J. V. Richards, Secretary

*Officers at time of the Conference

CONTENTS

PHYSICAL CONCEPTS

STRUCTURAL PRINCIPLES

ALLOY PHASE SURVEYS

SEARCH FOR A UNIFIED CRYSTAL CHEMISTRY OF METALS

W.B. Pearson

Division of Pure Physics, National Research Council of

Canada, Ottawa

ABSTRACT

New knowledge of the Fermi surface of metals and alloys is examined in relation to the crystal chemistry of metals, which depends primarily on an <u>energy-band factor</u> (involving crystal symmetry, electron concentration per primitive cell, and the symmetry of the energy bands at the Fermi level). The energy band factor is generally weak and is dominated by other factors, particularly when the effective electron concentration is relatively high (<u>bond factor</u>), and when the difference of potential between the components is relatively high (<u>electrochemical factor</u>). Critical assessment of geometrical effects in metals separates the trivial high coordination that may arise in the packing together of more or less spherical atoms (<u>coordination factor</u>), from a true <u>geometrical factor</u> which controls unit cell dimensions and imparts phase stability over a wide range of radius ratios of the component atoms. The existence of a geometrical factor is explicitly demonstrated, and as its only influence on E vs k relationships is structural, the energy-band factor may still control the choice of structure when the geometrical factor operates.

Factors controlling the layer stacking sequence of close packed arrays of atoms are discussed, particularly with reference to the lanthanons and their alloys, and the distribution of the $A1$, $A2$ and $A3$ structure types among the elements is considered.

1

INTRODUCTION

In the last decade we have seen the discovery of a very large number of new intermetallic phases, mostly belonging to already recognized structure types, and read many papers and reviews detailing the influence of relative size, space filling parameters, electron concentration and electronegativity on families of phases having the various structures; we now well appreciate many possible influences of Brillouin zones on the parameters of simple structures, and how the common metallurgical 12, 14, 15 and 16 coordination polyhedra are generated through the stacking of layer nets of atoms. In addition we have $\underline{A}$ and $\underline{B}$ series of metals (by no means always the same) which unite to give us the phases that are our current delight, but none of us have yet found a unified principle on which to base a crystal chemistry of metals and alloys! In contrast we see the simplicity of the crystal chemistry of ionic substances which is based on the relative sizes of the ions, and the crystal chemistry of semiconductors which depends on directed chemical bonds and valence rules.

The milestones in the crystal chemistry of metals include the Hume-Rothery[1] concept of electron phases in alloys of Cu, Ag and Au with the following B Group metals, followed by ideas of the electrochemical factor and the 15% size rule for extended solid solutions; Laves'[2] enunciation of the geometrical principles of the highest degree of space filling, symmetry and coordination for the structures of metals, when other dominant effects such as bond factors are absent; Frank and Kasper's[3,4] principles for the generation of the CN 12 icosahedron and the CN 14, 15 and 16 polyhedra by the appropriate stacking of various layer nets of atoms; and Pauling's[5] empirical bond number relationship $\underline{R}(1)-\underline{R}(\underline{n}) = 0.3 \log \underline{n}$, that appears to be generally valid regardless of the limitations of any system of atomic radii to which it is applied. Such is our heritage; the solution of crystal structure is the most important experiment of the solid state, but the determination of the Fermi surface is the second most important experiment that can be performed on metals, and now that we have direct experimental information on this, we must come to assess its meaning for crystal structure.

We presently see that no matter how many electrons are in the primitive cell of a particular structure, there are likely to be sheets of Fermi surface in only about four Brillouin zones and of these sheets, no more than two are likely to exert a dominant influence on the crystal structure. Of course, crystal structure prescribes the Brillouin zones, but the component atoms control the electron concentration and crystal potential and therefore influence the choice of structure to give the most favourable

electron energies. Since this is a property of all metals, we
must conclude that the crystal chemistry of metals is primarily
controlled by an electronic effect depending on the E vs k relationships, which involves the crystal symmetry, electron concentration per primitive cell, and the symmetry characteristics of
the energy bands and which we shall call the energy-band factor.
In practice we may expect to observe the operation of the energy-
band factor either in relation to electron concentration, or in
relation to the type of electrons, s, p, d or f participating in
the energy bands at the Fermi level, and the result of its operation may be either a change of crystal symmetry or structure, or
a change of crystal parameters without change of structure (eg.
such as have been associated with "Brillouin zone overlaps").
The energy band factor is generally a very weak effect which is
readily dominated by other factors such as the chemical-bond
factor, or the electrochemical factor which generally results in
the formation of phases with particular stoichiometries, so that
one of the components attains by electron sharing or transfer, a
filled $s^2 p^6$ valence subshell. The energy-band factor may even be
dominated by a geometrical factor, as we shall show. This concept conforms to our earlier knowledge of the crystal chemistry
of metals, from which we take two cogent examples in support of
the thesis.

ELECTRON CONCENTRATION AND THE ENERGY-BAND FACTOR

     First we look at the distribution of binary metallic phases
with the CsCl structure as a function of electron concentration
per atom (number of outer s, p and d valence electrons). Fig. 1
shows that there are maxima centred about 2.5 and 6 electrons per
atom (e/a). Note that there is not even an inflection on the
histogram at the limiting electron concentration of 1.5 e/a for
the Hume-Rothery β phases with this structure. The large peak in
the region of 2.5 e/a is due to the number of rare earths with
valency 3,   but even neglecting phases containing rare earths,
this situation remains unchanged, as shown by the dotted line in
Fig. 1. The empty lattice model for E vs k relationships gives
no clue why there should be a sharp cut-off of the CsCl structure
above an electron concentration of 3, and as it is apparent that
the CsCl structure forms mainly for chemical or valence reasons,
the cut-off above 3 e/a occurs because these give rise to the
FeB and CrB structures at about 3.5 e/a and to the rocksalt,
sphalerite or wurtzite structures at an electron concentration of
4 per atom (Fig. 1). When valence and electrochemical effects
are largely unimportant as in the Hume-Rothery alloys of the early
B Group elements with Cu, Ag and Au, then the energy-band factor
controls the occurrence of the β phase and sets the limiting
composition at an electron concentration of about 1.5.

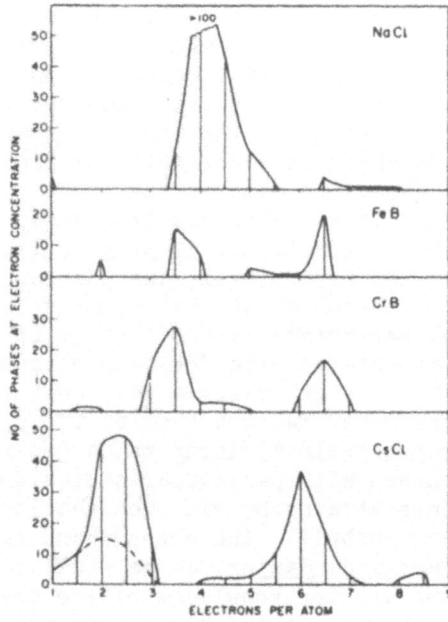

<u>Fig. 1</u>

Histogram showing the number of binary phases occurring with the CsCl, CrB, FeB and NaCl structures at the apparent valence electron (<u>s</u>, <u>p</u>, <u>d</u>) concentrations indicated.

The second large peak in the occurrence of CsCl phases which is centred about 6 e/a in Fig. 1, results from phases containing transition metal atoms. In accounting for this apparent electron concentration, we turn to our knowledge of the Fermi surface of metals and see that it results from several filled energy bands of electrons and that <u>the effective electron concentration in these alloys is about the same as in those of the lower electron concentration range</u>, which do not contain the transition metals.

In the structures of metals generally, it seems that we are effectively dealing with only a few electrons per atom, perhaps 1, 2 or 3 and the remaining apparent valence electrons merely occupy filled energy bands below these partially occupied bands. We expect these features also to apply to metallic phases such as the σ, μ, P and R phases which generally contain only transition metals. Thus we can consider that the effective electron concentration is perhaps 1 or 2 and much less than the apparent electron concentration of 6 or 7, which in part appears to influence the composition of such phases. When the effective valence electron concentration is 3 to 4 per atom, there is a probability of atoms attaining a filled valence subshell of 8 $\underline{s}^2\underline{p}^6$ electrons, and

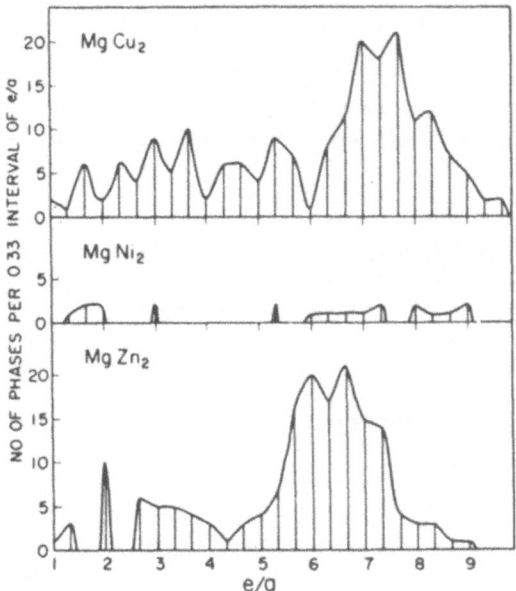

Fig. 2

Histogram showing the number of binary Laves phases ($MgZn_2$, $MgNi_2$
or $MgCu_2$ structures) occurring per 0.33 interval of apparent valence electron (s, p, d) concentration.

structures depending on bond factors, rather than on the more
metallurgical geometrical features, are obtained.

    In the second example we see that the structures of certain
Laves phases AB_2 are controlled by electron concentration[6,7,8].
For instance, Laves and Witte[6] find in ternary alloys of Mg with
B components such as Cu, Zn, Al, Si, Co or Ag, that the $MgCu_2$
structure occurs at electron concentrations from 1.33 to about
1.8, and the $MgZn_2$ structure is found in the range from 1.85 to
about 2.2 electrons per atom. If the $MgNi_2$ structure occurs, it
is found in the intermediate range between the $MgCu_2$ and $MgZn_2$
structures. When, however, we examine the distribution of binary
Laves phases as a function of average valence electron concentration per atom (Fig. 2), we see that they cover the whole range of
apparent electron concentrations, and that the limitations mentioned above are scarcely noticable on the histogram. Thus we
see another example of the operation of the energy-band factor
that may be recognized when other dominant factors are absent from
the alloys. The energy-band factor may also play a role in the
choice of the $MgCu_2$ or $MgZn_2$ structure for the transition metal
alloys, as Elliott and Rostoker[7] have attempted to demonstrate,
and as is apparent from Fig. 2, where the $MgZn_2$ structure occurs

preferably in the range of apparent electron concentrations about 5.5 to 6.5, whereas the $MgCu_2$ structure is preferred at larger apparent electron concentrations. In order to make further progress with such analyses we must know how many valence electrons occupy filled energy bands and thus what is the effective electron concentration.

We now turn to examine critically some of the other features and problems of metallurgical chemistry, such as space filling, the geometrical factor and the question of structures generated by the stacking of close-packed or other layer nets of atoms, in order to see how they may be fitted into a general framework.

GEOMETRICAL EFFECTS: COORDINATION, GEOMETRICAL AND BOND FACTORS

Although Laves'[2] geometrical principles for structure formation have been discussed for a decade, the relative importance of geometrical effects in the structures of metals still requires critical appraisal. The high coordination which may result from the packing together of more or less spherical atoms is a triviality which we shall call the coordination factor to distinguish it from a true geometrical factor that controls cell dimensions and the occurrence of phases with a given structure over a wide range of radius ratios of the component atoms. The operation of such a factor still has to be explicitly demonstrated.

The space filling concept introduced by Laves[2] and extended by Parthé[9] is of little use in examining geometrical effects in the structures of metals, because it is based on a hard-sphere model of the atom, and one of the main properties of metals is the compressibility of the atoms. It applies well to ionic compounds since the hard-sphere model is a good approximation for ions because the tails of the ionic wave functions cannot overlap much without strong ionic repulsion occurring; however the wave functions for metals overlap considerably. Furthermore, the space filling parameter (ϕ) which compares the volumes of the atoms in the unit cell, calculated from their radii, to the volume of the unit cell, depends on the hard-sphere model and so has no fundamental meaning for metals.

These considerations have led us[10,11] to adopt a model which permits the compression of the component atoms, A and B, in order that successively A-A, A-B and B-B contacts may be made in any given structure. These contacts are assumed to occur when the interatomic distances d_A, d_{AB} and d_B equal the sums of the atomic radii, or the atomic diameters: $2r_A = D_A = d_A$, $r_A + r_B = \frac{1}{2}(D_A + D_B = d_{AB}$ and $2r_B = D_B = d_B$. The "near neighbour diagrams" (n.n.d.) so derived have the property that the lines for A-A, etc. contacts

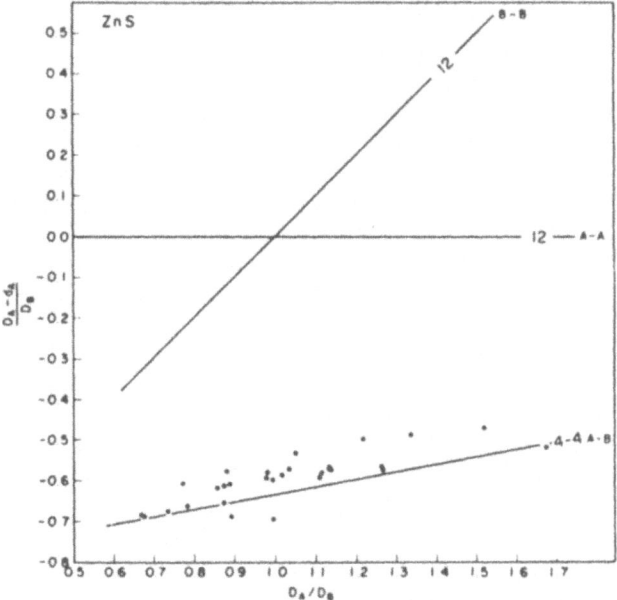

Near neighbour diagram for the sphalerite (ZnS, AB) structure[11]
(see text). Points represent the occurrence of phases with the
structure. Atomic radii for components in phases in Figures 3 to
11 are taken from Teatum et al.[24], and suitably corrected for coordination using Pauling's[5] bond number equation.

are a linear function of a reduced strain parameter, such as
$(D_A-d_A)/D_B$, versus the radius ratio, $r_A/r_B = D_A/D_B$ (See eg. Fig. 3).
It is immaterial whether, say, $(D_A-d_A)/D_B$ or $(D_A-d_B)/D_B$ is chosen as
the reduced strain parameter, since the latter only rotates the
diagram making the value of the line for B-B contacts zero instead
of that for the chosen A-A contacts; the relative geometry of the
diagram remains the same. Since $(D_A-d_A)/D_B$ and D_A/D_B values can be
determined for phases with a given structure, points can be recorded on the n.n.d. which represent the occurrence of phases, and
the distribution of these in relation to the lines for the various
contacts gives information on the factors which control the structure. Thus the n.n.d. can show the operation of bond factors, the
radius ratios (and unit cell axial ratios and atomic parameters if
these are involved) at which a coordination factor is to be found,
and the operation of a geometrical factor if one occurs.

Consider for instance the n.n.d. for the sphalerite structure
shown in Fig. 3; the well-known bond factor is clearly demonstrated
for phases with this structure since they lie along the line for
A-B contacts. A bond factor is also apparent in phases with the

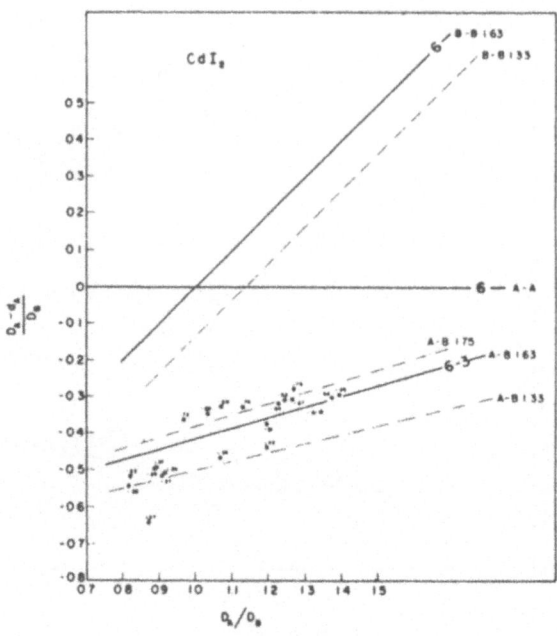

Fig. 4

Near neighbour diagram for the CdI_2 ($\underline{A}\underline{B}_2$) type structure[11] (hexagonal). The coordination of the various contacts is indicated and also the axial ratio of the unit cell to which the contact lines refer. Points represent phases with the structure and the numbers give the axial ratio for the unit cell.

CdI_2 structure where data follow the line for $\underline{A}$-$\underline{B}$ contacts according to the axial ratio of the hexagonal cell (Fig. 4). The n.n.d. for phases with the λ-$ThSi_2$ structure (Fig. 5) shows the importance of an $\underline{A}$-$\underline{B}$ bond factor since data for phases lie across the lines for 8-4 and 4-2 $\underline{A}$-$\underline{B}$ contacts. The data, however, have a mean slope parallel to the lines for $\underline{B}$-$\underline{B}$ contacts rather than those for $\underline{A}$-$\underline{B}$ contacts, indicating a secondary bond factor importance of the three-dimensional three-connected net of B atoms. The diagram also shows that in the range of radius ratios from about 1.75 to 2.1, the lines for $\underline{A}$-$\underline{B}$, $\underline{B}$-$\underline{B}$ and $\underline{A}$-$\underline{A}$ contacts meet in a small triangle, and phases located there would have a high CN of 20-9. If therefore the coordination factor controlled phases with this structure, they would be expected to form from components having a radius ratio of 1.9±0.1. Since known phases with the λ-$ThSi_2$ structure have radius ratios from 1.15 to 1.45 and are located according to bond factor requirements, it is clear that the coordination factor is of no influence in controlling phases with this structure.

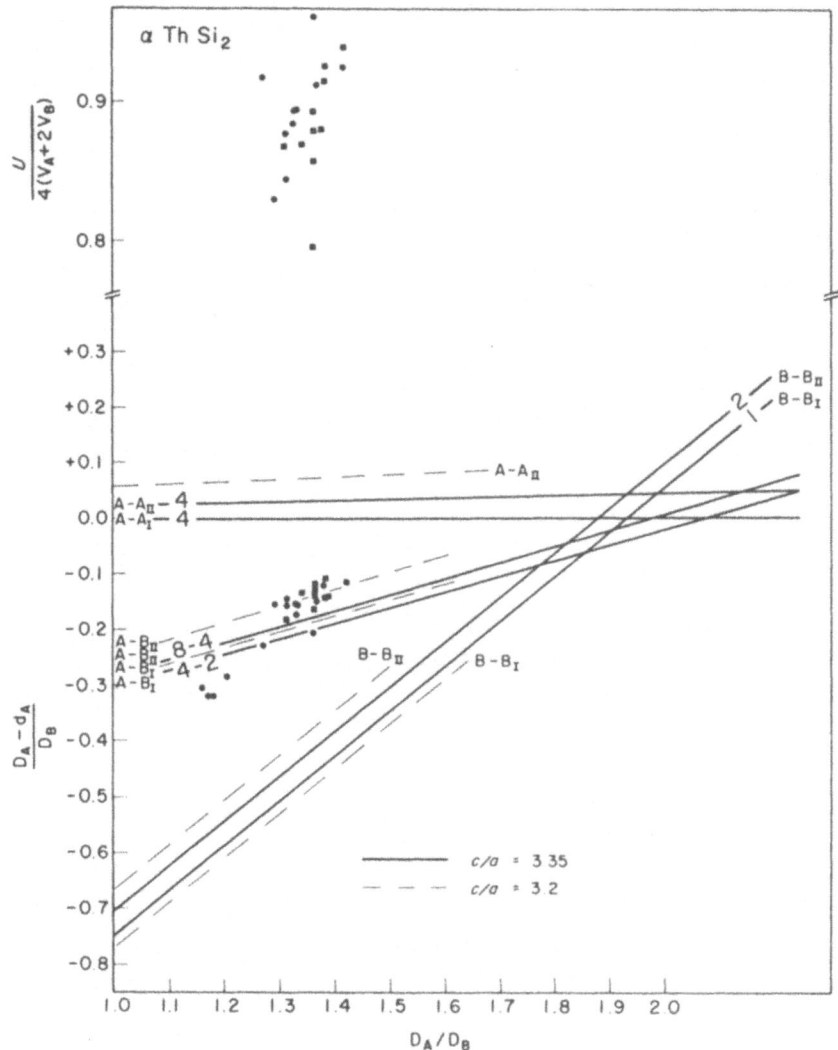

<u>Fig. 5</u>

Near neighbour diagram for the α-ThSi$_2$ (<u>AB</u>$_2$) structure[11] (tetragonal). The coordination for the contact lines is indicated.
These are drawn for axial ratios, <u>c</u>/<u>a</u> = 3.35——— and <u>c</u>/<u>a</u> = 3.2-----
and the ideal value of <u>z</u>$_{Si}$ = 5/12. Points ● represent phases
with normal α-ThSi$_2$ structures and points ■ phases with defect
structures.

The upper part of the diagram compares the unit cell volume
with the sum of the elemental atomic volumes of the same number
of atoms.

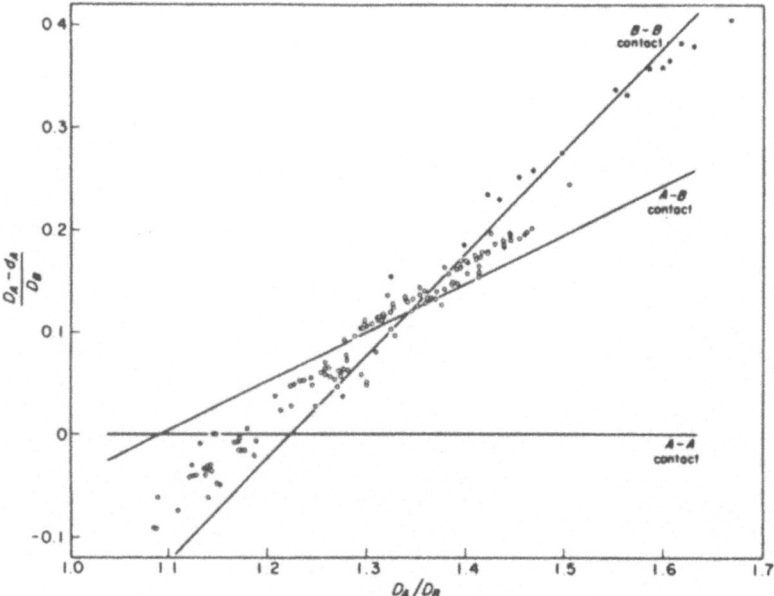

<u>Fig. 6</u>

Near neighbour diagram for the $MgCu_2$ (AB_2) structure (cubic).[10]
Points O represent phases with the structure and points ● repre-
sent phases with electronegativity difference $|x_A - x_B| \geq 1$.

        Such is the information regarding bond and coordination
factors that can be obtained from near neighbour diagrams. We
now turn to the n.n.d. for the $MgCu_2$ and $MgZn_2$ Laves phases (Figs.
6 and 7) to demonstrate the existence of a true <u>geometrical factor</u>
which controls the unit cell dimensions and occurrence of these
phases over a wide range of radius ratios. It is generally accepted that the Laves phases form mainly because of the geometrical
facility with which two components $\underline{A}$ and $\underline{B}$ pack together in the
proportion $\underline{AB}_2$ when the ratio of their radii is 1.225. Under
these conditions the ideal structure contains $\underline{A}$-$\underline{A}$ and $\underline{B}$-$\underline{B}$, but
no $\underline{A}$-$\underline{B}$ contacts (CN 4-6). However, the n.n.d. shows that at a
radius ratio of 1.347 and the appropriate $(D_A-d_A)/D_B$ value given
by the intersection of the lines for $\underline{A}$-$\underline{B}$ and $\underline{B}$-$\underline{B}$ contacts, both
$\underline{A}$-$\underline{B}$ and $\underline{B}$-$\underline{B}$ contacts occur without strain, together with $\underline{A}$-$\underline{A}$
contacts which are compressed somewhat (CN 16-12). This is clear-
ly the most favourable coordination factor for the structure,
provided that the $\underline{A}$ atoms are compressible, and the "ideal" radius
ratio of 1.225 ($=d_A/d_B$) is not the most favourable condition for
the occurrence of the Laves phases. A similar situation is also
found at a radius ratio of 1.093 where $\underline{A}$-$\underline{B}$ and $\underline{A}$-$\underline{A}$ contacts can
occur without strain and $\underline{B}$-$\underline{B}$ contacts are somewhat compressed.

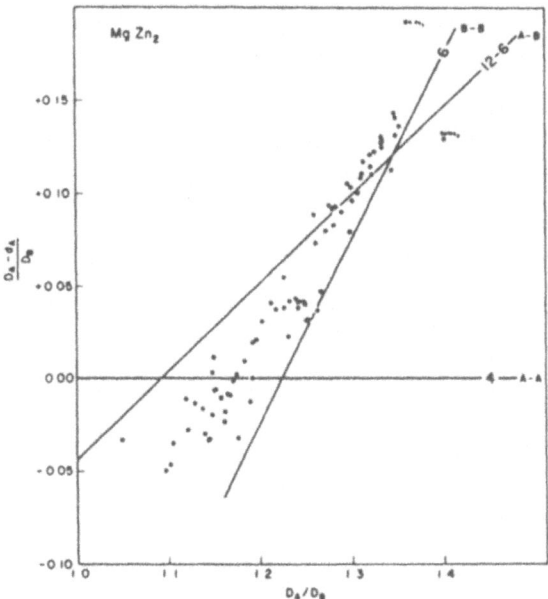

<u>Fig. 7</u>
Near neighbour diagram for the $MgZn_2$ (AB_2) structure[11] (hexagonal)
constructed for the ideal axial ratio c/a = 1.63 and the ideal
atomic parameters: z_{Mg} = 1/16, x_{Zn} = 5/6. The coordination for
the contact lines is indicated. Points ■ indicate some AMn_2
compounds with Mn radius for valency six.

In fact, the Laves phases are found over a wide range of radius
ratios from 1.04 to 1.67 and this results from the operation of a
true geometrical factor which involves the formation of A-B, B-B
and A-A contacts. The establishment of A-B and B-B contacts presumably provides the energy for the compression of the A-A contacts at radius ratios greater than 1.225. There is considerable
asymmetry in the distribution of the Laves phases about the D_A/D_B
value of 1.225, far more phases occurring at radius ratios greater
than this value (Fig. 6). This does not result from lack of components able to give radius ratios smaller than 1.225, but arises
from the fact that only 1/3 of the atoms in the structure (the A
atoms) have to be compressed in order to give A-B and B-B contacts
at radius ratios greater than 1.225, whereas below this value 2/3
of the atoms in the structure (the B atoms) would have to be compressed in order to give A-A and A-B contacts. The energy balance
does not permit this, and Figs. 6 and 7 show that below a D_A/D_B
value of about 1.15, the B-B contacts are no longer compressed
sufficiently to give A-A, let alone A-B contacts, and so the
phases become unstable and do not form at radius ratios much lower.
In fact, the geometrical factor ceases to maintain structural

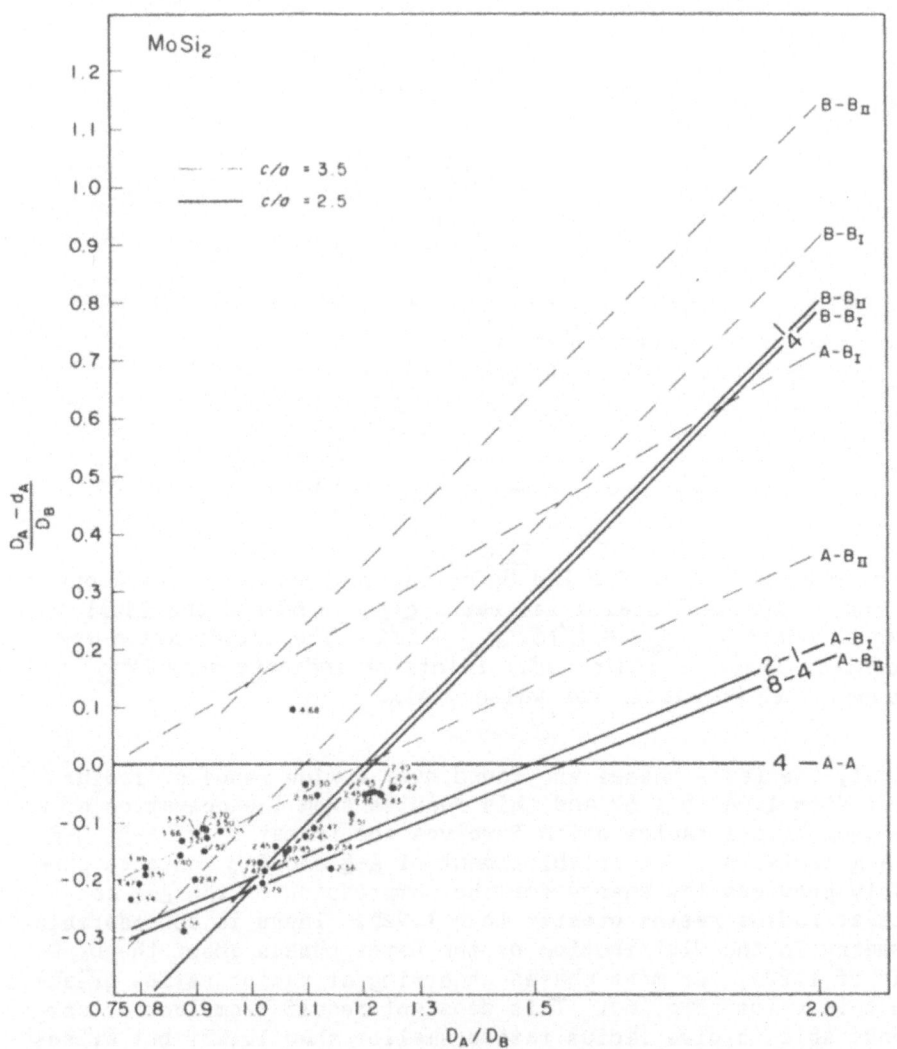

<u>Fig. 8</u>
Near neighbour diagram for the MoSi$_2$ (AB$_2$) structure[11] (tetragonal)
constructed for the ideal Si parameter, z_{Si} = 1/3. The coordination
for the contact lines is indicated; lines for c/a = 2.5 are drawn
full, those for c/a = 3.5 are broken. The axial ratio of the unit
cell is recorded beside points indicating phases with the structure.

stability at radius ratios lower than about 1.05. The characteristic of a true geometrical factor to be recognized on the n.n.d.
is the distribution of the phases over a wide range of radius
ratios between two or more intersecting lines of contacts giving
high coordination. If phases are located only closely about the
intersection of such sets of contact lines, then only the coordination factor is apparent, and there is no true geometrical factor.

Thus we have demonstrated the existence of a true geometrical
factor which can govern the unit cell dimensions and stability of
a structure over a wide range of radius ratios, and we have separated it from the purely trivial coordination factor. Fig. 8 shows
another example of the geometrical factor in the structure of the
$MoSi_2$ phase (tetragonal). Phases with this structure either have
an axial ratio, c/a, of about 2.5 or about 3.5. In either case
the phases are distributed between the lines for $A-B$ and $B-B$ contacts (whose locations depends on the c/a value) and we find the
phases over a wide range of radius ratios from about 0.75 to 1.25.

The rather more complex n.n.d. for the AlB_2 structure (hexagonal) shows the power of these diagrams for analysing the factors
controlling a structure, and shows that these factors may change
with changing structural parameters such as axial ratios. In
Fig. 9 it can be seen that the lines for 6 $A-A$, 12-6 $A-B$ and
3 $B-B_I$ contacts intersect at $D_A/D_B = 1.732$ if the axial ratio
is 1.075, and lines for 2 $A-A_{II}$, $A-B$ and $B-B$ contacts intersect
at $D_A/D_B = 1.5$ when the axial ratio is 0.866. Thus the coordination factor suggests that phases should be found between these
radius ratios and have axial ratios between 0.87 and 1.07. Instead, phases occur between radius ratios of about 0.95 and 1.80
and have axial ratios from 0.59 to 1.27. The n.n.d. shows that
there are two groups of AlB_2 type phases, and this is confirmed
by a diagram of axial ratio versus radius ratio. In one group
axial ratios vary from 0.95 to 1.27 and show only a very general
dependence on radius ratio. These phases lie between the lines
for $A-B$ and $B-B_I$ contacts and cross that for $A-A_I$ contacts, and
occur over a range of radius ratios from about 1.1 to 1.8, indicating that the structure and cell dimensions result from the operation of a geometrical factor. In the other group of phases there
is a more definite dependence of axial ratio on radius ratio which
results from the phases lying within, or close to, the triangle
formed by the lines for $A-A_{II}$, $A-B$ and $B-B_I$ contacts, starting with
the intersection of these contact lines at $D_A/D_B = 1.5$ and $c/a =$
0.866 (Fig. 9). Indeed, the phases follow the position of the line
for $A-A_{II}$ contacts as it varies with axial ratio of the unit cell,
showing in addition to the geometrical factor, the strong influence
of a bond factor involving the $A-A_{II}$ contacts. These represent
lines of A atoms lying in the c direction of the structure. The
n.n.d. also indicates that the graphite-like net of B atoms
(represented by the 3 $B-B_I$ contacts) is not a dominant feature of

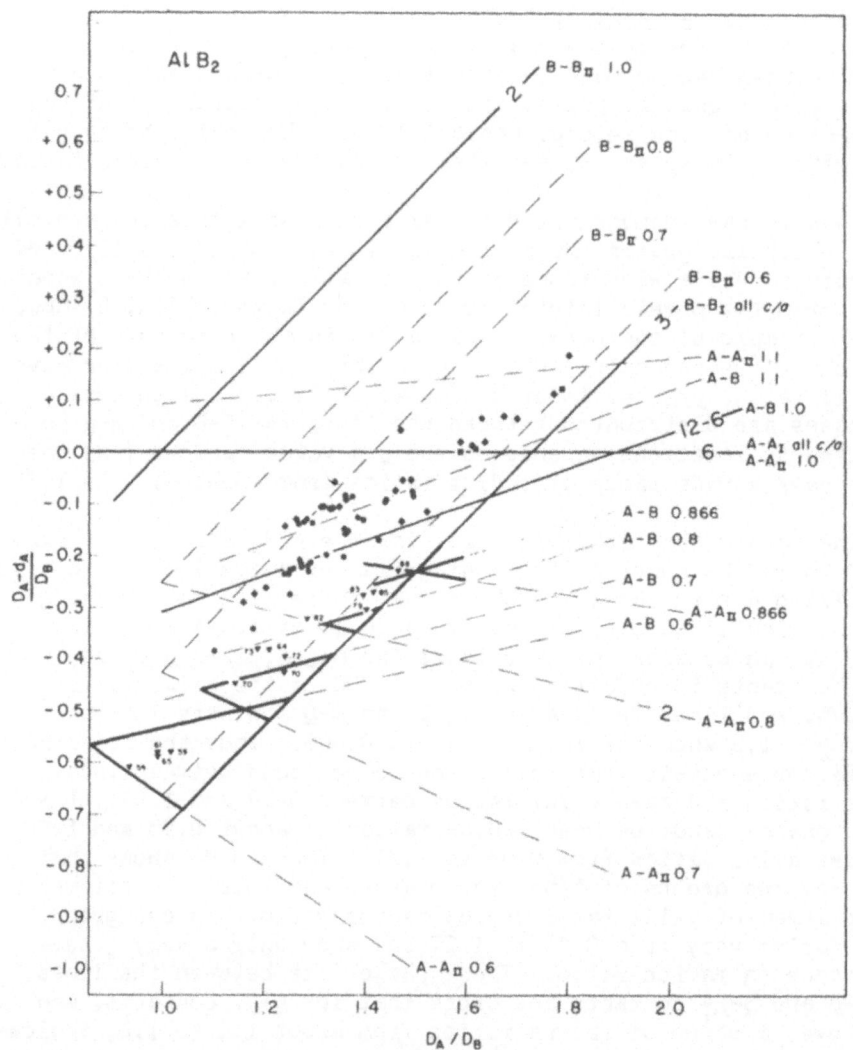

<u>Fig. 9</u>

Near neighbour diagram for the AlB_2 ($\underline{AB}_2$) structure[11] (hexagonal).
The coordination for the contact lines is given and also the axial
ratio of the unit cell to which they refer. Points represent
phases with the structure:

■ Phases with <u>c/a</u> values greater than 1.15
◆   "   "   "   "   from 1.05 to 1.15
●   "   "   "   "   " 0.95 to 1.05
▼   "   "   "   "   less than 0.9 (The actual values are
                                        recorded by these points).

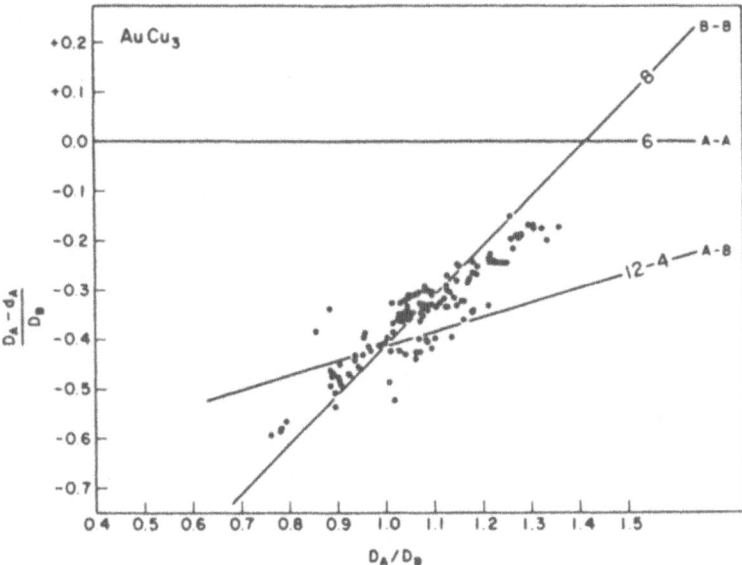

<u>Fig. 10</u>

Near neighbour diagram for the $AuCu_3$ ($\underline{AB}_3$) structure[11] (cubic).
The coordination for the contact lines is given. Points represent
phases with the $AuCu_3$ structure.

the structure; it only exerts such influence as is inherent in
the geometrical factor.

    We have constructed n.n.d. for many of the common metallic
structures, and the cases discussed above are representative
examples of our findings. Near neighbour diagrams give no information on the reasons for the Hume-Rothery 15% rule for extended
solid solutions, or on the reasons for superstructure ordering of
close packed solid solutions. They only show the occurrence of
the geometrical factor which is expected because of the close packing (cf. Fig. 10). Superstructure ordering and the extent of solid
solutions depend mainly on the energy-band factor (cf. eg.[12,13])
and/or elastic effects. In the latter case the results may be
non-intuitive because of the several parameters involved in the
elastic properties of the two components. Sometimes the n.n.d.
may indicate evidence of an energy-band factor. For example
Parthé and co-workers[14,15] have analysed the occurrence of phases
with the FeB and CrB structures finding two distinct groups; in
one formed by $\underline{B}$ components from Groups III and IV, the trigonal
prism of $\underline{A}$ atoms surrounding a $\underline{B}$ atom is elongated so that the
height exceeds the edge of the base, whereas in the other group
formed by $\underline{B}$ components from the Ni and Cu Groups, these prisms

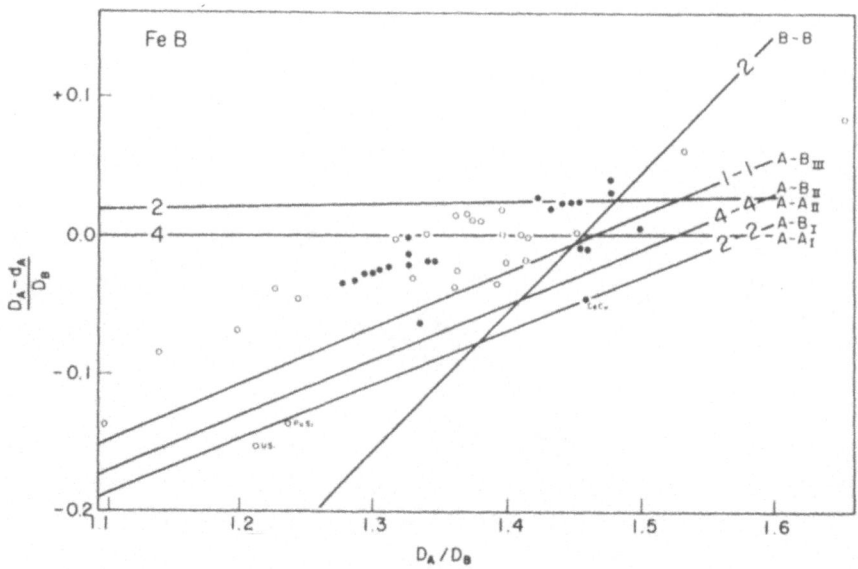

Fig. 11

Near neighbour diagram for the FeB (**AB**) structure[11] (orthorhombic) constructed for the axial ratios and atomic parameters of the FeB phase. The coordination for the contacts is indicated. Points represent phases with the structure, open circles being group I phases[14,15] with elongated trigonal prisms of **A** atoms, filled circles being group II phases with squashed trigonal prisms of **A** atoms. See text.

are squashed so that the height is less than the edge of the base. The n.n.d. (eg. Fig. 11) shows the dependence of these structures on an **A-B** bond factor, and indicates that there is no difference as regards bond or geometrical factors between the groups with elongated and squashed trigonal prisms. It might therefore be presumed that this difference results from the energy-band factor, and the squashed prisms found in the Ni and Cu Group compounds may result from the influence of the **d** bands in these metals.

STRUCTURES CONTAINING STACKED LAYERS OF ATOMS

Many metallurgical structures can be regarded as generated by the stacking of layers of atoms, two examples being the close-packed layers of atoms giving a variety of structures in which tetrahedral, octahedral, or twelve-fold coordination is important, and secondly various networks of atoms to give the CN 12 (icosa-hedron) 14, 15 and 16 polyhedra considered by Frank and Kasper[3,4] where coordination or geometrical factors are found to be important.

Close packed stacking sequences are frequently influenced by bond
factors, and they may depend on whether the sharing of faces, edges
or corners of the coordination octahedra or tetrahedra is favoured
by the component atoms. Thus the cubic sequence ABC of the "anions"
results in the sharing of corners of their octahedra and separation
of atoms located at the centres of the octahedra, whereas the hexagonal stacking sequence AB results in the octahedra sharing faces,
bringing atoms located at their centres closer together.

In the absence of strong bond factors, since close packing
is the geometrical ideal for the packing of spheres of equal size,
we may expect to find evidence of an energy-band factor in controlling the stacking sequence. Although our examination of this is
just beginning, we have already found certain evidence for it in
the structures of lanthanons and their phases. In this case the
energy-band factor does not involve electron concentration, but
apparently the relative degree of participation in the bonding
or energy bands of the $4f$ electrons compared to the d, s and p
electrons. Gschneidner and Valletta[16] first noticed that the
structures of the lanthanide elements and of inter-rare-earth
alloys change from c.p. cubic to double hexagonal c.p. (50% cubic
stacking), to rhombohedral Sm type (33% cubic stacking), to hexagonal close-packed as the ratio of the average interatomic distance
to the radius of the lanthanon $4f$ shell increases. Furthermore the
application of high pressures which decreases the interatomic distances, but presumably has little effect on the diameter of the
$4f$ shell, leads to structural changes in the opposite direction,
as expected. These structural sequences also follow the differences between the closest distances of approach (d_1 interlayer
and d_2 intralayer) and the diameter of the lanthanon $4f$ shells. A
similar situation appears to hold in lanthanide compounds which
involve c.p. arrangements. For example the $LnAl_3$ compounds have
the Ni_3Sn type structure (analogue of c.p. hexagonal) at the light
(La) end of the series, and this changes with increasing atomic
number of the rare-earth to a nine-layer rhombohedral type (analogue of the Sm structure, 33% cubic stacking), to the Ni_3Ti type
structure (analogue of the double hexagonal c.p. structure, 50%
cubic stacking), to a 15 layer rhombohedral structure (60% cubic
stacking), to the $AuCu_3$ structure (analogue of c.p. cubic) at the
heavy (Lu) end of the series. This change, which appears to be in
the opposite sequence to that found for the elements, can in fact
be shown to be in the same sequence if $r_{Ln}-r_{4f}{}_{Ln}$ values are calculated from the known lattice parameters of the phases on the assumption that the Al volume (taken from the elemental structure) remains
constant throughout the series of phases; then, exactly as with the
elements, an increase of $r_{Ln}-r_{4f}{}_{Ln}$ values leads to structural
changes from cubic close-packed in the heavy lanthanons to hexagonal close-packed structures in the light lanthanons as shown in Fig.
12[17]. Similar features can be detailed for other lanthanide

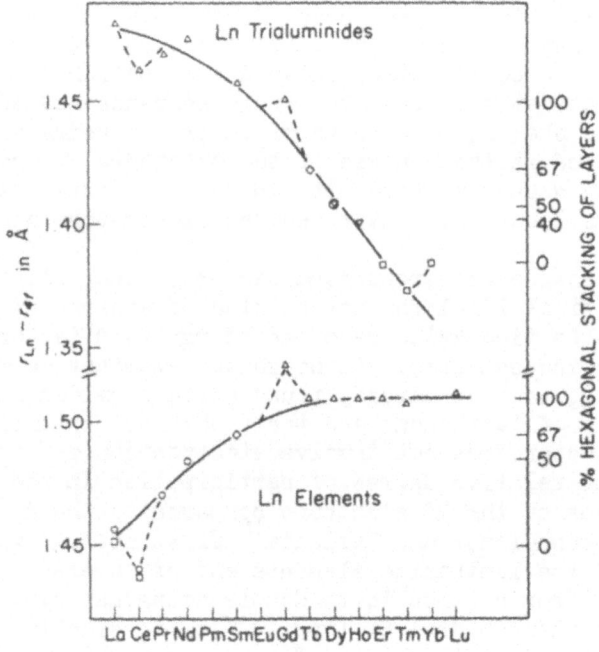

Fig. 12

$r - r_{4f}$ values (see text) for Ln elements and Ln trialuminides[17].
The percentage hexagonal stacking of the layers in the various
structures, is indicated on the right-hand side.

Symbol	Stacking of layers in relation to two neighbouring layers	Structure	
		Ln	LnAl₃
◻	c	Al, cubic	Cu₃Au, cubic
▽	cchchcchchcchch	-	Al₃Ho, rhombohedral
○	hchc	double hexagonal c.p.	Ni₃Ti, hexagonal
◇	chhchhchh	δ-Sm, rhombohedral	BaPb₃, rhombohedral
△	h	A3, c.p. hexagonal	Ni₃Sn, hexagonal

compounds such as those of LnTl₃ with the AuCu₃ structure and
LnHg₃ with the SnNi₃ structure.

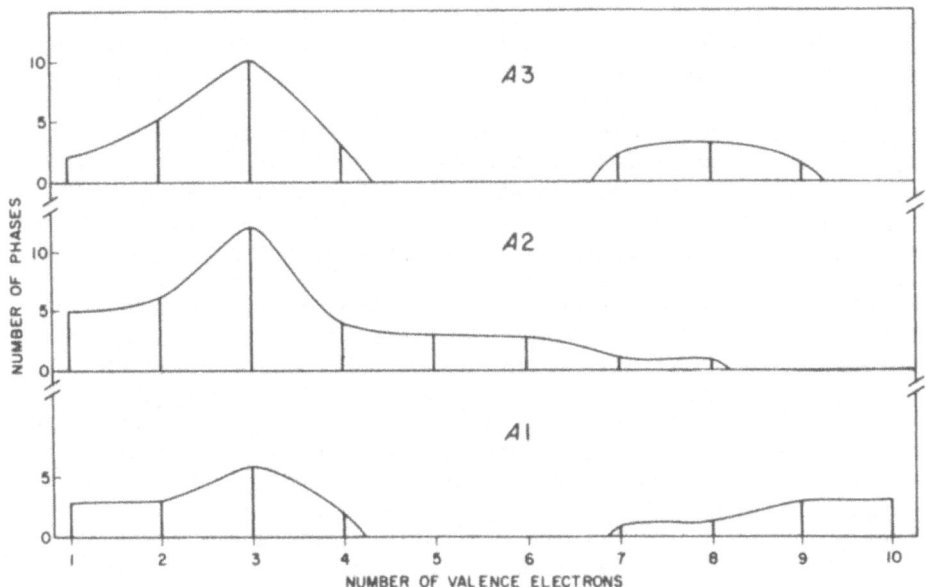

<u>Fig. 13</u>

Histogram for the occurrence of the A1, A2 and A3 structures among
the elements (excluding the actinides), showing the number of phases occurring at the valence electron (s, p, d) concentrations
indicated.

DISTRIBUTION OF A1, A2 AND A3 STRUCTURE TYPES

No crystal chemistry of metals would be complete without some
attempt to discuss the distribution of the A1 (f.c.c.), A2 (b.c.c.)
and A3 (h.c.p.) structures in elements and alloys. Fig. 13 shows
the frequency of occurrence of these in the elements as a feature
of apparent electron concentration. The allotropy that is observed
in the elemental structures suggests that there can be little
difference in energy between one modification and another. Engel[18],
followed by Brewer[19], has correlated the occurrence of these
structure types with the number of outer s and p electrons: b.c.c.
with one, h.c.p. with two, and f.c.c. with three. The known Fermi
surfaces of the Ni and Cu Group metals show that this cannot be
correct for the f.c.c. metals of these Groups. There may, however,
be good justification for the correlation in the case of the b.c.c.
and h.c.p. structures, but it should be noted that the stable
structures of Li and Na at the absolute zero of temperature are

h.c.p. rather than b.c.c. In the transition metal series band theorists would attribute the structure distribution to an energy-band factor associating the b.c.c. structure with a $d^{n-1}s$ configura-tion, h.c.p. with $d^{n-2}sp$ and f.c.c. with $d^{n-0.5}s^{0.5}$ to $d^{n-1}s$ [20]. As yet we do not know with any certainty what controls the distri-bution of these structures but the following thoughts may be perti-nent. 1). Cubic close packing with CN 12 (A1 structure) is likely to be found when filled energy bands occur just below the Fermi lev-el giving the atom a large core and rendering it relatively incom-pressible as for example in the case of Al, Pb, Ni and Cu. In this case the atom approximates most closely among metals to Laves' hard sphere model. 2). Because of the 8+6 coordination, the b.c.c. structure occurs preferably when there are no filled energy bands lying close below the Fermi level, and particularly when the atoms are compressible as in the case of the alkali metals. Experience with near neighbour diagrams suggests that the "size" of the alkali metals is not that given by the closest distance of approach (CN 8), but somewhere between this and the distance to the six further neighbours, so that the 8 closest contacts are somewhat compressed. In the smaller alkalis, Li and Na, we suggest that this situation may well be tolerated at high temperatures where thermal vibrations are active, but not at the lowest temperatures where transition occurs to the h.c.p. structure with ideal axial ratio and 12 coordination (cf. also Ti and Zr), so that the com-pression of the atoms in certain directions (as in the b.c.c. structure) is avoided. Such reasoning also accounts for the fact that there is little if any decrease in volume in the change from f.c.c. to the hexagonal structure, although space filling models indicate that the degree of space filling in the b.c.c. structure is several percent less than in the ideal h.c.p. structure; atomic compression in the b.c.c. structure with CN 8+6 and its absence in the ideal h.c.p. structure with CN 12 removes the expected difference of atomic volume of the Li and Na allotropes with the two structures. Applying this reasoning to b.c.c. transition metals, their low compressibility is noted, although the band structure of W, for example[21,22], shows that the two filled energy bands with predominantly d character lie well below the Fermi level 3). Finally, we suggest that the h.c.p. structure is to be found when a small overlap of the penultimate Brillouin zone occurs, such as leads to structural expansion in certain directions[23]. The h.c.p. structure, which has an extra degree of freedom in the relative lengths of the a and c axes compared to the cubic struct-ures, may accommodate this situation with a lower overall crystal energy than a cubic structure.

SUMMARY

This study of the crystal chemistry of metals draws on our new experimental knowledge of the Fermi surfaces of metals on the

one hand, and critically examines the concept of geometrical
effects in metals on the other. Since the Fermi surface is an
experimentally measurable property which exists for all metals,
we conclude that the crystal chemistry of metals depends primarily
on an energy-band factor, which may influence the crystal structure
and cell dimensions, either as a result of the electron concentration per primitive cell, or through the symmetry of the energy bands
at the Fermi level. The energy-band factor is generally a weak
effect which is likely to be dominated by several stronger factors
that arise when the effective electron concentration, or the difference in potential between the component atoms become relatively
large: 1) A bond factor, which is found when the required valence
and coordination conditions permit the formation of directed chemical bonds. Generally speaking, bond factors become stronger as
the number of valence electrons per atom (excluding electrons in
filled $\underline{d}$ bands) becomes larger (ie. from 2 to 4). 2) An electrochemical factor which arises when the electronegativity difference
of the component atoms is large, and which results in the formation of phases with particular stoichiometries that allow one or
more of the components to achieve a filled $\underline{s}^2\underline{p}^6$ valence subshell.
This situation which is associated with a strong chemical bond
factor, arises when the difference of potential of the component
atoms is large. 3) A geometrical factor resulting from the formation of a large number of atomic contacts, which controls the
cell dimensions and permits phases to occur over a wide range of
radius ratios of the component atoms. Since this factor has no
direct influence on the energy-band relationships other than
structural, the energy-band factor may still influence the choice
of structure when a geometrical factor operates, as in the case of
the superstructures formed from c.p. solid solutions, and as is
possible in the case of Laves phases. 4) In transition metal compounds with the Group V and VI elements P, As, Sb, S, Se and Te,
whether they are metallic or semiconducting, both crystal field
and bond effects may influence structures and cell dimensions,
and the resulting structures are generally those which give a
filled subshell of $\underline{d}$ electrons on the transition metal atoms,
the $\underline{t}_{2g}^{6}$ subshell being the most prominent.

Analysis of the geometrical effects in metals with the aid
of near neighbour diagrams leads to the separation of a trivial
coordination factor from the true geometrical factor which controls cell dimensions and structural stability.

Examination of factors determining the stacking sequence in
structures involving close packed layers of atoms, recognizes
the influence of bond factors when some layers fill tetrahedral
or octahedral holes of the others and of energy band factors
when the close packed layers are themselves stacked in close
packed spacing without intervening layers.

Finally, some reasons based on energy-band structures are advanced for the distribution of f.c.c., b.c.c. and h.c.p., A1, A2 and A3 structures among the elements and their alloys.

The analysis confirms previous observations that the crystal chemistry of metals is a complex phenomenon depending on the competing influence of numerous factors, yet by drawing on our knowledge of Fermi surfaces, it relates these to each other in a more satisfactory manner, particularly as a result of an explicit study of geometrical effects in the structures of metals.

I am indebted to Dr. J.-P. Jan for calculating the Fermi surface for the CsCl structure at an electron concentration of 6 per primitive cell using the empty lattice model.

REFERENCES

[1] W. Hume-Rothery, G.W. Mabbott and K.M. Channel Evans, Phil. Trans. Roy. Soc., A233, 1 (1934).

[2] F. Laves, Theory of Alloy Phases (Amer. Soc. Met., Cleveland, Ohio, 1956), p. 124.

[3] F.C. Frank and J.S. Kasper, Acta Cryst., 11, 184 (1958).

[4] F.C. Frank and J.S. Kasper, Acta Cryst., 12, 483 (1959).

[5] L. Pauling, J. Ammer. Chem. Soc., 69, 542 (1947).

[6] F. Laves and H. Witte, Metallwirtschaft, 15, 840 (1936).

[7] R.P. Elliott and W. Rostoker, Trans. Amer. Soc. Met., 50, 617 (1958).

[8] H. Oesterreicher and W.E. Wallace, J. Less-Common Metals, 13, 91 (1967).

[9] E. Parthé, Zeit. Krist., 115, 52 (1961).

[10] W.B. Pearson, Acta Cryst., B24, 7 (1968).

[11] W.B. Pearson, Acta Cryst., B24, 1415 (1968).

[12] J.F. Nicholas, Proc. Phys. Soc., 66A, 201 (1953).

[13] H. Sato and R.S. Toth, Phys. Rev., 124, 1833 (1961); 127, 469 (1962). Phys. Rev. Letters, 8, 239 (1962).

[14] O. Schob and E. Parthé, Acta Cryst., 19, 214 (1965).

[15] D. Hohnke and E. Parthé, Acta Cryst., 20, 572 (1966).

[16] K.A. Gschneidner and R.M. Valletta, Acta Met., 16, 477 (1968).

[17] W.B. Pearson, J. Less-Common Metals, 13, 626 (1967).

[18] N. Engel, Kem. Maanedsblad, 30, 53, 97 (1949); Trans. Quart. Amer. Soc. Met., 57, 610 (1964).

[19] L. Brewer, Phase Stability in Metals and Alloys, Rudman, Stringer and Jaffee, Ed. (McGraw-Hill, New York, 1967), p. 39. High Strength Materials, Zackay, Ed. (John Wiley, New York, 1965), p. 12.

[20] W.M. Lomer, Phase Stability in Metals and Alloys, Rudman, Stringer and Jaffee, Ed. (McGraw-Hill, New York, 1967), p. 241.

[21] L.F. Mattheiss and R.E. Watson, Phys. Rev. Letters, 13, 526 (1964).

[22] L.F. Mattheiss, Phys. Rev., 139, A1893 (1965).

[23] H. Jones, Proc. Roy. Soc., A147, 400 (1934); C.S. Barrett and T.B. Massalski, Structure of Metals, 3rd Ed. (McGraw-Hill, New York, 1966), p. 360.

[24] E. Teatum, K.A. Gschneidner and J. Waber, LA 2345 (U.S. Dept. of Commerce, Washington, D.C., U.S.A., 1960).

ALLOY PHASE STABILITY CRITERIA

Niels N. Engel

Division of Metallurgy
School of Chemical Engineering
Georgia Institute of Technology
Atlanta, Georgia 30332

INTRODUCTION

The electron pairing theory[1-3] combined with the postulate
that the simple, most common metallic structures are controlled by
one, two or three outer electrons, as proposed by the author,[4] enables the metallurgist to understand and predict phase formation
in many alloy systems on an empirical basis. Brewer[5,6] has used
this approach to make extensive surveys of transition metal systems, and Brewer type diagrams may be helpful in predicting unknown
alloy phases and phase diagrams. In this paper the theories mentioned above will be applied to phase diagrams of the transition
elements with elements of the second period, and to compound
formation in general.

For the elements of the two short periods of the periodic
chart, the electron pairing theory explains why the bonding energy
first increases and then decreases.[7] The bonding energy and phase
stability increase from sodium with one bonding electron to silicon which utilizes four electrons for bonding by promoting one s-
electron to the p-electron level. On passing from silicon to chlorine, the bonding energy decreases in spite of the increasing numbers of electrons in the outer shell because the electrons pair off
internally and become unavailable for bonding. In this way the
number of bonding electrons decreases from four for silicon to one
for chlorine and the energy for atomizing the elements at room
temperature drops from 109 to 29 kcal/mole.

Besides the normal elements which have only outer bonding
electrons, the three long periods contain the transition element
series in which d-electrons also participate in bonding. There are
nine transition elements in each series beginning with Sc, Y and La;

25

the noble metals Cu, Ag, and Au are included here as a ninth transition group since there is appreciable d-bonding in these metals.
In the transition elements up to the sixth column d-electrons are
added, yielding the maximum d-contribution for the elements Cr, Mo
and W. The stability of the elements increases accordingly (with
the exception of Cr). If further electrons are added, the d-electrons pair internally; consequently, the bonding strength and phase
stability decrease. Superimposed over the increase and decrease
of stability due to the increasing number of d-electrons described
are the tendencies of outer electrons to enter d-orbitals (in the
first half of the transition period) and of d-electrons to be ex-
cited to outer orbitals (in the second half). Both electronic transitions lower the energy of the metal and thus increase the bonding
strength.

CLASSIFICATION OF ALLOY SYSTEMS

When different elements are mixed to make alloys it may be re-
membered that almost 5,000 binary systems and more than 150,000
ternary systems can be formed by combining the hundred known elements. Even after excluding noble gases, etc., a very large number
of systems remains. A broad survey of alloy forming systems is possible if the elements are subdivided according to their "bonding
pattern", i.e. the type and number of bonding electrons, into three
main categories: the normal elements, the transition metals and
the rare earth, actinide metals.

Those elements bonded only by outer s- and p-electrons will be
called normal. All elements in the first three periods are normal.
Further normal elements are found in columns 1, 2, and 12 to 17 in
the three long periods. As stated, the transition elements have
d-electrons participating in bonding; they are found in columns 3
to 11 (both inclusive) in the three long periods. The rare earth
and actinide elements further have unpaired f-electrons which are
not assumed to actively participate in bonding;[8] however, the unfilled f-shell has a great impact on the bonding pattern, since it can
act as an electron sink. Six main groups of element combinations
can be distinguished, each with the potential of forming systems
with characteristic properties. Some of these will be discussed
here.

The first group consists of combinations of normal elements
with normal elements. A very important and interesting subgroup
contains the compounds formed by hydrogen, boron, carbon, nitrogen,
and oxygen, with each other, e.g., short organic compounds and
polymers, which will not be dealt with in this paper. The strongest bonding and the most stable phases are obtained when each atom
participates with four bonding electrons which is the maximum possible for normal element combinations. Column 14 elements such as
carbon in the form of diamond or compounds of early column 14 elements with each other such as silicon carbide show this. Four bonding electrons per atom also occur in equiatomic mixtures of column

13 with column 15 elements, the so-called III-V semiconductor compounds. If a three-electron element is mixed with a five-electron element to obtain diamond bonding, one electron must be transferred from the latter to the former; therefore, an ionic bond is superimposed over the diamond bond. Depending on the ionization potentials and the energy liberated in forming the electron pair, the ionization may be more or less complete. It is interesting to note that the melting point of GaAs, e.g., is about $300^{\circ}C$ higher than that of Ge and that AlP melts almost $800^{\circ}C$ higher than Si. These examples illustrate that electron transfer may result in bonding which is stronger than the covalent bonding produced by it.

Extended solid solubilities develop if the size factor[9] and bonding patterns are favorable. Among normal metals, extended solubilities are known in the systems Li-Mg, Zn-Al, Mg-In, Cd-Mg, and Tl-Pb where the numbers of bonding electrons of the elements do not differ by more than one (for normal elements). If the size factor is favorable and the bonding pattern is unfavorable, i.e., if the number of bonding electrons differs by two or more, no solubility occurs. Examples are the systems Be-Si and Al-Hg; here Hg is considered as a metal with less than one bonding electron because of its low melting and boiling points. In the language of thermodynamics, internal pressure levels are too far apart in systems such as Be-Si or Al-Hg.

In alloys of transition metals with transition metals, size factors are generally more favorable and bonding patterns, as far as structure-controlling outer electrons are concerned, vary only slightly from column to column. Solubilities, therefore, are often rather extended.

In alloys of transition metals with normal elements, extended terminal solid solubilities are often found on the transition metal side of the phase diagram because addition of normal elements will often enhance the d-electron bonding or at least not reduce it too severely. This is especially true for normal metals with many electrons. On the normal element side, terminal solid solubilities are generally limited because transition metal atoms in dilute solution are not able to develop d-electron bonds except when clustering or ordering takes place. Such solid solutions are therefore weakly bonded and will not prevail in competition with the transition metal phases or compounds in which d-bonding occurs.

Compound Diagrams

The electron pair bonding concept is helpful in predicting element combinations in which compounds should or should not be expected. To illustrate this, compound surveys for several elements are presented. The aluminum-normal element compounds are shown in Figure 1. (Information for compound diagrams and melting point diagrams is taken from References 10-14). On the abscissa the alloying elements are arranged with increasing number of electrons and increasing period numbers. On the ordinate, the known aluminum compounds are arranged with increasing amounts of alloying element going downward.

Aluminum forms compounds with all second period elements except Be. There are stoichiometric compounds with elements in columns 5 and 6; some additional compounds are formed with Mg, Zn, S, Se and Te.

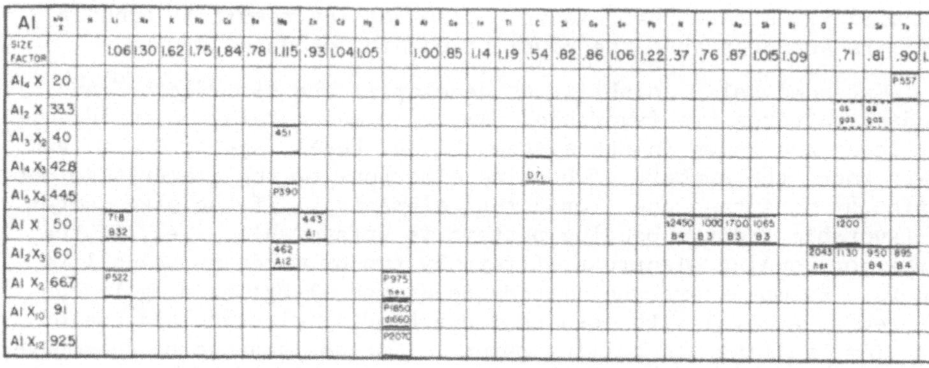

Al	at% X	H	Li	Na	K	Rb	Ca	Ba	Mg	Zn	Cd	Hg	B	Al	Ga	In	Tl	C	Si	Ge	Sn	Pb	N	P	As	Sb	Bi	O	S	Se	Te	Po
SIZE FACTOR			1.06	1.30	1.62	1.75	1.84	.78	1.115	.93	1.04	1.05		1.00	.85	1.14	1.19	.54	.82	.86	1.06	1.22	.37	.76	.87	1.05	1.09		.71	.81	.90	1.17
Al₄X	20																															P557
Al₂X	33.3																								.05 905	.08 905						
Al₃X₂	40						451																									
Al₄X₃	42.8																		D.7													
Al₅X₄	44.5						P390																									
AlX	50		718 B32				443 Al																2450 B4	1000 B3	1700 B3	1065 B3			1200			
Al₂X₃	60						462 Al2																					2043 hex	1130 B4	950 B4	895	
AlX₂	66.7		P522								P975 hex																					
AlX₁₀	91										P1850 1660																					
AlX₁₂	92.5										P2070																					

Figure 1. The aluminum-normal element compound diagram. All known stable compounds are marked by two horizontal lines at the appropriate composition. The upper number indicates the melting point of the alloy. Where "P" or "p" is added, the compound forms peritectically or peritectoidally; "e" indicates eutectoid decomposition. The structure type is presented at the bottom of each compound square.

A number of metals do not form compounds with Al because of unfavorable size factors; this is the case for the alkali metals Na, K, Rb, and Cs, and for Tl and Pb. Monotectics are formed and the elements do not even mix in the liquid state. Another group of metals does not form compounds despite favorable size factors because no state of lower energy can be reached by the relocation of electrons into different quantum states. In this case, mixtures of different atoms will not be able to develop as favorable a bonding pattern as the pure elements. Solid solutions will be less stable than pure elements, and monotectics or eutectics are formed; this is the case for combinations of Al with Cd, Hg, Ga, In, Si, Ge, and Sn. Only Zn, exhibiting an unusually favorable size factor with Al, forms extended solid solutions. When the melting points of the alloying element and of aluminum are almost equal and when the difference in electron concentration is small, a composition range in the center of the phase diagram may occur where the stability of phases with ordered atomic arrangements surpasses that of the weakened solid solutions; that is the case in the Al-Mg diagram. Such intermetallic phases always exhibit lower melting or decomposition temperatures than the components.

When aluminum is mixed with normal elements from the right
side of the periodic chart, it becomes possible to obtain more stable phases by electron transfer from atoms with more than four electrons to the aluminum atoms. As described above, the most stable
state will develop when both kinds of atoms acquire four unpaired
electrons to form the maximum number of outer electron bonds, with
further strengthening by ionic bonding due to electron transfer.
The fifth column elements form ordered (B3 type) diamond phases
with aluminum, and the sixth column elements form wurtzite (B4 type)
phases, both four-coordinated compounds. The enhanced stability of
these phases is expressed by the melting points which are much higher than those of aluminum or the alloying element.

The LiAl phase may be considered as an electron transfer phase
with the lithium atoms giving off one electron, resulting in four
electrons per aluminum atom. This enables the aluminum atoms to
form a diamond lattice with the lithium ions as interstitials. The
compounds with small atoms from the second period will not be treated.

The situation is entirely different when aluminum is alloyed
with transition metals, as demonstrated in Figure 2. The transition metals have less than three outer electrons per atom. Therefore, in alloys with aluminum the average outer electron concentra-

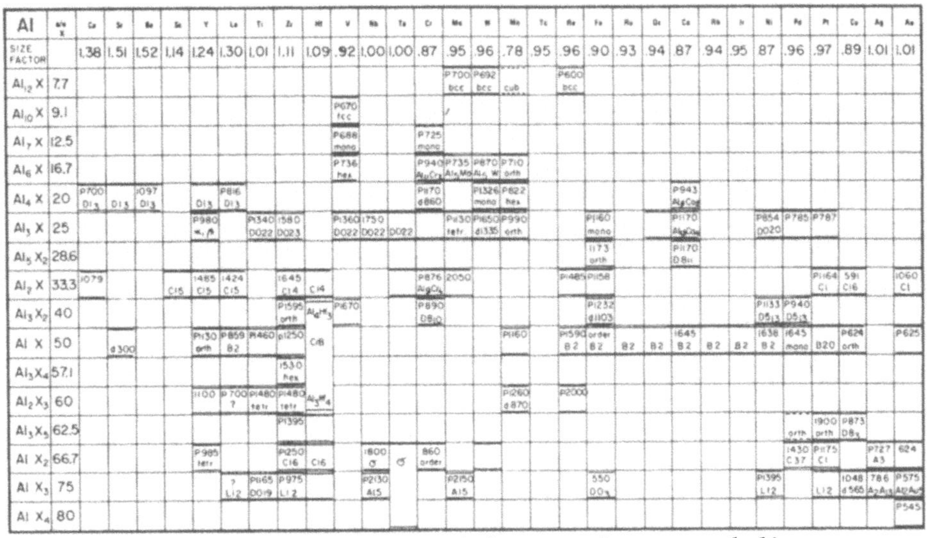

Figure 2. The aluminum-transition metal compound diagram.

tion is numerically increased. Because of the equilibrium between
outer electrons and d-electrons in the transition metal atoms, some
of the transferred electrons go into the d-shell and strengthen the
d-bonding. This is especially true for the third and fourth column
elements where compound formation flourishes. Extended terminal

solid solubility on the transition metal side, especially at elevated temperature, is also prevalent among these systems. It may be
pointed out that for Mo the Al_2X and AlX_3 phases seem to have maximum melting points (or peritectic formation temperatures) whereas
for Co and Ni the equiatomic compounds are most stable. Further
details will not be discussed here.

Alloys in which both components are transition metals are different from alloys of transition metals with normal elements for two
reasons. First, the atomic sizes for most transition elements are
reasonably close to each other, making extended solid solubility
ranges possible. Second, the stability largely depends on d-electron bonding which, according to the author's viewpoint, does not
influence the crystal structure. The phases are therefore controlled by the number of outer electrons as emphasized by Brewer[5] and
demonstrated in Figure 3 which is taken from Reference 5. Curved
lines connecting identical electron concentrations can be drawn

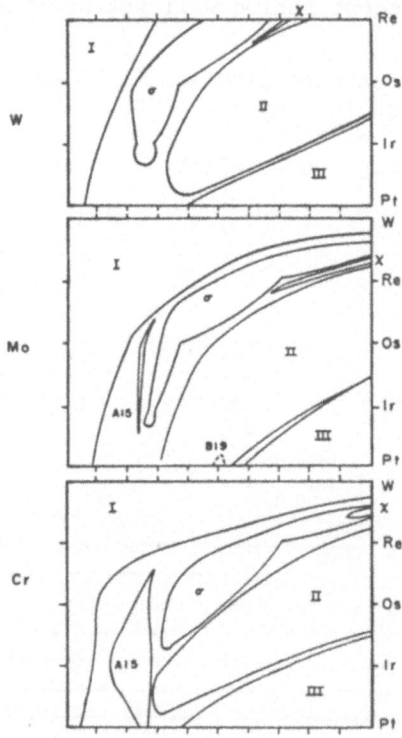

Figure 3. Brewer type diagrams
for sixth period transition metals. Taken from Reference 5.

through the alloy diagrams. The phase boundaries follow these
lines closely. Since bonding cannot be improved very much by mixing of neighboring or somewhat more distant transition metals, few
compounds are formed. The transition metals thus form extended
fields of electron concentration controlled phases with other transition metals. The phases constituting these fields often have ex-

tended ranges of solid solubility. Both fields with integral numbers of outer bonding electrons (indicated as I, II, III) and fields with nonintegral numbers of structure controlling electrons (σ, A15 types) exist. The Roman numeral I refers to phases with body centered cubic (A2 type) structures postulated to be controlled by one outer bonding electron, II refers to hexagonal close packed (A3 type) phases controlled by two outer electrons, and III refers to face centered cubic structures (A1 type) controlled by three outer bonding electrons.

The elements at the end of the transition period may form more stable phases by excitation or ionization of electrons. Such electronic changes can be initiated by elements located far away in the periodic chart, but not by neighboring elements and elements located nearby. As a result, additions of the two last mentioned groups of elements lead to the formation of solid solutions and phases determined by the electron concentration, whereas elements located far away produce compounds of different structures, some of which may be very stable. This is demonstrated for titanium in Figure 4; compound formation starts with Mn and continues to Au. The compound formation of copper is shown in Figure 5. Here no compounds

Ti	a/o R	Ca	Sr	Ba	Sc	Y	La	Ti	Zr	Hf	V	Nb	Ta	Cr	Mo	W	Mn	Tc	Re	Fe	Ru	Os	Co	Rh	Ir	Ni	Pd	Pt	Cu	Ag	Au
size factor		1.36	1.49	1.50	1.13	1.23	1.29	1.00	1.10	1.09	.91	.99	.99	.87	.94	.94	.78	.95	.95	.90	.92	.93	.87	.93	.94	.87	.95	.96	.89	1.00	1.00
Ti₆X	25																									1370 A15		P1520 A15		Li₂	1400 A15
Ti₃X	33																							P1055 E9₃	P1015 E9₃	1015 C11b					
Ti X	50														P950 B2				P1317 B2	B2	B2	1300 B2	2140 mono B2	1240 B2	1630 B2	982 B11	Li₂		1440
Ti₂X₃	60																									P935				
Ti X₂	66.7							1350 C15						1325					640 C14			P250 C3Cl6				P892		1480
Ti X₃	75													P250					640			P200 L1₂	L1₂	L1₂ 1378	DO24 DO24	1950	908 A3		
Ti X₄	80													P1230															Pt90 Ni₄Mo
Ti₅X₂₄	83															Pt750 Al2													

Figure 4. The titanium-transition metal compound diagram.

Cu	a/o R	Ca	Sr	Ba	Sc	Y	La	Ti	Zr	Hf	V	Nb	Ta	Cr	Mo	W	Mn	Tc	Re	Fe	Ru	Os	Co	Rh	Ir	Ni	Pd	Pt	Ag	Au
size factor		1.54	1.69	1.70	1.27	1.39	1.46	1.13	1.24	1.23	1.03	1.12	1.12	.98	1.07	1.07	.88	1.08	1.08	.97	1.04	1.05	.98	1.05	1.06	.98	1.08	1.09	1.00	1.13	1.13
Cu₁₃X	7		P675																											
Cu₈X	14.3				P95 hex																									
Cu₅X	16.7	935 hex																												
Cu₄X	20				985 hex	P785 hex																								
Cu₃X	25							908 hex	1115																					
Cu₅X₂	28.5									P1070																				
Cu₂X	33.3				935 hex	834 orth	P892																							
Cu₃X₂	40								P920	895																				
Cu X	50						B2	935 B11	P551 B2	975 B11	940																			
Cu X₂	66.7							1015 C11b	1000 C11b	C11b																				
Cu X₄	80	P710																												

Figure 5. The copper-transition metal compound diagram.

are formed with transition elements after Hf; only the first two column transition elements form compounds with copper.

Transition Metal-Interstitial Element Systems

From a phase stability viewpoint, the most interesting group of alloy systems is composed of systems in which one component is a transition metal and the other is a normal element. In these systems, phases of extreme stability can form, especially when the normal element belongs to the second period where atoms are small enough for interstitial solid solutions to be formed.[15] It is characteristic that, for example, carbon dissolves in solid transition metals with unfilled d-shells such as Ti and Nb, but not in normal metals such as Mg, Al, Zn or Sn. This fact supports the assumption that the small second row elements give off their electrons to the transition metal atoms when taken into interstitial solution. The necessary ionization energy will be largely compensated for or will be slightly overcompensated by the ionic bonding energy. Stable phases occur if the transferred electrons enter transition metal quantum states and form strong bonds. If the transferred electrons enter d-orbitals, as is possible in the early transition metals, some of the most stable phases known are formed. If they enter outer electron orbitals, the bonding will be weaker than if it is in second row normal elements, and the interstitial phase will be relatively unstable.

ZrC represents an ideal case for the explanation of the formation of a particularly stable phase. The zirconium atoms have four bonding electrons with the following substitution according to the Engel correlation:[4] for α-Zr, there are essentially two electrons in the d-level and two in the outer (s+p) level; for β-Zr there are three d-electrons and one s-electron. Adding an equiatomic amount of carbon means adding four electrons to each zirconium atom if complete ionization is assumed to occur. This results in eight electrons per zirconium atom of which slightly more than half will be in the d-level and slightly less than half in the outer electron levels. Since this phase exhibits the NaCl (B1 type) lattice (where the Zr atoms form a face centered cubic (A1 type) lattice in which C occupies the octahedral holes) it may have three outer electrons, according to the Engel correlation, leaving five electrons for the d-level.[16] This produces a bonding pattern which is stronger than that of tungsten, i.e., Zr (1-, 2-, 3-, $4s^2p^6d^5$, $5sp^2)^{4-}$ atoms form five d-bonds, three outer bonds and four ionic bonds. In Ti, Zr and Hf carbides, the bonding type is identical. The same B1 type structure is found in a series of transition metal carbides, nitrides and some oxides and borides. If the number of electrons per transition metal atom is increased by one, either by substituting transition metals from column 5 for transition metals from column 4 or by substituting nitrogen for carbon, the number of d-electrons is increased by one and an internal d-electron pair is formed. Bonding is thereby reduced by one d-bond. The stoichiometric compounds become less stable; the most stable compound is formed with a certain amount of interstitial atom vacancies as demonstrated for NbC in Figure 6.[11]

As far as melting points represent stabilities, the decrease in stability described can be demonstrated by a comparison of melting points of the fourth and fifth column metals and monocarbides, as shown in Figure 7. (The free energies of formation are correct

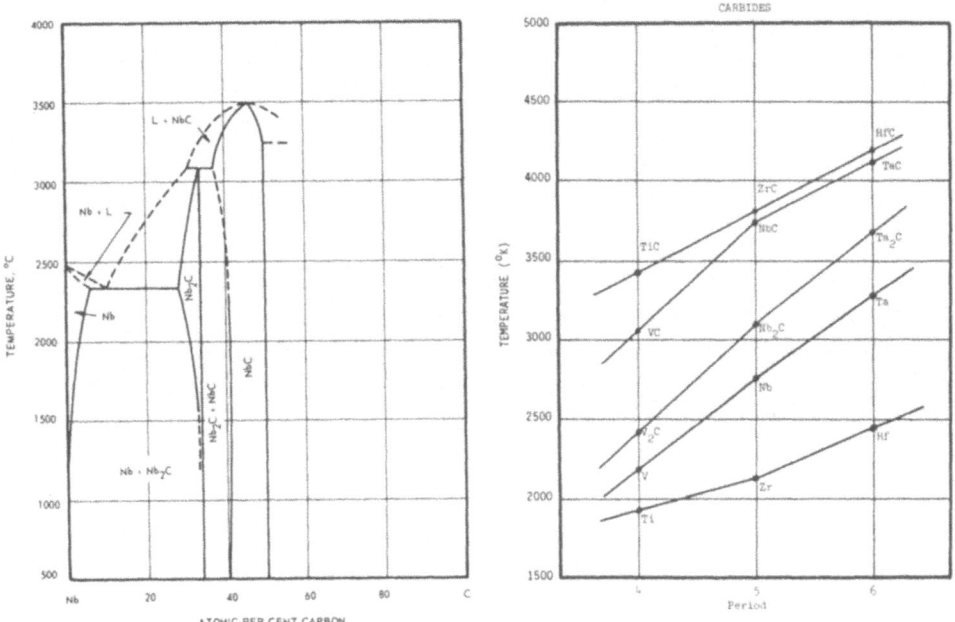

Figure 6. The niobium-carbon phase diagram. Taken from Reference 15.

Figure 7. Melting points of columns 4 and 5 transition metals, semicarbides and monocarbides. Taken from Reference 15.

quantities to use; however, for many carbides, nitrides, etc., these are not available. Therefore, the melting points are used as the next best and available criterion.) It can be seen that the melting points of the fifth column metals are 300 to 800°K higher than those of the fourth column metals. The monocarbides of the fourth column metals melt about 1000°K higher than the fifth column metals, and the maximum melting points of the fifth column monocarbides are only slightly lower. If it is assumed that each d-bond contributes 650°K and each outer electron contributes 300°K to the melting point, melting points can be calculated and indicate phase stability. In Figure 8, the calculated and actual melting points of the fourth column borides, carbides, nitrides, and oxides are compared and the electron configurations assumed above are shown. Figure 9

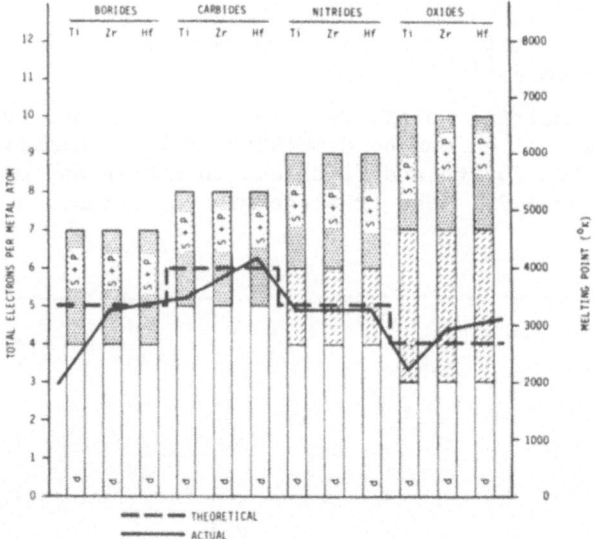

Figure 8. Theoretical electron configuration and comparison of calculated and actual melting points of equiatomic borides, carbides, nitrides, and oxides of fourth column transition metals. Taken from Reference 13.

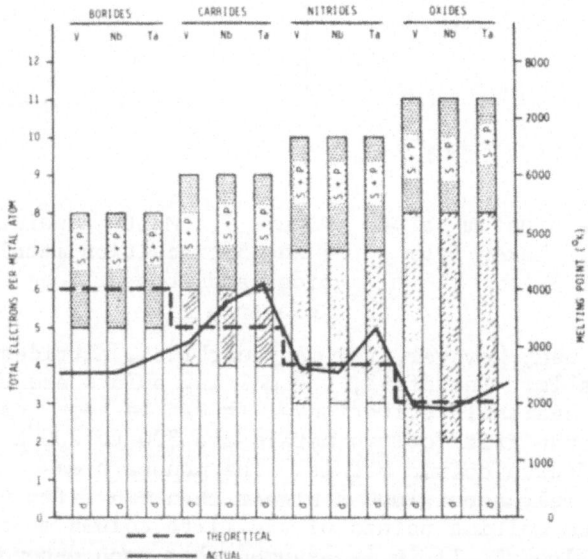

Figure 9. Theoretical electron configuration and comparison of calculated and actual melting points of equiatomic borides, carbides, nitrides, and oxides of fifth column transition metals. Taken from Reference 15.

gives the same comparison for the fifth column compounds.

As can be seen from Figure 8, for fourth column transition metals the carbides are the most stable phases. The monoborides are less stable because the number of electrons suffice only for four d-electrons, while the nitride is less stable because there are six d-electrons of which two form an internal pair leaving four to form bonds. If the same bonding mechanism is maintained, the third column elements would be expected to exhibit maximum melting points for the nitrides. This is actually the case. The increase in atomic size on going from column 4 to column 3 transition elements weakens the ionic bond considerably, thus making the stability peak less pronounced.

As stated, the stability of the NaCl structure is reduced when the total number of bonding electrons is decreased or increased from the optimal value of 8 bonding electrons per metal atom by substituting other atoms. A stability decrease is found for transition metals to the right of column 4 which contribute so many d-electrons that these become internally paired and thereby inactive in bonding. Carbides with two, three and approximately four metal atoms per carbon atom become stable besides or instead of the equiatomic carbide. In these carbides, the excess electrons from the carbon atoms are divided among several metal atoms and thereby reduce phase stability less. Even so, carbides cease to be stable between manganese and iron; the cobalt, nickel and copper column elements do not form stable carbides, although the iron carbide,

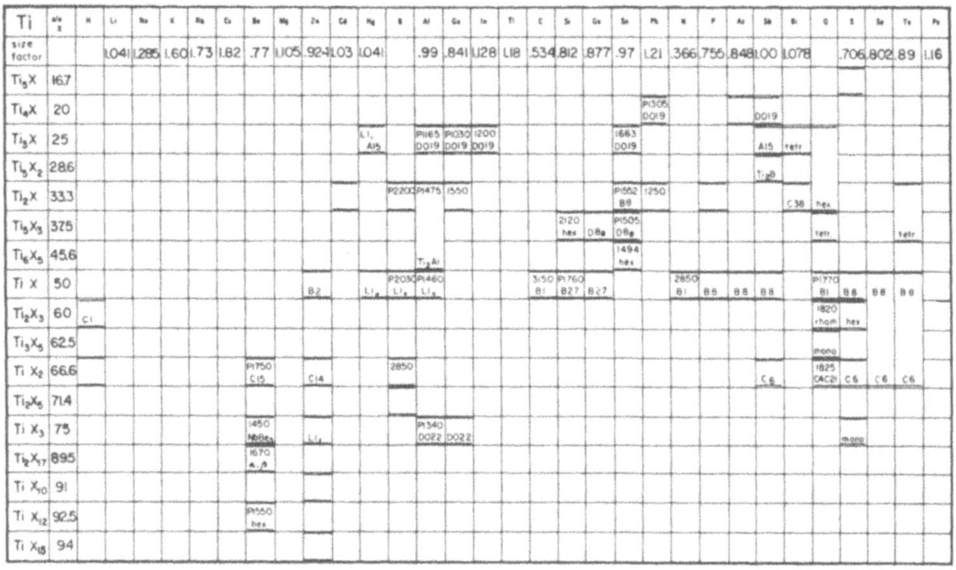

Figure 10.   Titanium-normal element compound diagram.

Fe_3C, can be readily retained. Transition metal nitrides exhibit a similar pattern of stability, whereas borides and oxides which are less stable are represented only by a few late transition metal compounds. The systematic stability variations in titanium alloys and the stability maximum for the carbide can be seen in the titanium-normal metal compound diagram, Figure 10.

The monocarbides and nitrides of uranium and plutonium are of interest and will be treated next. Figure 11 gives a survey of

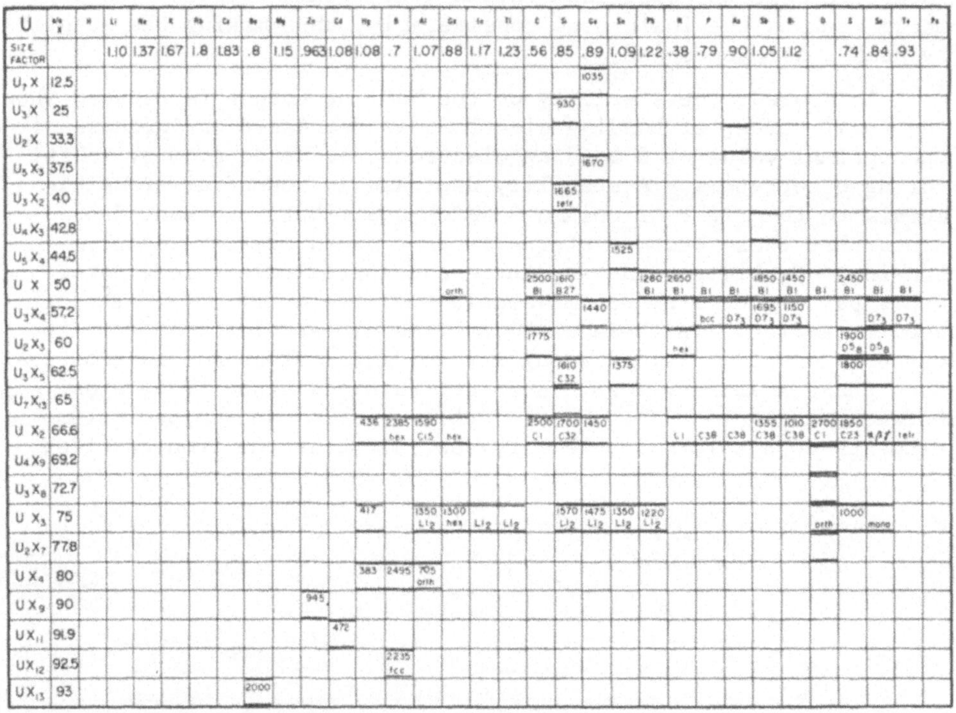

U / N/R		H	Li	Be	K	Rb	Ca	Sr	Mg	Zn	Cd	Hg	B	Al	Ga	In	Tl	C	Si	Ge	Sn	Pb	N	P	As	Sb	Bi	O	S	Se	Te	Po
SIZE FACTOR		1.10	1.37	1.67	1.8	1.83	.8	1.15	.96	1.08	1.08	.7	1.07	.88	1.17	1.23	.56	.85	.89	1.09	1.22	.38	.79	.90	1.05	1.12		.74	.84	.93		
U_7X	12.5																			1035												
U_3X	25																	930														
U_2X	33.3																															
U_5X_3	37.5																	1670														
U_3X_2	40																	1665 tetr														
U_4X_3	42.8																															
U_5X_4	44.5																		1525													
$U X$	50								orth				2500 B1	1610 B27					1280 B1	2650 B1	B1	B1	1650 B1	1450 B1			2450 B1	B1	B1			
U_3X_4	57.2														1440				bcc	D7₃	1695 D7₃	1150 D7₃				D7₃	D7₃					
U_2X_3	60												1775						hex				1900 D5₈	D5₈								
U_3X_5	62.5												1610 C32		1375								1800									
U_7X_{13}	65																															
$U X_2$	66.6				436 hex	2385 C15	1590 hex						3500 C1	1700 C32	1450				L1	C38	1355 C38	1010 C38	2700 C1	1850 C23			α,β,γ tetr					
U_4X_9	69.2																															
U_3X_8	72.7																															
$U X_3$	75				417	1350 L1₂	1300 hex	L1₂	L1₂				1570 L1₂	1475 L1₂	1350 L1₂	1220 L1₂							1000 orth	mono								
U_2X_7	77.8																															
$U X_4$	80				383	2495	705 orth																									
$U X_9$	90			945																												
$U X_{11}$	91.9			472																												
$U X_{12}$	92.5				2235 fcc																											
$U X_{13}$	93		2000																													

Figure 11. Uranium-normal element compound diagram. Taken from Reference 8.

compounds of uranium with normal elements.[8] The f-shell is so small that f-electrons do not contribute to bonding;[8] therefore, electrons transferred to the f-level become nonactive when the f-shell acts as an electron sink. At the melting point, uranium atoms have slightly more than 1.5 bonding d-electrons and about 1.2 outer bonding electrons; as detailed by the author,[8] this is in agreement with the melting point, neighboring element properties, alloy behavior and compound formation. Addition of carbon to form a monocarbide will almost double the number of electrons in bonding p- and d- levels while 1.3 electrons are lost to the f-electron sink, yielding a monocarbide and a correctly calculated melting point. In the nitride, one more electron is available. Most of

this electron goes into the f-shell, leaving a small addition to
the bonding levels; this increases the melting point by 150°K. The
uranium monoxide melting point is not known, but it has been extrapolated from the carbide, nitride, and sulfide; as predicted from
the bonding pattern presented, it must be higher than that of the
nitride.

Due to the filling of the f-shell the actinides are placed
between the third and fourth columns. Electron additions will increase the stability of the monocarbide phase by only a fraction of
the stability increase due to the addition of an electron in transition metals because the f-electron sink absorbs a great part of the
added electrons. The varying stability of the uranium alloys and
the stability maximum for the carbide can be seen from the compound
diagram, Figure 11.

As stated above, composite stability diagrams can be drawn
(Brewer[5]) for inter-transition metal systems in which phases indicated as I, II, and III types are correlated with outer electron
concentrations of 1, 2, and 3, respectively, as proposed by the
author.[4] The same is true for alloys of the fourth and fifth column
transition metals with normal elements from the second period.
Brewer type diagrams of these alloys are presented in Figures 12
and 13. The fourth column elements exhibit the hexagonal close
packed structure (type II) at low temperatures. This phase is
stabilized by addition of electrons, due to the increased number of
bonding d-electrons. However, the range of existence is determined
by the stability of neighboring phases, in this case the face

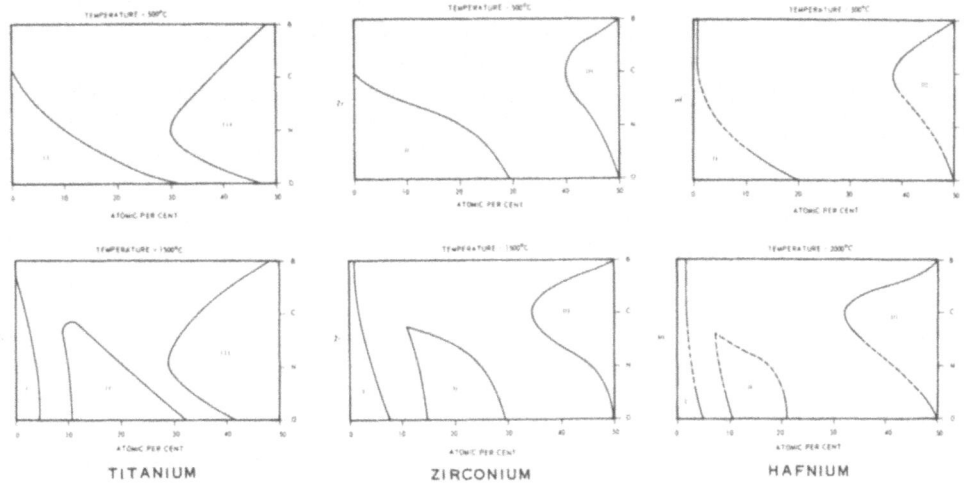

Figure 12. Brewer type diagrams for combinations of fourth column
transition metals with second period elements.

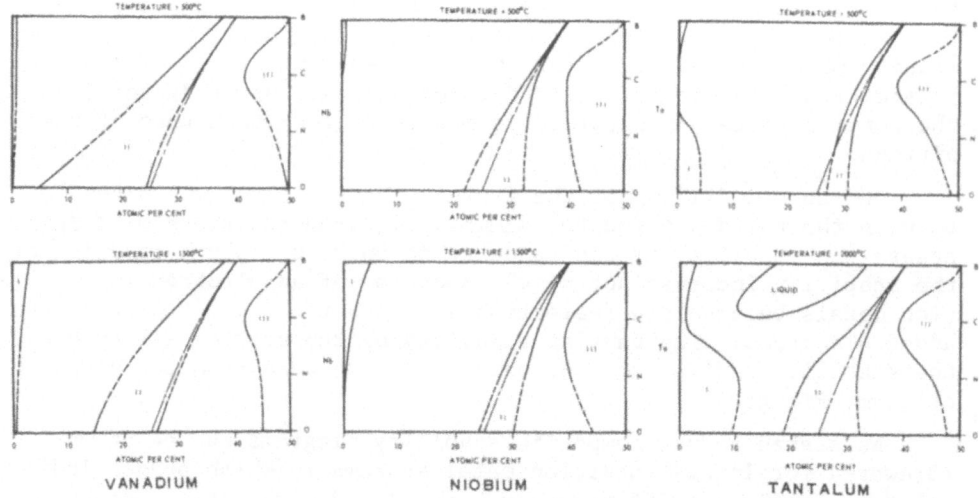

VANADIUM   NIOBIUM   TANTALUM

Figure 13. Brewer type diagrams for combinations of fifth column transition metals with second period elements. Calculated compositions for type II phases are shown by dot-dash lines.

centered cubic (type III) phase indicated at the right side of the diagrams. As discussed above, the carbide is the most stable compound; the nitride and the oxide become increasingly less stable due to internal pairing of d-electrons. Because of this destabilization of the type III phase with increasing number of electrons, the type II phase region expands as one passes from the top to the bottom of the diagrams in Figure 12. At high temperatures the pure metals assume the one electron structure, type I. The ranges of existence of the type I and type II phases are both determined by the stability decrease of the type III phase with an increasing number of electrons.

In the corresponding diagram for fifth column elements (Figure 13) all three phases are present in the sequence expected for increasing number of electrons at both low and high temperatures. The larger number of electrons causes the d-shell to be already approximately half filled in the hexagonal M_2X phases. Compositions with the electron distribution d^5sp are indicated by a dot-dash line. This electron distribution and the theoretical distribution d^6sp^2 in the type III phase agree with the composition ranges in the phase diagrams but they do not agree with the melting points shown in Figure 7, unless it is assumed that the specimens used for the melting point determinations had a high concentration of interstitial vacancies. The closeness of the melting points of ZrC and NbC and of HfC and TaC indicates this possibility for the fifth column carbides (see also Figure 6).

CONCLUSION

Mixing of atoms with the same type of electron distribution will generally not lead to electronic changes which cause compound formation. This has been shown for systems of copper and the nearest transition metal neighbors, titanium and the nearest transitional metal neighbors, and aluminum and thirteenth and fourteenth column elements. Mixing of atoms with different electron distributions will often cause electronic changes in either of the atoms whereby new and more stable bonding patterns may develop, as found in copper-early transition metal systems, titanium-normal metal systems, titanium-late transition metal or aluminum-transition metal systems.

Compound diagrams have been given in which each type of compound corresponds to a horizontal line which can often be traced through several compound diagrams. Such an arrangement of compounds with the same structure and bonding pattern generally exhibits a stability maximum. For the case of transition metal-second period element compounds, a bonding pattern has been presented which explains the main features of these very hard, high melting phases.

REFERENCES

1. G. N. Lewis, J. Am. Chem. Soc. 38, 762 (1916); J. Chem. Phys. 1, 17 (1933); Die Valenz und der Bau der Atome and Moleküle, Vieweg and Sohn A/G, Braunschweig (1927).

2. Linus Pauling, The Nature of the Chemical Bond, Cornell University Press, Ithaca, N. Y. (1960).

3. J. W. Linnett, Wave Mechanics and Valency, Methuen, London (1966).

4. N. Engel, "Metals and Alloys as Electron Concentration Phases," Kemisk Maanedsblad 30, 53, 97 (1949). Available in English translation from the author.

5. L. Brewer, in High Strength Materials (V. F. Zackay, Ed.), J. Wiley, New York (1965) p. 12.

6. L. Brewer, Science 161, 115 (1968).

7. L. Brewer, in Electronic Structure and Alloy Chemistry of the Transition Elements, P. Beck, Ed., Interscience, New York (1963) p. 221.

8. N. Engel, "Search for Metallic Uranium and Plutonium Containing Fuel Materials," Report to Oak Ridge National Laboratories (in print).

9. W. Hume-Rothery and G. V. Raynor, The Structure of Metals and Alloys, The Institute of Metals, London (1962).

10. M. Hansen, Constitution of Binary Alloys, Second Edition, McGraw-Hill Book Co., New York (1958).

11. R. P. Elliott, Constitution of Binary Alloys, First Supplement, McGraw-Hill Book Co., New York (1965).

12. G. V. Samsonov, Refractory Transition Metal Compounds, Academic Press, New York (1964).

13. F. H. Ellinger, C. C. Land and K. A. Gschneider, Jr., Alloying Behavior of Plutonium, Plutonium Handbook, Ed., O. J. Wick, Gordon and Breach Science Publishers, New York (1967).

14. William Burton Pearson, A Handbook of Lattice Spacings and Structures of Metals and Alloys, Pergamon Press, New York, I (1958), II (1967).

15. W. H. Elliott, "Application of the Electron Concentration Concept to Group IV-B and V-B Element Interstitial Compounds," M.Sc. Thesis, Georgia Institute of Technology, Atlanta, Georgia (1966).

16. N. Engel, Trans. ASM 57, 610 (1964).

EDITOR'S NOTE: The editor is obliged to point out that the validity of a number of the ideas expressed in the preceding article has been the subject of considerable discussion in the literature. This applies especially to the arguments linking the crystal structure and electronic configuration of noble metals and transition metals[1-8] and the arguments on charge transfer in NaTl and related phases (see the following two discussions). References 7 and 8 are in defense of the correlations mentioned above. The reader may wish to consult these references.

1. W. Hume-Rothery, Acta Met. 13, 1039 (1965).

2. W. Hume-Rothery, Acta Met. 15, 567 (1967).

3. W. Hume-Rothery, Progr. in Mat. Science 13(5) (1968).

4. Discussions in: High Strength Materials, V. F. Zackay, Ed., J. Wiley and Sons, New York, Chapter 2, p. 70 (1965).

5. Discussion in: Phase Stability in Metals and Alloys, P. S. Rudman, J. Stringer, and R. I. Jaffee, Eds., McGraw-Hill Book Co., New York, p. 239-249 (1967).

6. W. B. Pearson, "Search for a Unified Crystal Chemistry of Metals", this volume.

7. L. Brewer, Acta Met. 15, 553 (1967).

8. N. N. Engel, Acta Met. 15, 565 (1967).

DISCUSSION

I

"CHARGE TRANSFER" IN INTERMETALLIC COMPOUNDS:

EXAMPLE - THE SODIUM THALLIDE STRUCTURE

Lawrence H. Bennett

Metallurgy Division, Institute for Materials Research

National Bureau of Standards, Gaithersburg, Maryland[*]

Many properties of intermetallic compounds are intermediate between those for an ionic compound and those for a metallic alloy. It has therefore been usual to speak of an ionic bond co-existing with a metallic bond in these compounds.[1-3] Since metals are characterized by conduction ("free") electrons having high mobility with wave functions extending over the entire crystal, it is to be expected that charge neutrality would be maintained on an atomic scale throughout the metal. It is thus puzzling to consider how ions can coexist with "free" electrons.

One of the earliest descriptions incorporating coexisting metallic and ionic bonds was for the NaTl structure.[1,4] In this structure, each of the two components form a diamond lattice, which are interpenetrating. Zintl, in 1932, rationalized this structure by suggesting[4] that the electropositive Na gives up its outer electron to the less electropositive Tl, forming Na^+ and Tl^- ions and giving rise to ionic bonding. Tl has three valence electrons (free atom configuration $6s^2$, 6p) which together with the electron donated by Na forms an sp^3 bond of the type familiar in the diamond structure, e.g. in Si. Zintl recognized a basic problem with this view - the sodium thallide structure is found not only for alloys of alkali metals with Group III elements, but also for alloys with Group II elements such as in LiCd. Here the Cd (with free atom configuration $4d^{10}$, $5s^2$) was assumed to promote a d-electron in order to have sufficient electrons for the sp^3 bonding. Since bright colors (reds, greens, blues) are often associated with unfilled d-bands in ionic materials,[5] the reddish color or LiCd was taken as proof of this point of view.

[*]Mailing address: Washington, D.C. 20234

Though this view of the origin of the Zintl structure persists in reviews,[1-3] it was actually disproven in 1955 by Klemm and Fricke[6] who showed no susceptibility attributable to an unfilled d-shell in LiCd. Further strong evidence against the promotion of a d-electron was found by nuclear magnetic resonance.[7] Taking together the experimental data on magnetic susceptibility and nuclear magnetism, an energy band structure was proposed[7] more in keeping with a modern understanding of metals. The proposed density of electronic states diagram is shown schematically in Fig. 1.

The four electrons per Na,Tl atom pair in NaTl fill 2 bands (to "A"). Actually because non-stoichiometric NaTl is a defect lattice, the Fermi level is depressed to B, and thus NaTl is not an insulator but a semi-metal. The same result is perhaps obtained for stoichiometric NaTl due to a slight overlap to a third band. That NaTl is a semi-metal is shown both by the absence of Pauli susceptibility and the absence of a normal paramagnetic Knight shift. The density of states of LiCd is similar, but now there are only three valence-conduction electrons per Li,Cd pair, hence the Fermi level is at C in Fig. 1. Here there is normal Pauli paramagnetism and a normal paramagnetic Knight shift.

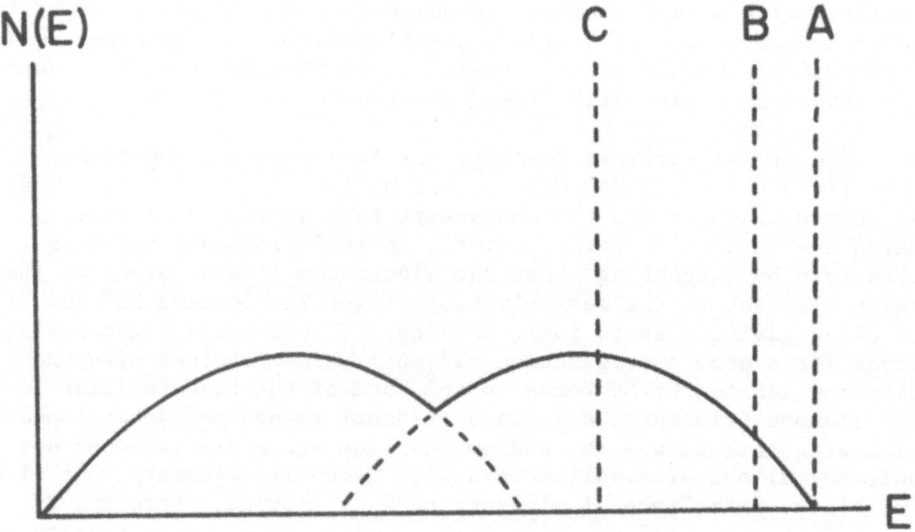

Figure 1.   Highly schematic energy band structure for the sodium thallide structure. NaTl has its Fermi level at B; LiCd has its Fermi level at C.

The lower energy band in Fig. 1 can be considered to consist of two electrons having wave-functions more or less centered on the alkali metal. Note that this would lead to a vanishing Knight shift at the alkali site, just as the alternative view of no valence electrons at the alkali site would. The upper energy band has one (for LiCd) or two (for NaTl) electrons with wave-functions centered on the heavy metal. Thus it would be possible to translate the energy band language of the physicist to the valence-bond language of the chemist by saying that the Tl or Cd have "transferred" one electron to the alkali metal. This charge transfer is in the opposite direction to the Zintl suggestion and is necessary to maintain approximate charge neutrality which would otherwise be upset by the electronegativity difference.[8] Pauling has emphasized the necessity of charge transfer to neutralize the charge of the atoms in metals, but he retains the nomenclature of partial ionic character of the bonds, which are derived from considerations of bond lengths and bond strengths.[8] More recently, Stern[9] has discussed the concept of "charging" in an alloy; he points out that the degree of charging in a given case is estimated by assuming approximate charge neutrality.

The color of LiCd in the proposed band picture is quite analagous to the color of Cu. Cd with one electron removed may be similar to Cu which has a filled d-band located about 2 eV below the Fermi surface. Neither the color of LiCd nor of Cu can be considered evidence of an unfilled d-band. In this connection it might be noted that the metals having unfilled d-bands, i.e. the transition elements, are not red!

In summary, metals are not ionic, and the puzzle of NaTl resolves itself into the problem of calculating the energy band structure and wave functions. A detailed calculation is well within the present ability of an A.P.W. calculation, but the complexity of the crystal structure would necessitate expensive computer time. Thus the highly schematic density of states shown in Fig. 1 will have to suffice for the present.

REFERENCES

1. N. N. Engel, preceding paper.

2. For example, L. Brewer, in High Strength Materials, V. F. Zackay, Ed. (John Wiley and Sons, New York, 1964), p. 37.

3. P. W. Anderson, Concepts in Solids (W. A. Benjamin, Inc., New York, 1964), p. 6.

4. E. Zintl and G. Bauer, Z. Phys. Chem. B20, 245 (1932).

5. See, for example, H. B. Gray, <u>Electrons and Chemical Bonding</u> (W. A. Benjamin, Inc., New York, 1965), p. 183.

6. W. Klemm and H. Fricke, Z. Anorg. Chem. <u>282</u>, 161 (1955).

7. L. H. Bennett, Phys. Rev. <u>150</u>, 418 (1966).

8. L. Pauling, <u>The Nature of the Chemical Bond</u> (Cornell Univ. Press, Ithaca, 1960), 3rd-ed., p. 431 ff.

9. E. A. Stern, in <u>Energy Bands in Metals and Alloys</u>, L. H. Bennett and J. Waber, Eds. (Gordon and Breach, New York, 1968)

THE NaTl STRUCTURE

W.B. Pearson

Division of Pure Physics, National Research Council of

Canada, Ottawa

The character of phases with the NaTl type of structure was questioned on several occasions with the suggestion that they are 'ionic', as first proposed by Zintl and Woltersdorf[1] and discussed by Laves[2]. The idea of charge transfer so that Al^-, Ga^-, In^-, or Tl^- with 4 electrons can form a strong diamond-type network[1] (apart from the difficulty of having also Group II components Zn and Cd) is objectionable in postulating strong covalent bonding between like atoms that are negatively charged. The situation here is quite different from that in phases with, say, the Li_3Bi, Na_3As or $Tl^ITl^{III}Se_2$ type structures where ionic charge transfer from one of the Li or Na atoms, or the Tl^I atom, has been invoked to establish, between the remaining unlike atoms, covalent bonding that gives filled valence subshells on the anions.

The Laves[2]-Parthé[3] spacefilling diagram (Fig. 1) shows that when $r_A/r_B < 1$, points for the actual phases with the NaTl structure lie along the line for A-A contacts rather than that for B-B contacts as expected. The near-neighbour diagram (Fig. 2, see Pearson[4] for description of symbols) shows this information in another form. Here A and B are chosen so that $D_A/D_B(r_A/r_B)$ is always greater than unity. This diagram (based on covalent sizes for CN 8 for the component atoms) emphasizes the well-known fact that, in phases with this structure with $D_A/D_B > 1$, both A-A and A-B contacts are compressed until B-B contacts are established. This observation does not result from using covalent sizes instead of allowing for "ionicity", A^+B^-, in phases where the alkali metal is the A component. Such charge transfer would reduce the size of the A atom, D_A, by δ_A and increase that of the B atom by δ_B, so that $(D_A - d_A - \delta_A)/D_B + \delta_B$ values should have been plotted against

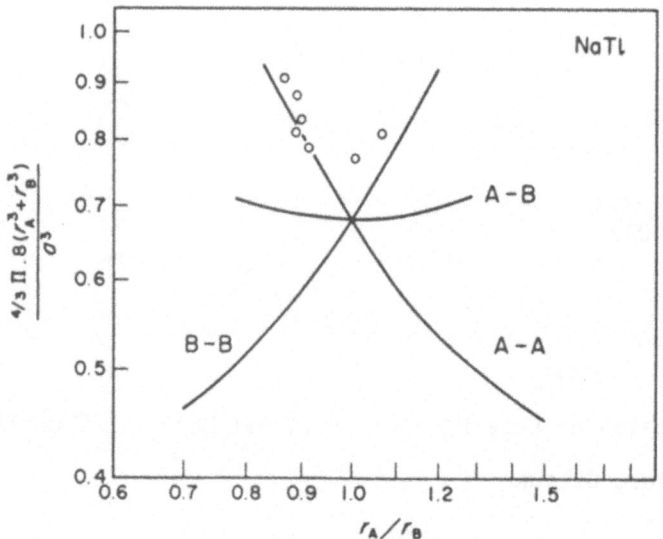

Fig. 1
Laves-Parthé[2,3] type spacefilling diagram for the NaTl structure.
Points represent actual phases with the structure.

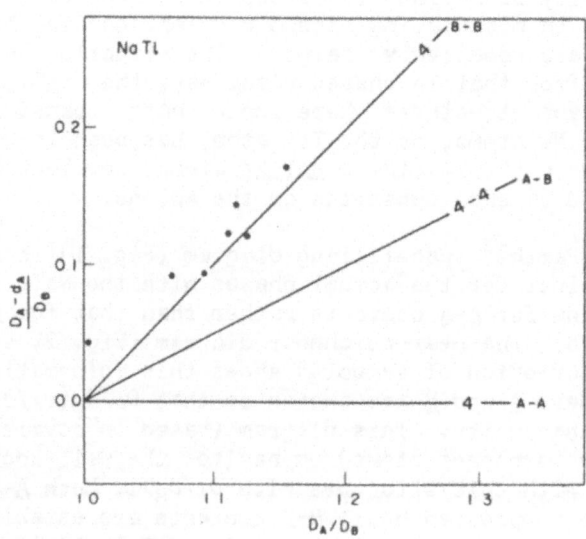

Fig. 2
Near-neighbour diagram[4] for the NaTl structure. Points represent
actual phases with the structure.

$(D_A-\delta_A)/D_B+\delta_B$ for these phases in Fig. 2. It can be demonstrated readily that these relative changes in atomic diameters can only move the points for the phases along the line for B-B contact (δ_B=0) or into the region above it (δ_B>0), confirming more strongly the conclusion that B-B, A-B and A-A contacts are all formed.

It is notable that in InLi (r_{In}/r_{Li}=1.067), where Li is the smaller B component, Li-Li (B-B) contacts are also established (Fig. 2). Hence it can scarcely be argued that the A components are compressed in order to make a strong diamond network of the Group III metals, since this compression occurs no matter whether a Group I or III element is the B component! Indeed, the following arguments indicate that the NaTl structure is a normal metallic type which is favoured over the CsCl structure when the component atoms have certain properties: The site occupation in the CsCl and NaTl structures is identical with interpenetrating cubes of atoms giving 8-8 nearest neighbour coordination in each case. This is achieved by 8-8 A-B contacts in the CsCl structure and these control the structural dimensions, whereas in the NaTl structure it is achieved by 4-4 A-B, 4 A-A and 4 B-B contacts. Therefore, if the compressibility of the A atoms (r_A/r_B>1) is such as to allow A-B and B-B contacts to be established, then 8-8 CN is achieved at a smaller overall atomic volume relative to the component atoms in the NaTl structure than in the CsCl structure. Other things being equal, a lower free energy may therefore be expected for a phase in the NaTl than the CsCl structure, provided that the A component is readily compressible and the radius ratio does not depart much from unity. These conditions are apparently satisfied in NaTl phases with the alkali metals as the A component, and radius ratios about 1.1. In the two NaTl phases where the alkali metal is the smaller B component, the radius ratio is closer to unity (1.004 and 1.067) requiring relatively small compression of the other component. LiTl does not have the NaTl structure because with Li as the B component, r_{Tl}/r_{Li}=1.10 is already too large for Tl to be compressed sufficiently for 4 Li-Li contacts to form. Potassium compounds do not have the NaTl structure since even for the largest Group III element, thallium, the radius ratio r_K/r_{Tl}=1.397, which is much too large to allow ready compression of the K atoms to permit the formation of 4 Tl-Tl contacts. Equiatomic phases of the alkali metals with Group IV elements generally have complex structures in which valence rules are satisfied[5].

The circumstances which we have discussed, suggest that phases with the NaTl structure are neither 'ionic', nor dominated mainly by strong diamond type covalent bonds, but characteristically metallic phases adopting 8-8 coordination like the metallic phases with the CsCl structure, but achieving this with a lower free energy when two favourable circumstances arise simultaneously: i) the larger A component is highly compressible and ii) the radius ratio is not much larger than unity.

DISCUSSION

REFERENCES

[1] E. Zintl and G. Woltersdorf, Z. Elektrochem., 41, 876 (1935).

[2] F. Laves, Theory of Alloy Phases (Amer. Soc. Met., Cleveland, Ohio, 1956), p. 124.

[3] E. Parthé, Z. Kristallogr., 115, 52 (1961).

[4] W.B. Pearson, this volume.

[5] W.B. Pearson, Acta Cryst., 17, 1 (1964).

THE CONCEPT OF A PARTIAL ELECTRON CONCENTRATION VALUE AND ITS APPLICATION TO PROBLEMS IN CRYSTAL CHEMISTRY

Erwin Parthé

School of Metallurgy and Materials Science, and
Laboratory for Research on the Structure of Matter,
University of Pennsylvania, Philadelphia, Pennsylvania

INTRODUCTION AND BASIC DEFINITIONS

Hume-Rothery in 1926 was the first to point out that in certain intermetallic systems there exists a correlation between composition, structure and electron/atom ratio of the intermetallic phases. Since that time many groups of compounds have become known where a connection between the stability range and the number of participating electrons has been assumed. To mention only a few one could list the group of transition metal disilicides (Nowotny, 1963), the group of ternary σ and related complex phases with high atom coordination (Nevitt, 1963), the group of shift structures ("Verwerfungsstrukturen") based on the Cu_3Au type (Schubert, 1964), and the group of the normal and defect tetrahedral structures (Parthé, 1964). In all these structures the parameter of interest is the so called valence electron concentration VEC which for a compound of composition $A_m B_n$ is given by

$$VEC = \frac{m \cdot e_A + n \cdot e_B}{m + n} \qquad (1)$$

where m and n are composition parameters and e_A and e_B the valence electron number of elements A and B respectively.

For the alkali and alkaline earth elements and all elements to the right of the Ni group the valence electron numbers are gener-

49

ally assumed to be identical to the group numbers of the periodic system. The valence electron numbers for the transition elements are open to controversy, since they may be different for different families of compounds; occasionally they have been assumed to be zero (Ekmann, 1931), frequently they are also approximated by the group number.

I wish to show here that there exist families of compounds for which the interesting parameter is not the VEC value but instead a partial VEC value. If one separates VEC into parts according to

$$VEC = \frac{m}{2(m+n)} \cdot \frac{m \cdot e_A + n \cdot e_B}{m} + \frac{n}{2(m+n)} \cdot \frac{m \cdot e_A + n \cdot e_B}{n} \tag{2}$$

it is possible to define

$$(VEC)_A = \frac{m \cdot e_A + n \cdot e_B}{m} \tag{3}$$

and

$$(VEC)_B = \frac{m \cdot e_A + n \cdot e_B}{n} \tag{4}$$

as partial VEC values.

The concept of a partial VEC value has been applied in the past to a number of different structural problems, however it was not always realized or stated that the quantity used was actually a partial VEC value. I wish to discuss here three applications of the partial VEC values which involve the same kind of electron algebra, but have different reasons why partial VEC values can be used profitably.

1) THE VALENCE COMPOUNDS AND THE CONCEPT OF THE PARTIAL ELECTRON CONCENTRATION

Valence compounds are compounds where the individual atoms in the structure obtain filled valence shells either by accepting, sharing or donating electrons. The term valence compounds was at first applied only to the normal valence compounds where the cations have the exact number of valence electrons to complete the octet shell of every individual anion present. In this case the total number of valence electrons per anion, the electron/anion ratio, is eight or using our notation $(VEC)_{anion} = 8$. Later the term valence compounds was extended to include also the poly-anionic valence compounds with $(VEC)_{anion} < 8$ and the polycation-ic valence compounds with $(VEC)_{anion} > 8$ which then permitted the inclusion of compounds having anion-anion or cation-cation bonds (Pearson, 1964). The whole concept of valence compounds has been treated recently by a number of authors in great detail (Hulliger & Mooser, 1965; Parthé, 1967; Goryunova & Parthé, 1967; and others) and it would serve no purpose to discuss it here. I wish, however, to state briefly the equation for valence com-pounds applicable to compounds having only two-electron bonds. If the composition of a valence compound is $C_m A_n$, the general equation for valence compounds (in the version given by Parthé, 1967) can be written as

$$(VEC)_{anion} = 8 + \frac{m}{n} \cdot CC - AA \qquad (5)$$

where CC is the average number of cation-cation bonds per cation (or, in a different interpretation, the average number of electrons per cation which are not involved in cation-anion bonds)
and AA is the average number of anion-anion bonds.

Without going into the details of the application of equation (5) we find that the partial VEC value relates to the number of cation-cation and anion-anion bonds in the structure. Provided certain conditions are fulfilled, compounds having the same partial VEC value will have the same structural features. For example, in the simplest case, polyanionic valence compounds will have anion-anion dumb-bells if $(VEC)_{anion} = 7$ or anion-anion chains if $(VEC)_{anion} = 6$. It might be mentioned here that for certain struc-ture families, for example, the tetrahedral structure compounds, the total VEC value and the partial VEC value can both be used to obtain complementary results concerning the structural features of the compounds. For more details on this and the proper appli-cation of equation (5) the above cited references should be consulted.

(2) PARTIAL VEC VALUES APPLIED TO TERNARY β- Mn PHASES

In a recent study on the extension of β-Mn structure phases in ternary systems containing Mn, another transition element and Si or Ge or Sn by Bardos, Malik, Spiegel & Beck (1966) it

was noted that the β-Mn phase field extension in alloys containing
silicon is quite different from the phase field extension for alloys
containing Ge or Sn. Particularly in certain ternary systems the
β-Mn phase field is extended with Si, while the corresponding
Ge and Sn phase fields are reduced. In other ternary systems it
is just the reverse. The explanation put forward by Beck and his
co-workers is as follows: silicon, germanium and tin all provide
their valence electrons to the pool; however, in Si the empty d
band is sufficiently broadened and lowered to overlap the Fermi
surface, allowing d states associated with silicon atoms to be
occupied by itinerant d electrons. In these compounds Si behaves
like a transition element. On the other hand in Ge and Sn the d
bands are occupied. The wave functions of the itinerant d electrons must be orthogonal with respect to the wave funcions of the
3d electrons of Ge or 4d electrons of Sn. This condition results
in a high repulsive potential which effectively excludes the itinerant d electrons from the d states of the germanium and tin atoms.

The condition for the occurrence of the β-Mn structure is the
availability of 7 valence electrons for every element in the compound which permits its d states to be occupied by itinerant d
electrons. As Si admits itinerant d electrons the composition of
Si containing β-Mn phases will correspond to VEC = 7. However,
Ge and Sn do not admit d electrons and for Ge- and Sn- containing
phases the condition will be $(VEC)_d = 7$ if we use $(VEC)_d$ as a
shorthand notation for a partial VEC value which indicates the
total number of valence electrons divided by the number of atoms
which admit in their d states itinerant d electrons. Beck and co-
workers have used these ideas to plot lines in the different ternary
diagrams either for VEC = 7 or for $(VEC)_d = 7$ which agree well
with the experimentally observed ternary β-Mn phase field extensions. There is however a small complication because the
valence electron contributions of Si and Ge or Sn atoms are not
equal. In order to obtain good agreement it had to be assumed
that Si contributes 4 electrons, while Ge and Sn provide only 2
electrons. For details the cited paper should be studied.

(3) STUFFED SKELETON STRUCTURES WITH CONSTANT PARTIAL VEC VALUE

The concept of a partial VEC value can be applied to certain
cases of "skeleton" structures where certain atoms form a rigid
framework where the holes are (partially) filled with other atoms.
It is assumed that a particular number of valence electrons is

necessary for the formation of a particular skeleton. If the skeleton forming elements don't have enough electrons some electron donors have to be inserted into the holes of the skeleton to increase the electron supply in order to reach the required value. It is clear that the number of the filler atoms depends not only on their electron contribution but also on the valence electron contribution of the skeleton forming elements. With different skeleton formers the overall composition of compounds with a particular skeleton may be different. A constant partial VEC value assures that in different compounds all having the same skeleton the number of electrons per skeleton site is constant.

We shall discuss the application of the partial VEC values for three groups of skeleton structures. In the chemical formulae given below the skeleton forming elements will be placed in square brackets. The symbol $(VEC)_{[\ ]}$ will denote the particular electron number which is required for the formation of a particular skeleton.

Stuffed SiO_2 structures with $(VEC)_{[\ ]} = 5.33$

Numerous examples of skeleton structures are known for minerals and other inorganic compounds. Many aluminosilicates, particularly the feldspars are based on a complete framework of linked tetrahedra similar to those in SiO_2. If some of the Si atoms (with 4 valence electrons) are replaced by Al atoms (with 3 valence electrons) some electron donors must be inserted into the holes of the framework to compensate for the valence electron loss which occurred with the Si-Al exchange. Examples of compounds which have a SiO_2 framework are given in Table 1.

Composition of compounds with stuffed SiO_2 framework	Type of SiO_2 framework formed by atoms in the bracket	References
$Li^{1+}\left[Al^{1-}SiO_4\right]$	high - quartz	Hahn, Behruzi & Lohre (1966)
$\gamma\text{-}Li^{1+}\left[Al^{1-}O_2\right]$	low-cristobalite	Bertaut, Delapalme, Bassi, Durif-Varambon & Joubert (1965); Marezio (1965)
$K^{1+}\left[Al^{1-}O_2\right]$	high-cristobalite	Barth (1935)
$K^{1+}\left[Fe^{1-}S_2\right]$	fibrous quartz; identical to SiS_2	Boon & MacGillavry (1942)

Table 1: Examples of compounds with stuffed SiO_2 structures.

In all these compounds with a SiO_2 framework 5.33 electrons are necessary per framework site and we can say that for these compounds the partial VEC value $(VEC)_{[\;]} = 5.33$. In this particular case the simple rule of electroneutrality leads immediately to the proper chemical formulae.

Problems with filled tetrahedral structures having $(VEC)_{[\;]} = 4$.

The problem becomes more interesting if we consider compounds which are not normal valence compounds. The first group of compounds for which a skeleton-filler scheme has been proposed are filled tetrahedral structures which are characterized by a diamond structure skeleton with additional atoms in the holes. According to Zintl and Woltersdorf (1935) the bonding in LiAl can be interpreted in a similar fashion as for stuffed SiO_2 structures, namely partly ionic, and partly covalent: the Li atoms donate one electron each to the Al atoms which then have the 4 valence electrons necessary to form four tetrahedral sp^3 orbitals. In Fig. 1, one may compare the diamond structure,

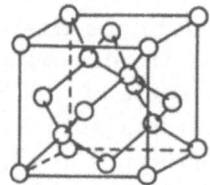   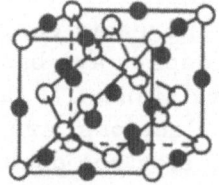

C, Si, Ge, α-Sn   LiAl, LiGa, LiIn, NaIn, NaTl
                              also LiZn, LiCd

Fig. 1: The diamond and a "stuffed" diamond structure.

formed only with elements having already 4 valence electrons, with the filled diamond structure observed with $Li^{1+}[Al^{1-}]$, $Li^{1+}[Ga^{1-}]$, $Li^{1+}[In^{1-}]$, $Na^{1+}[In^{1-}]$ and $Na^{1+}[Tl^{1-}]$. Unfortunately, at our present state of knowledge the interpretation of the filled tetrahedral structures given by Zintl and Woltersdorf cannot be considered valid anymore. The particular reason is the occurrence of the LiAl structure with compounds like LiZn and LiCd where the diamond framework forming elements Zn or Cd cannot obtain the necessary four valence electrons to form the sp^3 hybrids. This problem is still unsolved.

Stuffed white tin structures with $(VEC)_{[\ ]} = 4$

Recently a family of transition metal compounds has become known for which--as I believe--the concept of a partial VEC is justly applicable. Most of these new compounds have been struc- turally analyzed in Vienna by Nowotny and his co-workers (Völlenkle, Preisinger, Nowotny & Wittmann, 1967; Völlenkle, Wittmann & Nowotny, 1966). The structure of one compound, $Rh_{17}Ge_{22}$, was solved here in Philadelphia (Jeitschko & Parthé, 1967). The compounds belonging to this family are transition metal silicides, germanides, stannides and gallides of varying composition, all of which are characterized by a transition metal skeleton similar to the white tin structure. This frame- work has the rather unique property of permitting the insertion of filler atoms in a regular fashion for different amounts of filler atoms. Without wanting to go into the details, the filler atoms may be thought to be arranged in pairs. These pairs are now inserted into the white tin framework, one above the other, only rotated with respect to each other. If many filler atoms are necessary the filler atom pairs are close together, if less filler atoms are needed the filler atom pairs are wider spaced.

The crystal structures of white tin, Ir_4Ge_5, Ru_2Sn_3 and Tc_4Si_7 are shown in Fig. 2. Similarly as in Fig. 1, the

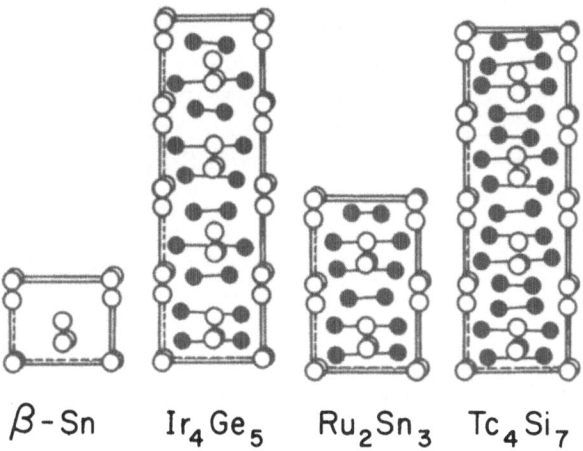

$$\beta\text{-Sn} \qquad Ir_4Ge_5 \qquad Ru_2Sn_3 \quad Tc_4Si_7$$

Fig. 2. The white tin and three "stuffed" white tin structures. The Si atom arrangement in Tc_4Si_7 and the Ge atom arrange- ment in Ir_4Ge_5 are only schematic. The filler atom pairs are properly spaced but actually may be rotated from the posi- tions shown in the figure.

framework is indicated with open circles, the filler atoms with smaller black circles. It is interesting to note that the stuffed white tin structures have tetragonal unit cells which are multiples of the white tin structure type cell. The reason for this lies in the particular way the filler atoms are inserted. In general, there is a relation between unit cell dimension and composition. If the composition is T_nX_m where T indicates a transition element and X either Si, Ge, Sn or Ga, the super cell will contain n identical white-tin cells stacked one on top of another. In this framework are inserted 2 m equally spaced filler atom pairs.

At present we know 28 binary and ternary compounds which have tetragonal stuffed white tin structures. Their compositions are given in the third column of Table 2. There are known also a number of orthorhombic phases which have stuffed white tin structures. However, as we don't know the electronic influence on the distortion, all these orthorhombic compounds shall be omitted from this discussion. We note from the formulae given in Table 2 that the ratio of T to X atoms can be quite complicated and (except for two pairs) all given over-all-compositions and crystal structures are different except that the latter always have a white tin skeleton of the transition metal atoms.

Let us assume that to form a white tin skeleton we need a certain number of electrons, let us say ξ or, expressed differently, $(VEC)_{[\ ]} = \xi$. If a transition element does not have ξ, but only $\xi - y$ electrons, then y electrons would have to be provided by the filler atoms. If we make the further assumption that the stuffed atoms contribute always all their valence electrons to the transition metal atoms, that means Si, Ge and Sn provide 4, Ga 3 and As 5 electrons, we can calculate how many electrons have been accepted by each transition element. These values are given in the fourth column of Table 2. We note that for all stuffed white tin structure compounds containing a particular transition element the number of the accepted electrons per transition metal atom is approximately constant. The electron contribution of the transition elements in these compounds is then ξ minus the values of column 4.

The complicated chemical formulae of the stuffed white tin structures can obviously be related to the unique property of the white tin skeleton of permitting the insertion of different quantities of filler atom pairs. Slight changes in the electron contribution of the transition elements can be compensated by changing

Group number of transition element	Chemical symbol of transition element	Observed compositions of compounds having stuffed white tin structures	Electrons per transition element accepted from the stuffed elements	Expected electron transfer to the transition elements assuming that the latter obtain filled d bands and 4 electrons for valence band	Difference of observed value minus expected value	Remarks
8_2A	Ir	Ir_4Ge_5	5.00 ·	5.0	±0.0	Calculated compositions
		$Ir_{17}(Ga_{0.15}Ge_{0.85})_{22}$	4.98	5.0	-0.02	$Ir_{17}(Ga_{0.137}Ge_{0.863})_{22}$
		$Ir_{11}(Ga_{0.35}Ge_{0.65})_{15}$	4.98	5.0	-0.02	$Ir_{11}(Ga_{0.333}Ge_{0.666})_{15}$
		$Ir_{19}(Ga_{0.80}Ge_{0.20})_{30}$	5.05	5.0	+0.05	$Ir_{19}(Ga_{0.833}Ge_{0.166})_{30}$
		Ir_3Ga_5	5.00	5.0	±0.0	
	Rh	$Rh_{17}Ge_{22}$	5.18	5.0	+0.18	
		$Rh_{43}(Ga_{0.10}Ge_{0.90})_{57}$	5.17	5.0	+0.17	
		$Rh_{23}(Ga_{0.25}Ge_{0.75})_{31}$	5.05	5.0	+0.05	
		$Rh_{12}(Ga_{0.35}Ge_{0.65})_{17}$	5.17	5.0	+0.17	
		$Rh_{39}(Ga_{0.50}Ge_{0.50})_{58}$	5.20	5.0	+0.20	
		$Rh_{43}(Ga_{0.75}Ge_{0.25})_{69}$	5.21	5.0	+0.21	
		$Rh_{10}Ga_{17}$	5.10	5.0	+0.10	
8_1A	Ru	Ru_2Sn_3	6.00	6.0	±0.0	
		$Ru_{69}(Ga_{0.05}Ge_{0.95})_{104}$	5.95	6.0	-0.05	
		$Ru_{11}(Ga_{0.15}Ge_{0.85})_{17}$	5.95	6.0	-0.05	
		$Ru_{23}(Ga_{0.25}Ge_{0.75})_{36}$	5.87	6.0	-0.13	
		$Ru_{19}(Ga_{0.35}Ge_{0.65})_{31}$	5.95	6.0	-0.05	
		$Ru_{13}(Ga_{0.50}Ge_{0.50})_{22}$	5.92	6.0	-0.08	
		$Ru_{23}(Ga_{0.75}Ge_{0.25})_{41}$	5.79	6.0	-0.21	Shift in composition
7A	Tc	Tc_4Si_7	7.00	7.0	±0.0	
	Mn	$Mn_{11}Si_{19}$	6.91	7.0	-0.09	
		$Mn_{26}Si_{45}$	6.92	7.0	-0.08	
		$Mn_{15}Si_{26}$	6.93	7.0	-0.07	
		$Mn_{27}Si_{47}$	6.96	7.0	-0.04	
6A	Mo	$Mo_{13}Ge_{23}$	7.08	8.0	-0.92	
		Mo_9Ge_{16}	7.11	8.0	-0.89	
		$Mo_{23}Ge_{41}$	7.13	8.0	-0.87	
	Cr	$Cr_{11}Ge_{19}$	6.91	8.0	-1.09	
		$Cr_{10}(Ge_{0.936}As_{0.063})_{17}$	6.91	8.0	-1.09	
5A	V	$V_{17}Ge_{31}$	7.29	9.0	-1.71	

For a summary of older literature references on the structure data see: Jeitschko & Parthé (1967) and Völlenkle, Preisinger, Nowotny & Wittmann (1967)

New references: $Rh_{10}Ga_{17}$, Ir_3Ga_5: Völlenkle, Wittmann & Nowotny (1967)

$Mn_{15}Si_{26}$: Knott, Mueller & Heaton (1967) also Flieher, Völlenkle & Nowotny (1967)

$Cr_{10}(Ge_{0.936}As_{0.063})_{17}$: Boller, Wolfsgruber & Nowotny (1967)

$Mn_{26}Si_{45}$, $Mn_{27}Si_{47}$: Flieher, Völlenkle & Nowotny (1967)

Ir_4Ge_5: Panday, Singh & Schubert (1967)

Table 2: List of tetragonal stuffed white tin structures

slightly the concentration of the electron donors. This involves
a change in the spacing of the filler atom pairs which can be
determined very accurately from the X-ray diffraction patterns
and in turn permits the fixing of the composition of these compounds with a precision which can not be obtained by chemical
analysis.

Let us now investigate in more detail the slight changes in the
electron contribution of one transition metal in different stuffed
white tin structures. Some of these variations are real, others
not necessarily. Iridium in Ir_4Ge_5 and Ir_3Ga_5 contributes
exactly ξ-5 electrons, however there are reported deviations
for the ternary alloys. While the X-ray diffraction experiments
allow to determine the ratio Ir/(Ga + Ge) with high precision
due to the formation of independent transition metal and filler
atom substructures, it is experimentally difficult to control
exactly the Ga/Ge ratio in these ternary compounds. It might
very well be that very slight shifts in the Ga/Ge ratio had
occurred during the preparation. To demonstrate this point in
the last column of Table 2 are given the compositions of
Ir-Ga-Ge alloys having the proper Ga/Ge ratio such that the
valence electrons provided by them permit each Ir atom to absorb
exactly 5 electrons. A comparison with the reported compositions given in column 3 shows that the differences in composition
are very small.

Variations in the electron contribution occur with the molybdenum germanides: $Mo_{13}Ge_{23}$, Mo_9Ge_{16} and $Mo_{23}Ge_{41}$. The
difference in the electron contribution amounts however only to
5/100 of an electron per Mo atom. This minute change in the
electron contribution could be caused easily by a variation of the
reaction temperature while forming the compound. However,
there are no detailed experimental data available to support or
to refute this hypothesis. A slightly larger variation of 7/100 of
an electron is observed for Rh in $Rh_{17}Ge_{22}$ and $Rh_{10}Ga_{17}$.

We can conclude that the differences for the transition elements
are at most 0.1 electrons/atom except for the ternary alloys
where changes in composition could have occurred during preparation.

In Fig. 3 are shown the y and indirectly also the ξ-y values as
function of the group number of the transition elements. The
electron contribution of the transition metals decreases linearly

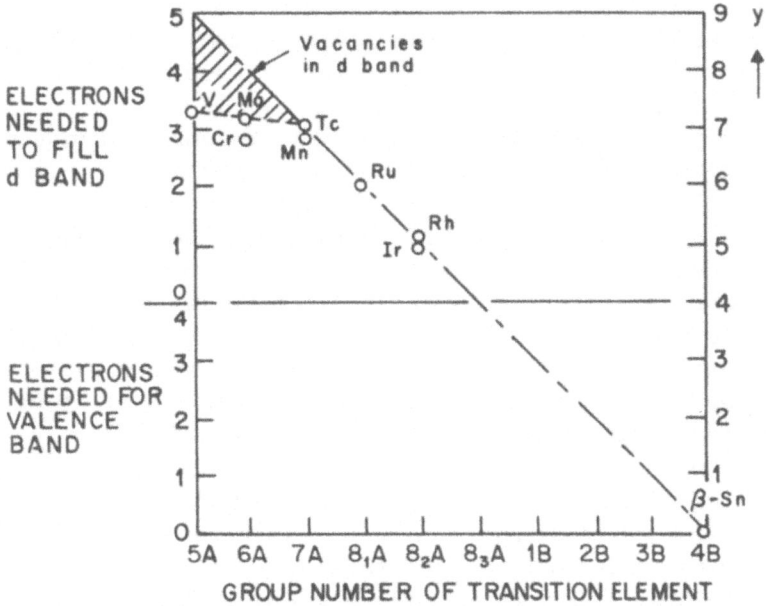

Fig. 3: The number of electrons accepted by the transition
          metal atoms in stuffed white tin structures and their
          possible distribution into d bands and valence bands

from Ir to Tc with the group number, however there is a change
of slope for the line between Mn and V.

So far the occurrence of the filled white tin structure has been
explained assuming a fixed number of ξ electrons which is necessary for a white tin framework to form. As a consequence the
change in the slope from Sn over Ir to Tc as compared to the slope
from Tc to V had to be related to a change in the electron contribution of the transition metal, in the first part being linearly
proportional to the group number, in the second part not. The
experimental data can however be explained by a modified bonding scheme. Let us at first investigate white tin itself. It has a
completely filled d shell and 4 electrons in the valence band.
There is sufficient other experimental evidence available to indicate that a white tin structure can form only if there are 4 or at
least approximately 4 valence electrons. In the stuffed white tin
structures the transition elements which form a white tin framework are assumed to have the same electron configuration as the
tin atoms in white tin, namely a filled d band and 4 electrons in
the valence band. To obtain this electron configuration the

transition elements have to absorb electrons from the filler atoms just as before, but we know now in advance how many it should, if we know the group number of the transition element.

Using this concept we can calculate what the compositions of the stuffed white tin structure compounds should be. To denote the composition we shall use here ionic formulae with square brackets. The large numerals in the brackets indicate the number of the s, p and d electrons (≡ group number except for Ir, Rh) of the elements that form the white tin skeleton, their superscripts denote the number of electrons which are being accepted from the donors. The large numerals outside the brackets stand for the valence electron numbers of the filler atom, the superscripts indicate the number of electrons which they have donated to the framework. The subscripts are composition parameters. Some possible formulae for an electron value of 14 per framework site (or to be more correct a value of 4 provided the d band has been filled with 10 electrons) together with the chemical formulae of known examples are given in Table 3.

Filler atoms provide each 4 electrons	$[10^{4-}]4^{4+}$ (Pt Ge)	$[9^{5-}]_4 4_5^{4+}$ Ir$_4$Ge$_5$	$[8^{6-}]_2 4_3^{4+}$ Ru$_2$Sn$_3$	$[7^{7-}]_4 4_7^{4+}$ Tc$_4$Si$_7$
Filler atoms provide each 3 electrons	$[11^{3-}]3^{3+}$ (AuGa) $\quad$ $[10^{4-}]_3 3_4^{3+}$?	$[9^{5-}]_3 3_5^{3+}$ Ir$_3$Ga$_5$	$[8^{6-}]3_2^{3+}$ (RuGa$_2$)	

Table 3: Some possible composition formulae and known examples of stuffed white tin structures.

In Table 3 some of the chemical formulae are in parentheses. These compounds (PtGe, AuGa with MnP type, RuGa$_2$ with TiSi$_2$ type) have white tin frameworks, however, they are not tetragonal but orthorhombically deformed and will not be further discussed.

In the fifth column of Table 2 are listed the number of electrons the transition metals are expected to absorb according to this hypothesis. In the sixth column are listed the differences between the observed and the expected values, which can be also obtained from Fig. 3. The differences between experimentally found and theoretically expected values are very small for compounds containing transition elements of groups 7, 8_1 and 8_2. The maximum deviation being ± 0.21 electrons per transition metal atom. But for Mo, Cr and V compounds we find that these transition elements accept fewer electrons than expected.

Cr and Mo atoms are supposed to accept 8 electrons and V even 9 electrons in order that each transition metal atom has a filled d band and the 4 necessary electrons in the valence band; however, we can conclude from the experimental data that Cr accepts only 6.9, Mo about 7.1 and V 7.3 electrons. As the white tin skeleton can form only if there are 4 electrons in the valence band we must conclude that the d shells of Cr, Mo and V are not completely occupied, in particular V having only 8.3, Cr 8.9 and Mo 9.1 electrons. It does not seem possible at this time to state why we have these vacancies in the d shells of the transition elements of the 5th and 6th group, but it should be possible to verify their existence by magnetic measurements.

We see thus that it is possible to explain the stuffed white tin structure in two ways. In the first method we assumed that the condition for the occurrence of a white tin structure is given by $(VEC)_{[ ]} = 5$ where we counted all the d electrons of the transition elements and the electrons transferred from the stuffed atoms. In the second method we formulated $(VEC)_{[ ]} = 4$ as the important condition, where we counted only the electrons in the valence band of the transition elements.

In conclusion we can say that the concept of a partial electron concentration has found useful applications for valence compounds, electron compounds with β-Mn structure and stuffed skeleton structures although the reasons for its applicability are quite different. The concept has permitted the explanation of known structure data, and the prediction of structural features of yet unknown compounds. There is no reason to assume that the compound groups discussed are the only ones where the concept can be applied.

This study is a contribution from the Laboratory for Research on the Structure of Matter, University of Pennsylvania supported by the Advanced Research Projects Agency, Office of the Secretary of Defense.

REFERENCES

D. I. Bardos, R. K. Malik, F. X. Spiegel & P. A. Beck, Trans. AIME, 236, 40, (1966).

T. F. W. Barth, J. Chem. Phys. 3, 323, (1935).

E. F. Bertaut, A. Delapalme, G. Bassi, A. Durif-Varambon & J. C. Joubert, Bull. Soc. franc. Minér. Crist. 88, 103, (1965).

H. Boller, H. Wolfsgruber & H. Nowotny, Mh. Chem. 98, 2356, (1967).

J. W. Boon & C. H. MacGillavry, Rec. Trav. Chim. 61, 910, (1942).

W. Ekmann, Z. Phys. Chem. 12B, 57, (1931).

G. Flieher, H. Völlenkle & H. Nowotny, Mh. Chem. 98, 2173, (1967).

N. Goryunova & E. Parthé, Mater. Sci. Eng. 2, 1, (1967).

Th. Hahn, M. Behruzi & G. Lohre, Acta Cryst. 21, part 7, A53, (1966).

F. Hulliger & E. Mooser, Progr. Solid State Chem. 2, 330, (1965).

W. Hume-Rothery, J. Inst. Metals 35, 295, 307, (1926).

W. Jeitschko & E. Parthé, Acta Cryst. 22, 417, (1967).

H. W. Knott, M. H. Mueller & L. Heaton, Acta Cryst. 23, 549, (1967).

M. Marezio, Acta Cryst. 19, 396, (1965).

M. V. Nevitt, in Electronic Structure and Alloy Chemistry of the Transition Elements, P. Beck, Ed. (John Wiley, New York, 1963), p. 101 ff.

H. Nowotny, in Electronic Structure and Alloy Chemistry of the Transition Elements, P. Beck, Ed. (John Wiley, New York, 1963), p. 179 ff.

P. K. Panday, G. S. P. Singh & K. Schubert, Z. Krist. 125, 274, (1967).

E. Parthé, Crystal Chemistry of Tetrahedral Structures (Gordon and Breach, New York, 1964).

E. Parthé, in Intermetallic Compounds, J. Westbrook, Ed. (John Wiley, New York, 1967) p. 180 ff.

W. B. Pearson, Acta Cryst. 17, 1, (1964).

K. Schubert, Kristallstrukturen zweikomponentiger Phasen (Springer, Berlin, 1964) p. 94 f.

H. Völlenkle, A. Preisinger, H. Nowotny & A. Wittmann, Z. Kristallogr. 124, 9, (1967).

H. Völlenkle, A. Wittmann & H. Nowotny, Mh. Chem. 97, 506, (1966).

H. Völlenkle, A. Wittmann & H. Nowotny, Mh. Chem. 98, 176, (1967).

E. Zintl & G. Woltersdorf, Z. Electrochem. 41, 876, (1935).

STRUCTURAL PRINCIPLES OF GIANT CELLS

Sten Samson

Gates and Crellin Laboratories of Chemistry,[*]

California Institute of Technology, Pasadena, California

INTRODUCTION

It is a curious coincidence that the paper by Linus Pauling,[1] which first described the crystal structure of an intermetallic compound, presented simultaneously one of the most complicated structure problems that lay ahead. In this paper Pauling discussed the atomic arrangement of Mg_2Sn, which has the calcium fluoride structure, and also gave a brief account of an X-ray diffraction study of crystals of $NaCd_2$. The diffraction patterns of these crystals were so complicated, however, that it was not then possible to assign indices with certainty to many of the spots. The unit of structure was later[2] reported to be a cube that has an edge length slightly over 30Å and contains about 384 sodium atoms and 768 cadmium atoms. The space group is $Fd3m$ (O_h^7). Thus, we see that the existence of this most complex compound has been known for 45 years, ever since the first structure determination of an intermetallic compound was published.

A compound with structure apparently similar to that of $NaCd_2$ is βMg_2Al_3. Cursory investigations[3,4] of small crystal fragments were found to represent a cubic structure, space group $Fd3m(O_h^7)$, with approximately 1166 atoms per unit cube of edge length $a_0 = 28.22$Å.

Observation of powder diffraction patterns led to the conclusion[5] that βMg_2Al_3 is isomorphous with Cu_4Cd_3. This phase can be

[*]Contribution No. 3585.

obtained only after prolonged annealing and forms through a solid-state reaction of a metastable eutectic mixture of the two phases $CdCu_2$ ($C36$ type[6,7]) and Cd_8Cu_5 ($D8_2$ type[8]), that always seem to precipitate first on solidification of the melt. It is, therefore, difficult to grow crystals of this compound. When single crystals were finally obtained after extremely long periods of annealing, they were found, indeed, to be cubic,[9] but the X-ray diffraction patterns turned out to be drastically different from those of $NaCd_2$ and βMg_2Al_3. A cursory investigation[9] showed that the probable space group is $F\bar{4}3m(T_d^2)$, $F432(O^5)$, or $Fm3m(O_h^5)$ with approximately 1116 atoms per unit cube of edge $a_0 = 25.87$Å. This work was completed recently.[6] The final results obtained correspond to one formula unit of $Cu_{640}Cd_{484}$ per unit cube of space group $F\bar{4}3m$.

Up to the present $NaCd_2$, βMg_2Al_3, and Cu_4Cd_3 have seemed to exhibit the largest structural units that have been observed in intermetallic compounds. These units, containing more than 1100 atoms each, are referred to here as the "giant cells".

It is possible that more compounds of comparable or even greater complexity exist, but that, as yet, these have escaped my detection. I should welcome any information or suggestions as to the likelihood of this.

The reason for interest in these complex structures is that they incorporate a large number of crystallographically independent coordination shells and hence represent a valuable source of information with regard to the atomic configurations that lead to maximum stability.

THE MOST IMPORTANT COORDINATION POLYHEDRA OBSERVED IN COMPLEX METALLIC STRUCTURES

The description of crystal structures, in which all the atoms are arranged so as to fill space, as is the case in intermetallic compounds, is virtually a portrayal of the configuration of atoms around single atoms. In some cases, the configurations can be described conveniently by reference to the five regular polyhedra. In the majority of cases, however, one has to resort to polyhedra of a more complicated geometrical nature. These are cumbersome to describe in the course of a structure study and are therefore discussed separately in the following sections. Some configurations have been given short names: each name refers to the metallic phase in which the configuration was first discovered. In the interest of brevity, the word ligancy (L) is used throughout instead of coordination number (CN).

The Friauf Polyhedron (Ligancy 16)

This polyhedron can be derived through relatively simple
modifications of the cubic closest-packed arrangement of spheres of
equal size (face-centered cube). A configuration of sixteen such
spheres is shown in Fig. la. Removal of a tetrahedron of four
contiguous spheres from this aggregate results in the framework of
twelve spheres (called B spheres), arranged about the corners of a
truncated tetrahedron, as shown in Fig. 1b. The central cavity is

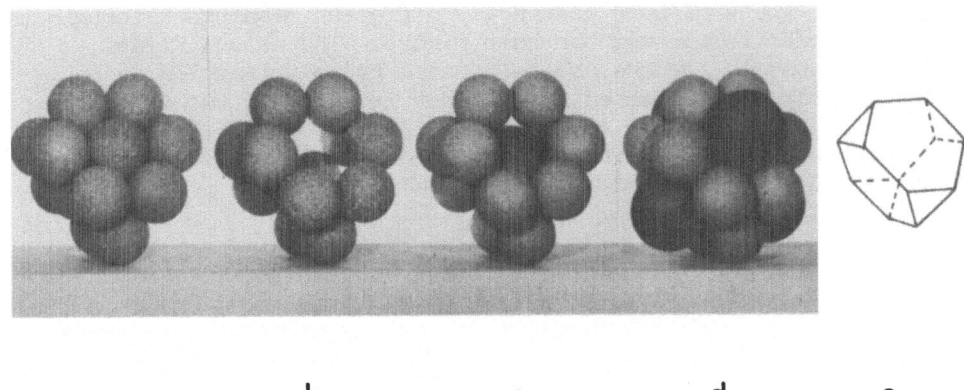

a   b   c   d   e

Figure 1.   To derive the Friauf polyhedron from an aggregate of
            sixteen spheres of equal size arranged in the cubic
            closest packing. (a) The group of sixteen spheres.
            (b) and (c) The truncated tetrahedron. (d) The
            aggregate of seventeen spheres referred to as the
            Friauf polyhedron. (e) A formal representation of the
            Friauf polyhedron. Here, the atoms out from the centers
            of the hexagons are not indicated for reasons explained
            in the text.

capable of accommodating a thirteenth sphere (called A sphere) with
a radius 1.35 times that of the surrounding spheres, as shown in
Fig. 1c. Out from the center of each of the four hexagons there is
an additional sphere of the large kind (up to 35 per cent larger in
radius), as shown in Fig. 1d. The central (large) sphere, accordingly, is surrounded by sixteen spheres, twelve small and four large
ones.

This group of seventeen spheres is called a Friauf polyhedron,
since it was first discovered in the Friauf phases $MgCu_2$[10] and
$MgZn_2$.[11] It consists, as we see, of two integral parts: (1) the
truncated (say, positive) tetrahedron bounded by four hexagons and

four triangles (twelve vertices); (2) the regular (negative) tetrahedron (four vertices, representing four large atoms).

In the formal representation of the Friauf polyhedron shown in
Fig. 1e, the spheres out from the centers of the hexagons forming
the regular, negative tetrahedron are not indicated, since, in most
instances, each such sphere is shared between two adjacent Friauf
polyhedra or between a Friauf polyhedron and a different kind of
coordination shell. The reader should be alerted to the fact that
in subsequent discussions reference will be made sometimes to the
Friauf polyhedron and sometimes to the truncated tetrahedron, and
that these two terms have distinctly different meanings although
they may refer to one and the same figure. The actual Friauf
polyhedron has 16 corners and is bounded by 28 triangular faces
although only the truncated tetrahedron is shown in the figures.

Figure 2 shows a layer of truncated tetrahedra arranged so as
to fill a plane. Each such tetrahedron shares three of its four

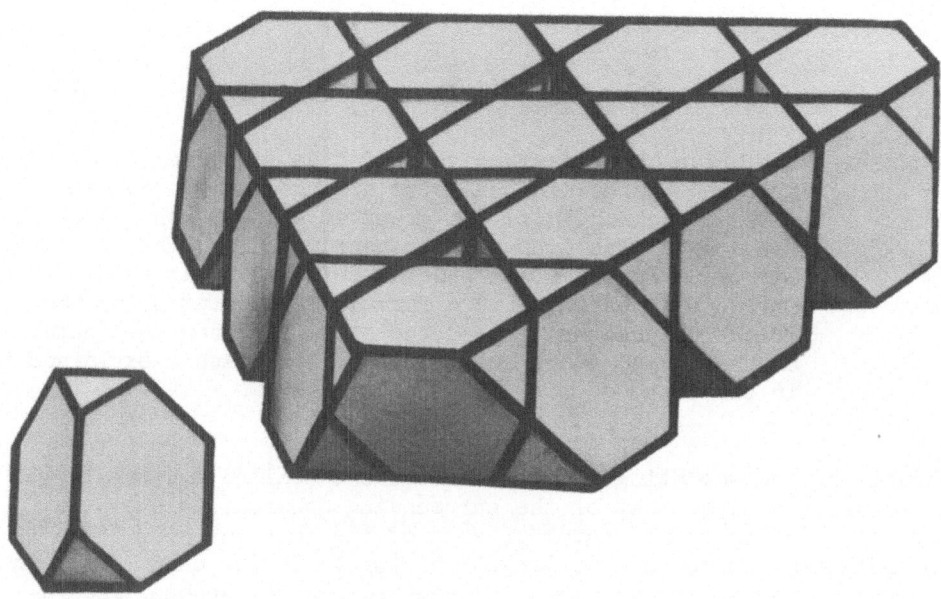

Figure 2. A close-packed layer of truncated tetrahedra forming
           Friauf polyhedra.

hexagons with three other polyhedra; the unshared fourth hexagon
may be shared by a truncated tetrahedron of the next layer that
may be superimposed on this one. It is seen that then each large
atom, called A, out from the center of a hexagon represents, in
turn, the center of an adjacent Friauf polyhedron.

A geometrical feature of close-packed, regular truncated
tetrahedra of edge length a is that the center-to-center distance
(A–A) becomes 1.23 × a, whereas the center-to-vertex distance (A–B)
is 1.17 × a. With twelve contiguous spheres of radius 0.5 a
(B spheres) at the vertices, the truncated tetrahedron can accommodate a central sphere (A sphere) of radius (1.17 − 0.5)a = 0.67a,
whereas in the close-packed layer the A–A distance corresponds to
a radius of only (0.5 × 1.23)a = 0.615a, which is 8 per cent shorter.

The Friauf polyhedra observed in the three structure types C14
($MgZn_2$[11]), C15 ($MgCu_2$[10]), and C36 ($MgNi_2$[12]) are very nearly of this
kind, and hence, exhibit metrical properties that are clearly
inconsistent with spherically shaped atoms. It seems that the A
atoms are elongated in the directions of the twelve A–B bonds and
shortened in the directions of the four tetrahedral A–A bonds.
In fact, the A atoms would seem to have a tetrahedral shape if the
B atoms were spherical. Probably, neither A nor B is spherical.

The μ-Phase Polyhedron (Ligancy 15)

The twelve spheres of equal size shown in Fig. 3a are arranged
about the vertices of a truncated trigonal prism bounded by eight
triangles (four above and four below; see also Figs. 3d,e) and
three hexagons. The central cavity of this framework is capable of
accommodating a thirteenth sphere of radius, about 1.31 times that
of the surrounding spheres, as shown in Fig. 3b. Out from the
center of each of the three hexagons there is an additional sphere
of the large kind, as shown in Fig. 3c. The central large sphere,
accordingly, is surrounded by fifteen spheres, twelve small (called
B) and three large ones (called A).

This group of atoms was first observed in the μ phases W_6Fe_7,
W_6Co_7, Mo_6Fe_7, and Mo_6Co_7[13] and is therefore referred to as the μ-
phase polyhedron.

A formal representation of this polyhedron is shown in Figs.
3d,e. Again, the atoms out from the centers of the hexagons are
not shown, since, in most cases, each such atom is shared between
two adjacent μ-phase polyhedra or between a μ-phase polyhedron and
a different kind of coordination shell with which it shares its
hexagon.

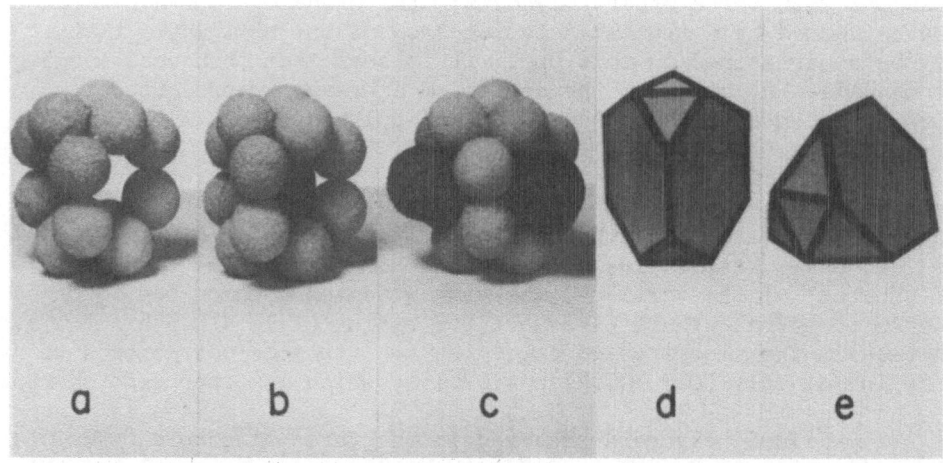

Figure 3.   (a) Twelve spheres of equal size are arranged about the
            vertices of a truncated trigonal prism. (b) The central
            cavity is appropriate for a thirteenth sphere of radius
            1.31 times that of the surrounding ones. (c) Three more
            large spheres are added to form the μ-phase polyhedron
            of L15. (d) and (e) A formal representation of the μ-
            phase polyhedron: here the atoms out from the hexagons
            are not indicated.

In the μ phases[13] these polyhedra are arranged so as to fill a
plane, as is shown in Fig. 4. It is of interest to note that here
the center-to-center distance between adjacent truncated trigonal
prisms (sharing hexagons) is exactly the same as the center-to-
vertex distance (if they were regular polyhedra), which is
$(2/\sqrt{3})\cdot a = 1.155a$. If, again, twelve contiguous B spheres of
radius 0.5a (one-half the edge length of the truncated prism) were
at the vertices, the center-to-vertex distance would correspond to
a central sphere of radius $(1.155 - 0.5)a = 0.655a$, whereas the
center-to-center distance (A–A distance) corresponds to a radius of
only $(0.5 \times 1.155)a = 0.578a$, which is nearly 12 per cent shorter.
Thus, it seems that each A atom (large atom at the center) is
elongated in the directions of the twelve A–B bonds and shortened
in the directions of the three trigonal A–A bonds. In fact, each
A atom appears to be trigonally deformed, even if the B atoms were
assumed to be spherical. However, in reality, it is highly unlikely
that either A or B is spherical.

Again, the reader should be alerted to the fact that only the
truncated trigonal prism will be shown in subsequent figures,
although the term μ-phase polyhedron refers to the coordination
shell that has 15 corners and 24 triangles.

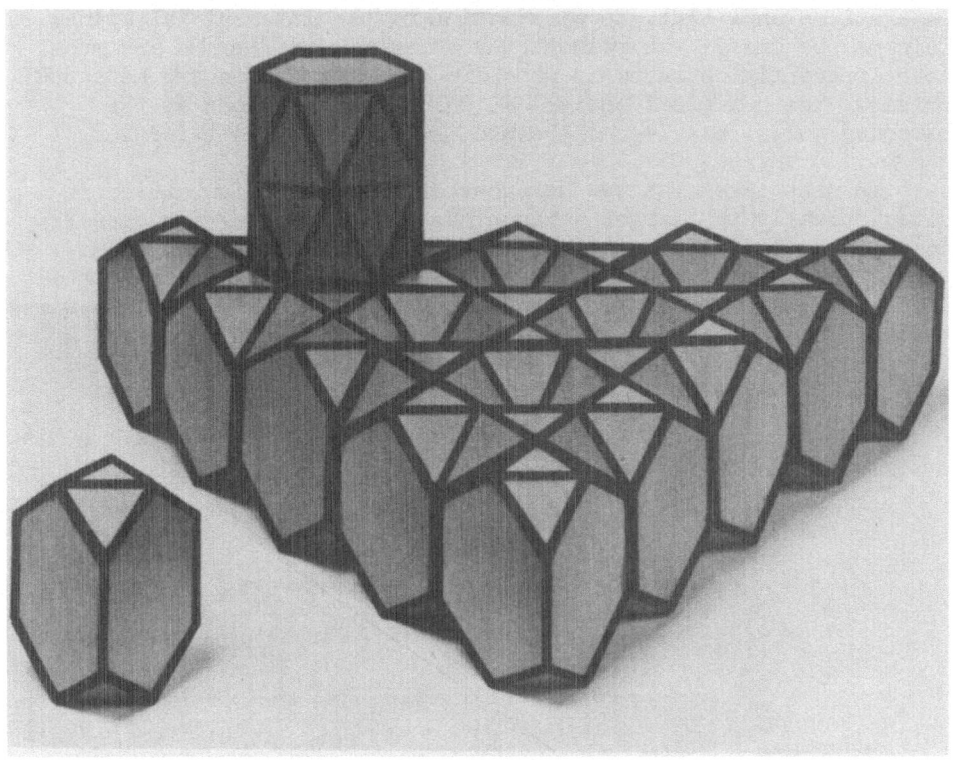

Figure 4. Arrangement of μ-phase polyhedra so as to fill a plane.

The Hexagonal Prism and Antiprism with Two Atoms
at the Extended Poles (Ligancy 14)

A considerable number of the coordination shells of ligancy
14 observed in intermetallic compounds belong to this group.

The regular hexagonal prism with square prism faces, having
eighteen edges (excluding poles) of length a corresponds to a
center-to-vertex distance of $a \cdot \sqrt{5}/2 = 1.118a$ and, accordingly, is
appropriate for the accommodation of a central sphere which has a
radius of 0.618a, which is 1.236 times that of the twelve
surrounding spheres of radius 0.5a.

Conversion of a hexagonal prism (plus two atoms at the extended
poles, L14) into an antiprism leads to an increase in the number of
edges from 30 to 36, and the six square prism faces are replaced by
twelve triangles. Addition of spheres out from the centers of
square prism faces gives rise to octahedral interstices, whereas

the triangles of the antiprisms lead to the creation of tetrahedral interstices and, thus, to a reduction of the interstitial space. This may be one of the reasons why hexagonal antiprisms are much more frequently observed in metallic structures than are hexagonal prisms. The hexagonal antiprism, which has two atoms at the extended poles, has 14 corners and is bounded by 24 triangles.

In most cases the two hexagons of each antiprism differ in size; usually the larger hexagon has all or part of its corners occupied by large atoms, the smaller by small atoms. In cases where these antiprisms are arranged so as to form a close-packed layer (Fig. 5) the side of each large hexagon is $2/\sqrt{3} = 1.155$ times that of the small one.

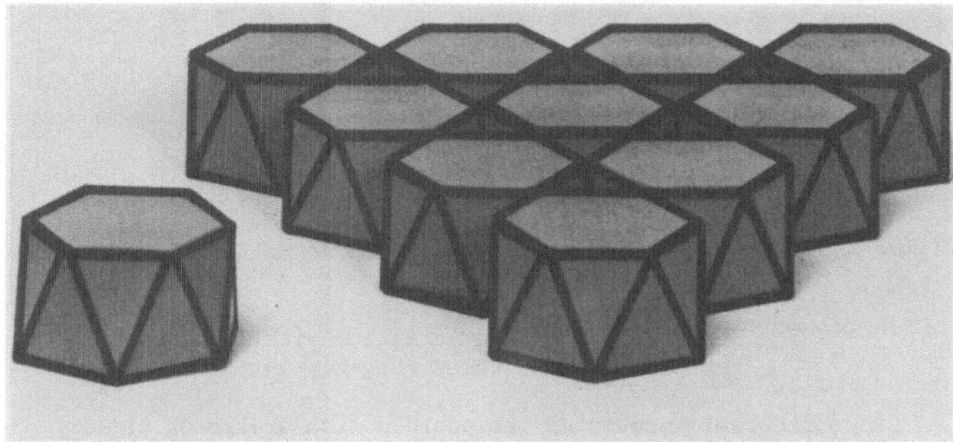

Figure 5. Hexagonal antiprisms arranged so as to form a close-
           packed layer. The side of each large hexagon is
           $2/\sqrt{3}$ times that of each small hexagon.

Frequently, two antiprisms share a large hexagon and thus form a bi-antiprism (μ phase[13], P phase[14]); see also Figs. 7 and 8. A close-packed layer of such complexes is observed in the μ-phase structure; see Fig. 6. The hexagonal bi-antiprism referred to here is not to be confused with the coordination shell of ligancy 18 observed in the $D2_d$ ($CaCu_5$) type of structure, in which the central atom is at the center of the large hexagon.

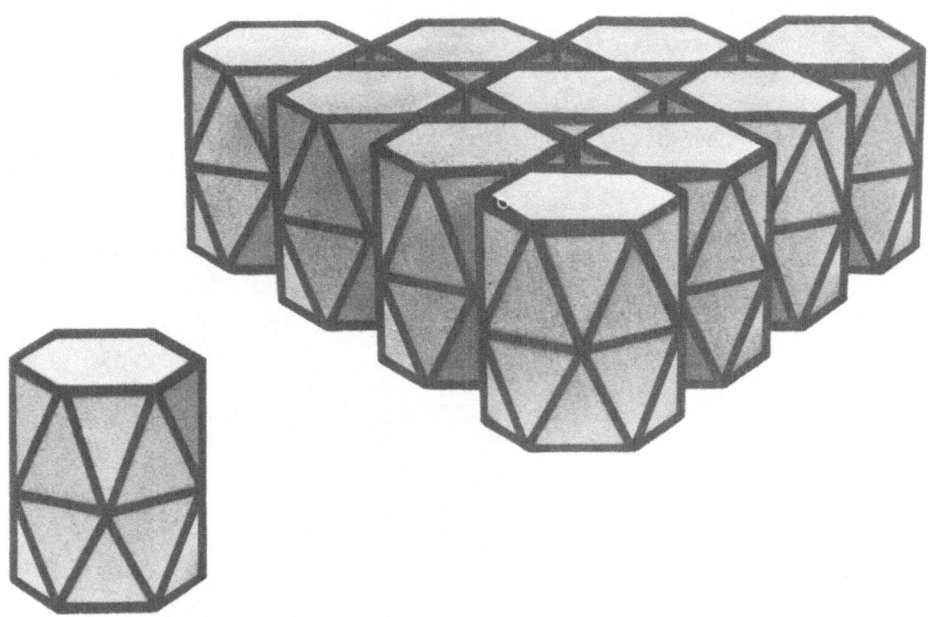

Figure 6.   Two close-packed layers of hexagonal antiprisms, in
            which each large hexagon is shared by the lower and
            the upper layers, as is the case in the μ-phase
            structure.

Each atom at the extended pole of an antiprism is surrounded
usually by a μ-phase polyhedron, a Friauf polyhedron, or, in turn,
by an antiprism (Figs. 6, 7, 8).

The Friauf polyhedron (L16), the μ-phase polyhedron (L15), and
the hexagonal antiprism (plus two atoms at the poles, L14) are
related to each other as follows: The Friauf polyhedron can be
described as a hexagonal antiprism which has one atom out from the
center of the small hexagon and three atoms that form a triangle
out from the center of the large hexagon. The μ-phase polyhedron
is obtained through replacement of these three atoms by two atoms.

The Cubo-Octahedron, the Icosahedron, and the Pentagonal Prism with Two Atoms at the Poles (Ligancy 12)

The icosahedron can be derived through relatively simple
modifications of the cubo-octahedron (which is observed in the
cubic closest-packed structures). Figure 9a shows fourteen
spheres arranged at the lattice points of a face-centered cube.

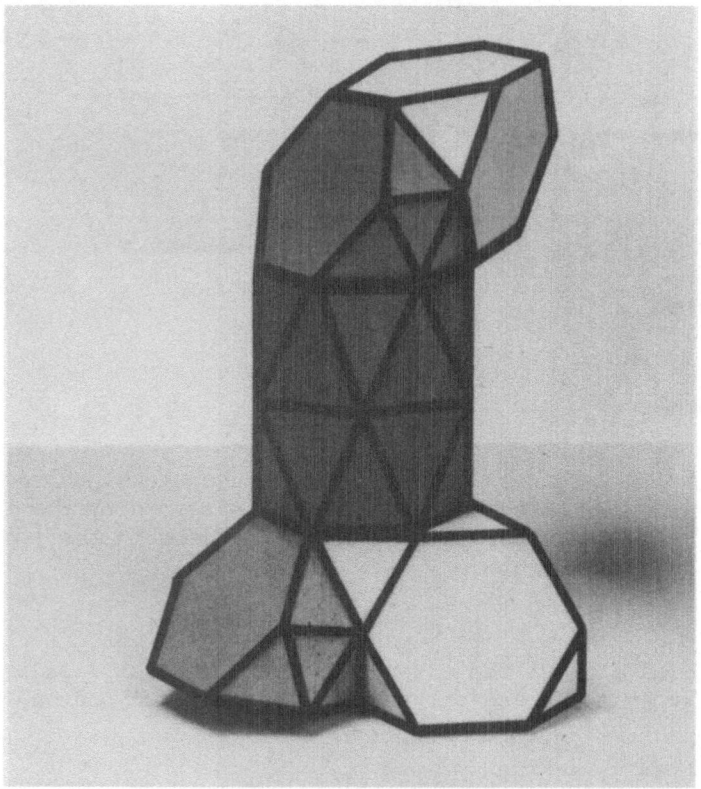

Figure 7. The sequence of contiguous polyhedra as observed in
the P(Mo-Ni-Cr) phase: μ-phase polyhedron, Friauf
polyhedron, hexagonal bi-antiprism, μ-phase
polyhedron, Friauf polyhedron ····. The zigzag chain
is of infinite length.

Figure 8. A layer of Friauf polyhedra in which the horizontal
           triangles are shared with μ-phase polyhedra, and the
           hexagons with hexagonal antiprisms. A set of six
           contiguous μ-phase polyhedra is shown to the left,
           a hexagonal bi-antiprism (right) is created by each
           such sixfold set.

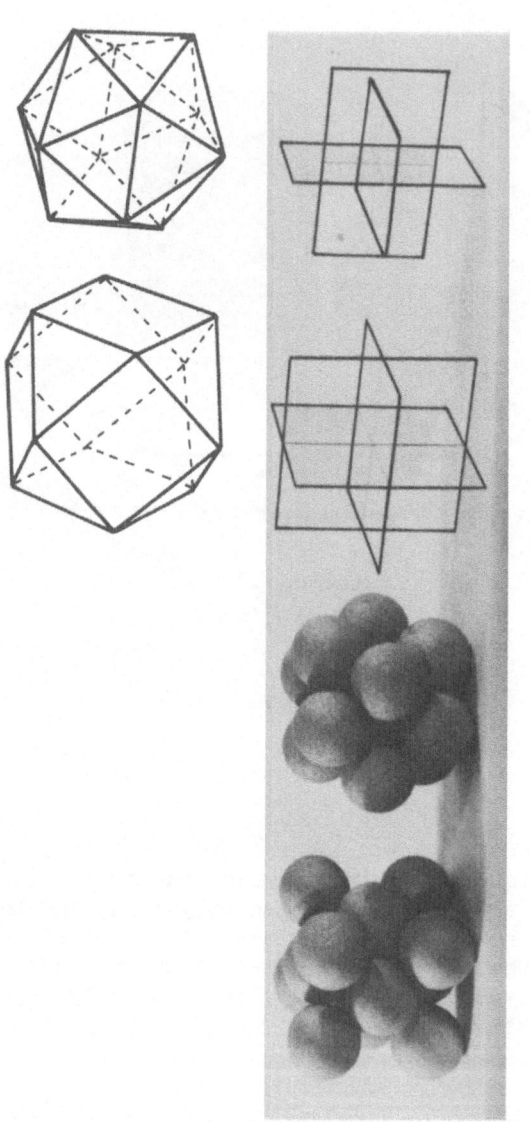

Figure 9.   The group of twelve spheres at the vertices of the cubo-octahedron
shown in (b) is brought into view by translation of the origin of
the face-centered cube shown in (a). The cubo-octahedron is
shown in (c) and the icosahedron in (d).

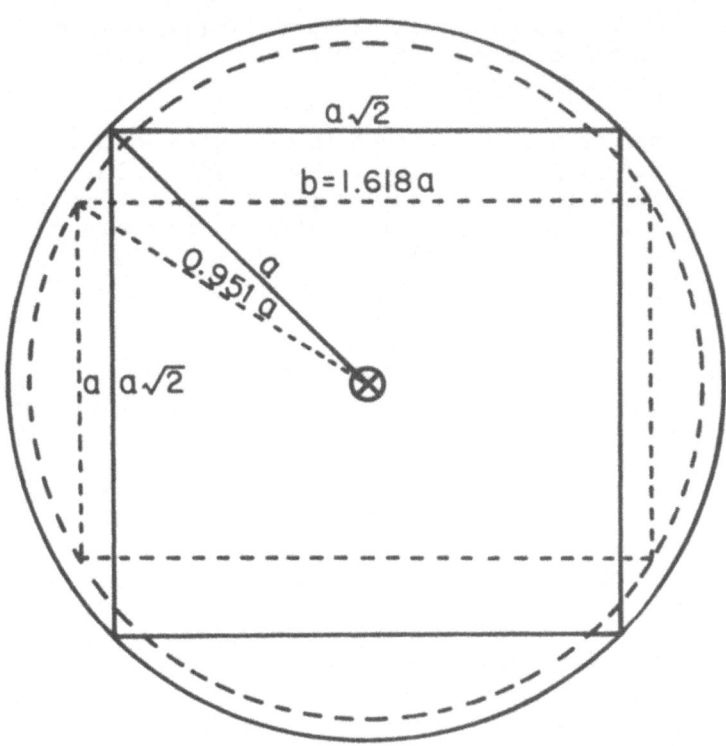

Figure 10. The metrical nature of the icosahedron (broken lines)
            as compared to that of the cubo-octahedron (solid
            lines). If a is the nearest vertex-to-vertex distance
            of each polyhedron, then the side of each square in
            Fig. 9c is a·$\sqrt{2}$ (here, solid lines), and the rectangles
            in Fig. 9d have the sides a and b = 1.618 a (here,
            broken lines). It is seen that the icosahedron has a
            shorter center-to-vertex distance than the cubo-
            octahedron.

The cubo-octahedron (Fig. 9b) is brought into view through translation of the origin of the cube shown in Fig. 9a by one-half of the length of the cube edge along any one of the three axes or one-half of the body diagonal. It is seen (Fig. 9c) that the vertices of the cubo-octahedron are located at the corners of three intersecting, mutually perpendicular squares. If the edge length of the cubo-octahedron (bond distance) is a, then the side of each square is $a \cdot \sqrt{2}$. Replacement of each square by a rectangle that has the sides a and b = 1.618 a results in the configuration shown in Fig. 9d. It is called the icosahedron since it is bounded by twenty equilateral triangles.

Each corner of this polyhedron is connected with five other corners that form a plane pentagon. Each pentagon has the side a, and the distance between each corner and the next nearest corner is 1.618 a; that is, equal to the long side of the rectangle. There are, accordingly, many different orientations in which the set of three mutually perpendicular rectangles can be fitted into the upper part of Fig. 9d.

The icosahedron has 15 twofold axes (30 edges), 10 threefold axes (20 triangles), and 6 fivefold axes (12 vertices) as well as 15 planes of symmetry and other elements of the second kind.

The conversion of a cubo-octahedron, which has 24 nearest-neighbor distances (or edges) of length a, into an icosahedron of 30 nearest-neighbor distances of the same edge length a (replacement of the squares of side $a \cdot \sqrt{2}$ by rectangles of sides a and 1.618 a, Figs. 9c,d) results in a shortening of the center-to-vertex distance of 4.9 per cent and a corresponding decrease in volume of slightly more than 7 per cent, see Fig. 10. Accordingly, with twelve contiguous spheres of equal size at the vertices, the central sphere of the icosahedron has to be nearly 10 per cent smaller in radius.

From Fig. 9d it becomes immediately apparent that the icosahedron can be described also as a pentagonal antiprism, which has two atoms at the extended poles. It thus provides a condition favorable for the formation of tetrahedral interstices, which, in turn, result in a minimum of interstitial space. In fact, each icosahedron may be regarded as consisting of twenty contiguous and slightly deformed tetrahedra that have one vertex in common at the icosahedron center.

The icosahedron (pentagonal antiprism) is much more frequently observed than is the pentagonal prism (with two atoms at the extended poles), which gives rise to octahedral interstices, one around each center of a square prism face. The pentagonal prism corresponds to a radius ratio of nearly unity.

Of the three coordination shells of ligancy 12 discussed here, surrounding a central sphere which may become 10 per cent smaller, the icosahedron is the smallest (see Fig. 10) and hence corresponds to lower energy or higher stability.

To my knowledge a cubo-octahedron has not yet been observed in a true intermetallic compound; however, it is not uncommon in carbides, silicides, and interstitial compounds.

Irregular Coordination Shells of Ligancy 11, 13, and 14

Of the large variety of irregular coordination shells that exist, the three types described below occur most frequently in the giant cells: (1) A shell of ligancy 11 is formed by replacing the pentagon of an icosahedron with a tetragon. (2) Polyhedra of ligancy 13 are created through the widening of part of an icosahedron, and addition of a thirteenth ligand, that can penetrate the opening. (3) Ligancy 14 is produced through penetration of two of the approximately square prism faces of a pentagonal prism (occupied poles) by two atoms.

Interpenetrating and Contiguous Polyhedra

In a space-filling structure, the atoms at the vertices of any polyhedron are, in turn, surrounded by coordination shells that penetrate each other and the central one, and the various kinds of interpenetrating polyhedra are often incommensurate. This, as well as other factors that will be discussed later, leads to distortions, which sometimes become considerable. Therefore, throughout the discussions presented later, it should be understood that such terms as Friauf polyhedron, μ-phase polyhedron, icosahedron, etc., do not necessarily refer to the regular configurations in which each triangle is equilateral.

Because of interpenetration it is difficult to show by means of a model all existing polyhedra simultaneously. The figures shown in the following sections exhibit only the contiguous polyhedron; that is, those that share faces, edges, or corners but do not interpenetrate. In each such figure, each corner of a facet then represents the center of another coordination shell that has to be described in the text or shown in a separate model.

One of the most striking examples of interpenetrating polyhedra is found in the μ phase. In the close-packed layer of hexagonal bi-antiprisms shown in Fig. 6, each corner of a large hexagon represents the center of a μ-phase polyhedron, and thus, the layer shown in Fig. 4 represents exactly the same atomic arrangement as the one shown in Fig. 6. In fact, each hexagon that can be made out in the horizontal plane in Fig. 4 belongs to a bi-antiprism, and the one shown at the rear, left-hand corner can be imagined as having been pulled out of that layer.

THE CRYSTAL STRUCTURES OF β Mg_2Al_3 AND $NaCd_2$

The structures of these two compounds are partially disordered but presumably in slightly different ways, as indicated by the difference between the two stoichiometric ratios. The fundamental structural features are the same, however. For βMg_2Al_3 the details of the disorder have been worked out,[9] but for $NaCd_2$ they are still uncertain.[15] $NaCd_2$ reacts with oxygen or moisture and gradually decomposes during X-ray examination. It is difficult, therefore, to obtain X-ray data of a quality that will suffice for the determination of all the structural details. Therefore, only βMg_2Al_3 is discussed below; the idealized ordered model is described first, and then some details of the disorder are given. The accurate length of the cube edge is $a_0 = 28.239Å$ for βMg_2Al_3 and $a_0 = 30.56Å$ for $NaCd_2$; the space group is in both cases $F_d3m(O_h^7)$.

The basic building block of the structure consists of five Friauf polyhedra that are arranged about an approximate fivefold axis of symmetry, as shown in Fig. 11. The five polyhedra are, in the crystallographic sense, of three different kinds (F1, F2, and F3). The group of polyhedra lies on a plane of symmetry; therefore, the left half of each figure is a mirror image of the right half. The dihedral angles of a tetrahedron are 70°32' and hence correspond to nearly one-fifth of a complete rotation. The aggregate (Fig. 11) consists of 47 atoms and is called the VF polyhedron. Each atom out from the center of a hexagon is not indicated for reasons explained earlier; see also Fig. 1.

Six VF polyhedra are arranged about the vertices of an octahedron in such a way as to produce four additional Friauf polyhedra, F4, located at the vertices of a regular tetrahedron and sharing hexagons with polyhedra F1. F4 is dark and only one is seen in Fig. 12. The resulting complex consists of 234 atoms and comprises 34 Friauf polyhedra. It has symmetry T_d (Fig. 12), and hence, each two diametrically opposed VF polyhedra are turned 90° with respect to one another. The twelve outer Friauf polyhedra are of the type F3.

A second such T_d complex is meshed with the first one, as shown in Fig. 13. These two complexes share hexagonal faces of the F2 polyhedra and are related to one another by a diamond glide. Three more T_d complexes can be added in a similar fashion. Each T_d complex is accordingly connected with four others that are

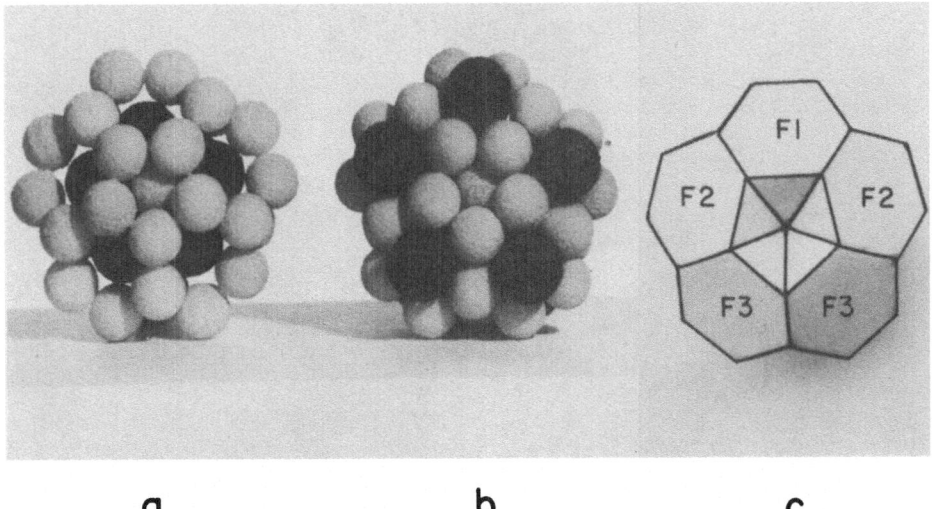

a   b   c

Figure 11.   (a) Five contiguous truncated tetrahedra about a
             fivefold axis of symmetry. (b) The VF polyhedron.
             (c) A formal representation of the VF polyhedron.
             The atoms out from the centers of the hexagons are
             not indicated.

arranged about the vertices of a regular tetrahedron (Figs. 14 and
15). The atom out from the center of each dark hexagon of an F4
polyhedron is shared between three truncated tetrahedra, each one
belonging to a Friauf polyhedron F3 of different T_d complex. A
similar kind of vertex sharing is also observed in $\gamma Mg_{17}Al_{12}$,[16] and
$\epsilon Mg_{23}Al_{30}$;[17] refer also to Fig. 25.

     Continued stacking of T_d complexes leads to an infinite three-
dimensional network (Fig. 16) in which each T_d complex of 234 atoms
shares atoms with four others so as to reduce the average number of
atoms per T_d complex to 144. The cubic unit of structure contains
eight T_d complexes; they account for 1152 atoms. Eight more atoms
(magnesium) have to be added, each of them at the center of a
Friauf polyhedron F5 that cannot be brought into view with the
opaque models used here. Each F5 polyhedron shares edges with
twelve F3 polyhedra (that is, six VF polyhedra) and lies at the
center of the sphere shown in Fig. 17. There are eight such
spheres in the cubic unit; each one is penetrated by four others

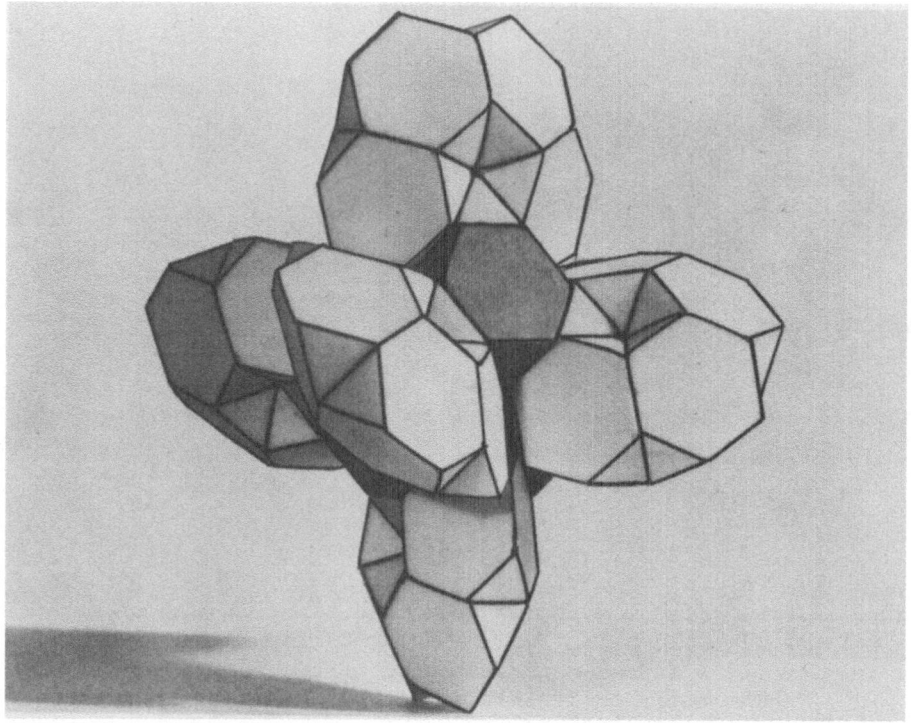

Figure 12. The complex of 234 atoms formed by six VF polyhedra
that are arranged around the vertices of an octahedron
of T_d symmetry. In the disordered model six of the
twelve outermost Friauf polyhedra are distorted.

of the same kind. With the addition of 32 more atoms (aluminum) out
from the centers of the triangles of eight such F5 polyhedra, the
entire complement of 1192 atoms in the ordered structure is
accounted for.

The unit of structure contains 280 Friauf polyhedra (L16), 96
μ-phase polyhedra (L15), 64 hexagonal antiprisms, each with two
atoms at the extended poles (L14), 128 coordination shells of
ligancy 13, and 624 icosahedra (L12). Each magnesium atom may be
assumed to have ligancy 14, 15, or 16, and each aluminum atom,
which is smaller (radius ratio Mg/Al ~ 1.14), ligancy 12.

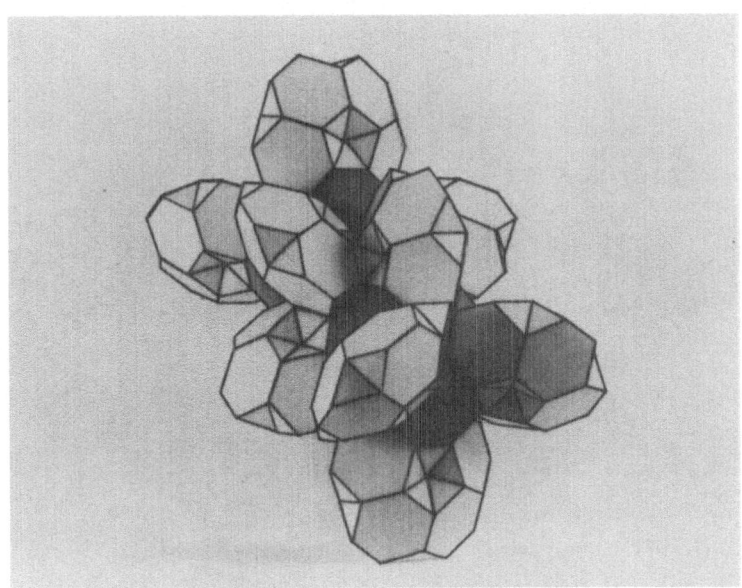

Figure 13.   A second complex of 234 atoms inserted into the one
             shown in Fig. 12. The two complexes are related to
             each other by a diamond glide.

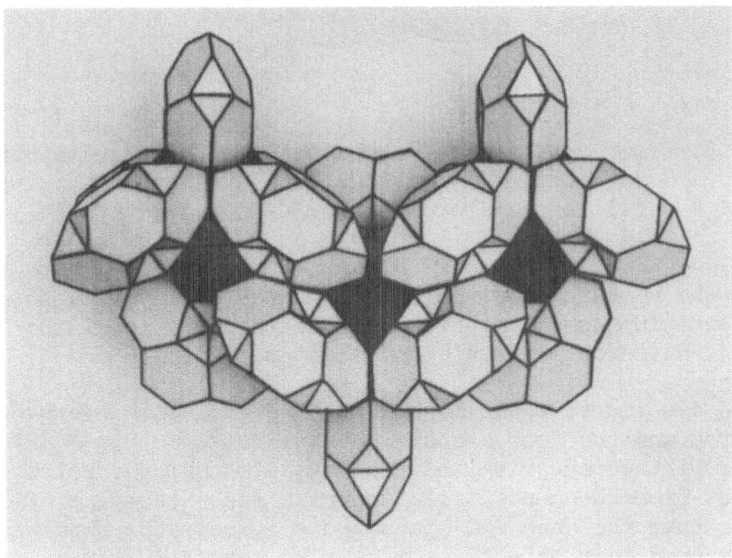

Figure 14.   Three complexes of 234 atoms each forming part of the
             aggregate shown in Fig. 15.

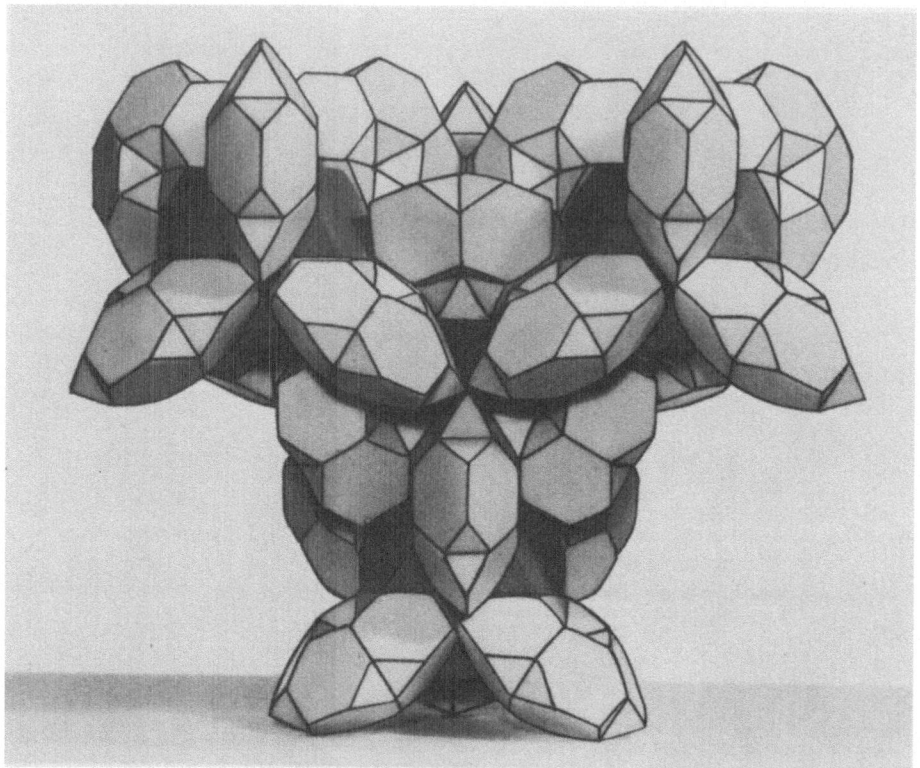

Figure 15.   Four complexes of 234 atoms each arranged about one
             such complex. The four complexes are at the vertices
             of a regular tetrahedron.

Some of the coordination shells of ligancy 13 may enclose magnesium,
and others aluminum. In this idealized ordered model the 1192 atoms
occupy 17 different sets of equivalent positions.

    The disordered model is obtained by replacing every other F5
polyhedron and the four associated aluminum atoms ($\frac{1}{8} \times 32$) with a
centered pentagonal prism that has two atoms at the poles and two
atoms out from the centers of two prism faces (complex of 15 atoms).
In order that the observed space-group symmetry be retained, it has
to be assumed that this 15-atom complex occurs in six orientations,
and that there is a random interchange of the set of four F5 polyhedra and the set of four 15-atom complexes in the individual unit
cells. The details of the disordered arrangement cannot be shown by

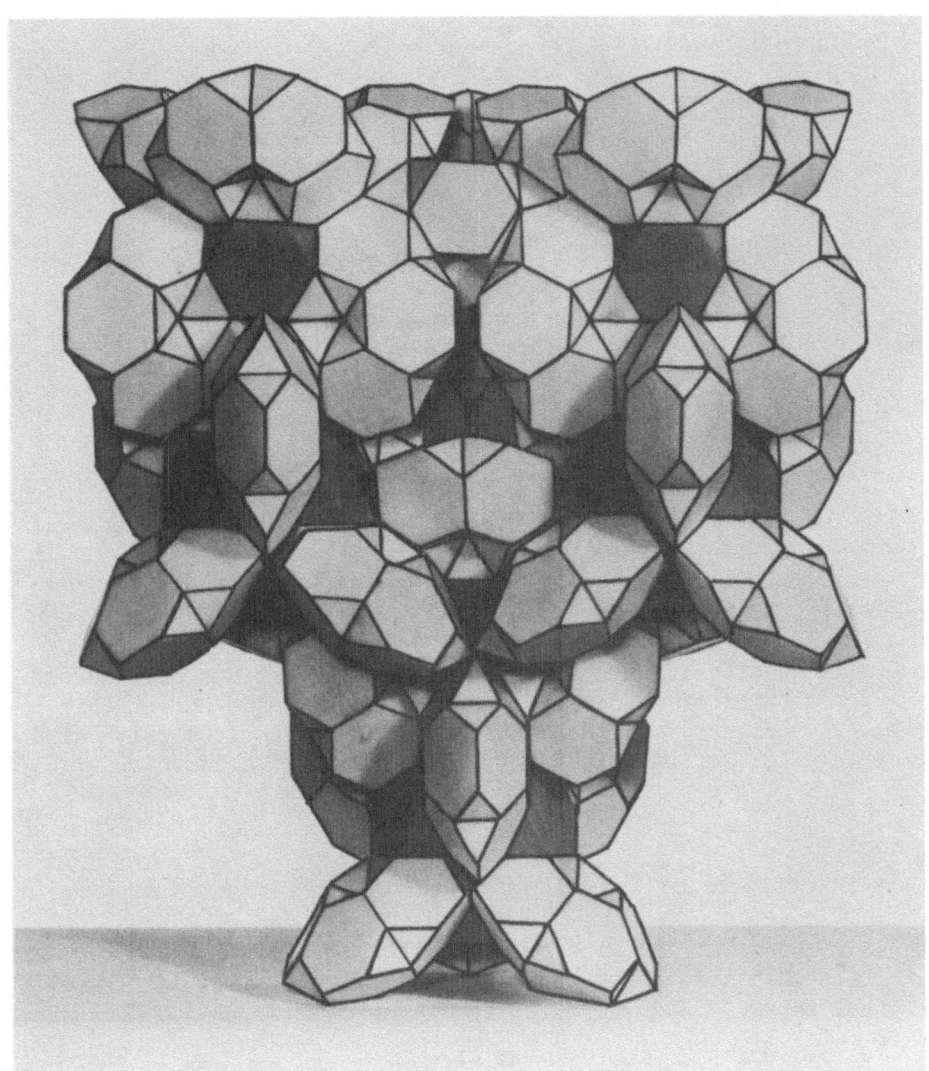

Figure 16. Continued stacking of the 234-atom complexes leads to the configuration shown above. The spherical arrangement of VF polyhedra shown in Fig. 17 can be recognized here. Its center is located about two-thirds up the vertical center line of this figure.

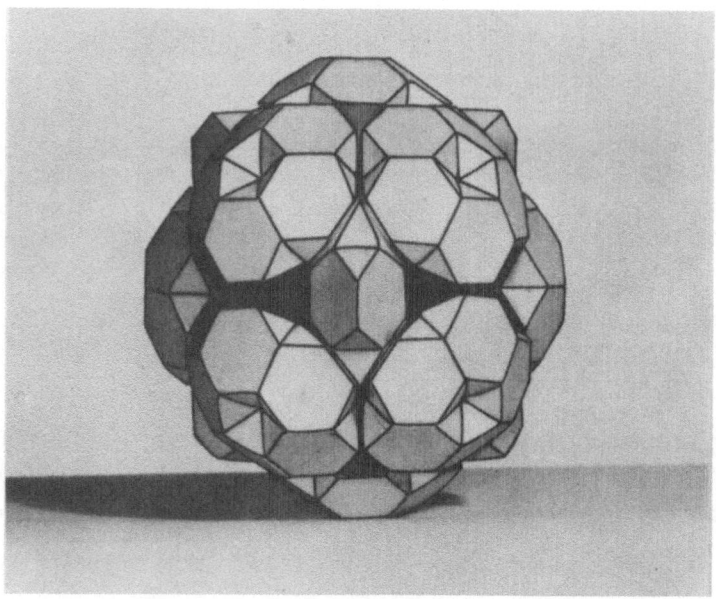

Figure 17.   Twelve VF polyhedra (4×3) form a sphere around each
             one of the eight points $00\frac{1}{2}$, etc. Six additional VF
             polyhedra are arranged about the vertices of a second
             kind of T_d octahedron which can be made out on this
             figure. At the center of this sphere is a fifth kind
             of Friauf polyhedron, called F5, which is shared
             between these six VF polyhedra.

means of models of the kind presented here, but packing maps[18] are
extremely well suited for this purpose as can be seen in the
original paper.[9]

       To account for this kind of disorder, the complex of 234 atoms
$(T_d$ complex) shown in Fig. 12 has now to be modified. Of any two
diametrically opposed VF polyhedra in this complex, one remains
unchanged, while the other contains, instead of two F3 polyhedra,
two modified Friauf polyhedra.

       In the three-dimensional network of the modified T_d complexes
each second sphere of the kind shown in Fig. 17 will then have a
15-atom complex at its center instead of the F5 polyhedron with the
associated four aluminum atoms (21-atom complex). Since there are
four spheres of each kind per unit of structure, the total number of

atoms is $1192 - 4$ $(21 - 15) = 1168$. The occurrence of the 15-atom complexes in six orientations results in the displacement of certain other atoms in the structure for part of the time.

The atoms of the disordered model occupy 23 different sets of equivalent positions. In an ordered structure, the number of different polyhedra is equal to the number of point sets; but here, the number of different polyhedra (in the crystallographic sense) is increased to 41. The disordered atomic arrangement corresponds to 252 Friauf polyhedra, 24 μ-phase polyhedra, 48 hexagonal anti-prisms (plus two atoms at the extended poles), 672 icosahedra, and 172 more-or-less irregular coordination shells of ligancy 10 to 16, forty-eight of which are modified Friauf polyhedra. The disorder, accordingly, results in a gain of 48 icosahedra per unit of structure.

Some Notes on the Derivation of this Structure

As is seen, the idealized ordered model has been completely accounted for with the use of only five crystallographically different Friauf polyhedra (except for the 32 atoms that had to be added out from the triangles of the eight F5 polyhedra; the eight extra magnesium atoms added belong to F5), whereas the (ordered) structural unit comprises seventeen crystallographically different kinds of polyhedra. The Friauf polyhedra dominate the structure, although they are by far outnumbered by the other kinds of coordination shells, especially by icosahedra; this is because the truncated tetrahedra do not penetrate one another.

In any structure so far known to incorporate Friauf polyhedra, the truncated tetrahedra are contiguous; thus far they have never been observed to penetrate each other. The Friauf-polyhedra framework, therefore, represents the most favorable starting point in the search for a trial structure in which it is anticipated.

This observation has been of extreme importance in the derivation of this structure as well as that of Cu_4Cd_3[6] and others,[19,20] in which packing maps[18] had to be used. Probably it would have been possible to deduce the trial structure by starting out with icosahedra. A few of them placed at appropriate points on the packing map[18] would have suggested the existence of Friauf polyhedra, and from here on there would not have been too much reasonable latitude for the positioning of the remaining atoms. This approach would have been considerably more tedious and difficult, however.

THE CRYSTAL STRUCTURE OF Cu_4Cd_3

Although the unit of structure of this compound contains somewhat fewer atoms than that of βMg_2Al_3 (and $NaCd_2$), 1124 as compared to 1168 per unit cube, it was considerably more difficult to determine. The reason for this is the lower symmetry and a resulting significant increase in the number of structural parameters. The space group is $F\bar{4}3m(T_d^2)$ and the more accurate length of the cube edge is $a_0 = 25.871\text{Å}$. In fact, this structure comprises two different substructures that penetrate one another, and each one is of considerable complexity. Each substructure represents a diamond arrangement, in one case of Friauf polyhedra, in the other case of icosahedra. The structure lacks a center of symmetry, and hence calculations of Fourier syntheses are meaningless unless the trial models of both substructures are nearly correct and are used simultaneously as a basis for the phase-angle determinations. Again, the structure was solved exclusively with the use of a packing map.

The metallic radius observed for cadmium in intermetallic compounds is almost identical to that observed for magnesium, whereas copper atoms are somewhat smaller than aluminum atoms. The difference in size is appropriate for the formation of Friauf polyhedra, each with cadmium at the center and copper at the vertices of the truncated tetrahedron.

The diamondlike arrangement of Friauf polyhedra consists of the three types of complexes shown in Figs. 18a, b, and c. The octahedron of T_d symmetry, Fig. 18a, comprises ten Friauf polyhedra (4 F1 + 6 F2). The tetrahedral arrangement shown in Fig. 18c consists of five Friauf polyhedra (F5 + 4 F6), and the four dark polyhedra in Fig. 18b are F3 + 3 F4. The cubic unit contains four octahedra (Fig. 18a) that are arranged about the points $\frac{1}{2},\frac{1}{2},\frac{1}{2}$; $\frac{1}{2}00$; $0\frac{1}{2}0$; $00\frac{1}{2}$ (point set 4b in $F\bar{4}3m$) and four tetrahedra (Fig. 18c) that are at the points $\frac{1}{4},\frac{1}{4},\frac{1}{4}$, etc. (point set 4c). The layer of the dark Friauf polyhedra serves as a link between the octahedra and the tetrahedra. The infinite, three-dimensional framework of Friauf polyhedra thus formed is shown in Fig. 19. It is seen that the tetrahedra, the dark layers, and the octahedra alternate in a zigzag fashion.

The diamondlike arrangement of icosahedra may be described in terms of two kinds of complexes. One of them is shown in Fig. 20. It is seen (Fig. 20a) that five icosahedra, sharing vertices, are arranged about an approximate fivefold axis of symmetry, thus enclosing a pentagonal prism; each shared vertex represents also the center of a pentagonal prism. A set of six such fivefold rings is arranged at the vertices of an octahedron of T_d symmetry. All six rings interpenetrate and share icosahedra in such a way that the aggregate (Fig. 20c) consists of fourteen icosahedra that enclose six pentagonal prisms of the kind shown in Fig. 20a. Figure 20b shows two such fivefold rings interpenetrating at right angles. It is now seen that the pentagonal prism (in Fig. 20a) is shared by two icosahedra, one above and the other below the plane of the paper.

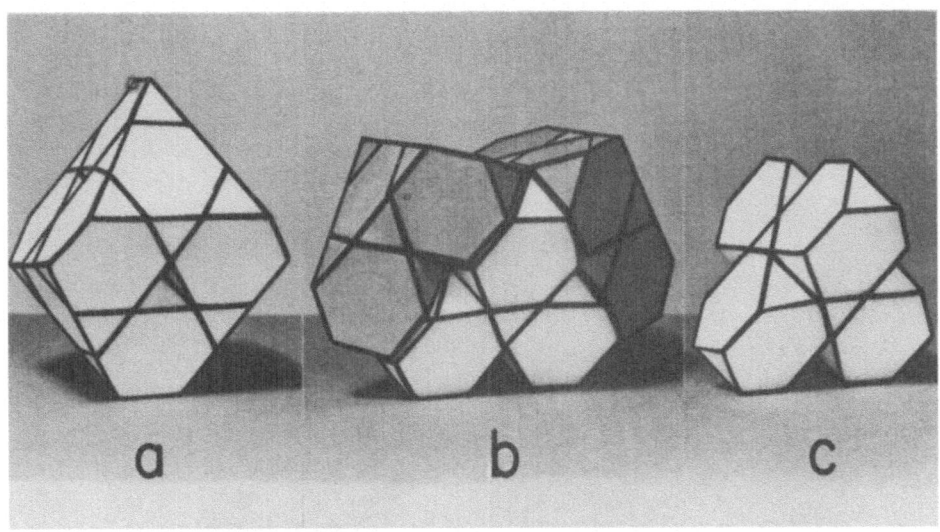

Figure 18.   The three types of atomic groupings that form the
             infinite, three-dimensional framework of Friauf
             polyhedra shown in Fig. 19. (a) A complex of ten
             Friauf polyhedra forming an octahedron of T_d symmetry.
             (b) The dark Friauf polyhedra attached to the complex
             shown in (c) serve as a link between (a) and (c).
             (c) Five Friauf polyhedra arranged to form a
             tetrahedron.

The two icosahedra have one vertex in common at the center of the
pentagonal prism, and each icosahedron center is at an extended
pole of that prism. Each additional vertex that is shared between
two icosahedra represents the center of a pentagonal prism (which
has two atoms at its extended poles), as can be made out on the
figures. Accordingly, thirty-six more pentagonal prisms are
created. In twelve of these prisms, two prism faces are deformed
in such a way that two more atoms are added as ligands (to provide
ligancy 14).

    Accordingly, the aggregate shown in Fig. 20c represents
fourteen icosahedra, thirty pentagonal prisms, each one with two
atoms at the poles (ligancy 12), and twelve pentagonal prisms, each
of which has two more atoms penetrating two prism faces (ligancy 14).

    The second icosahedral complex is shown in Fig. 21c. It
consists of a set of six pairs of interpenetrating icosahedra. The
center of each icosahedron of such a pair represents a vertex of the

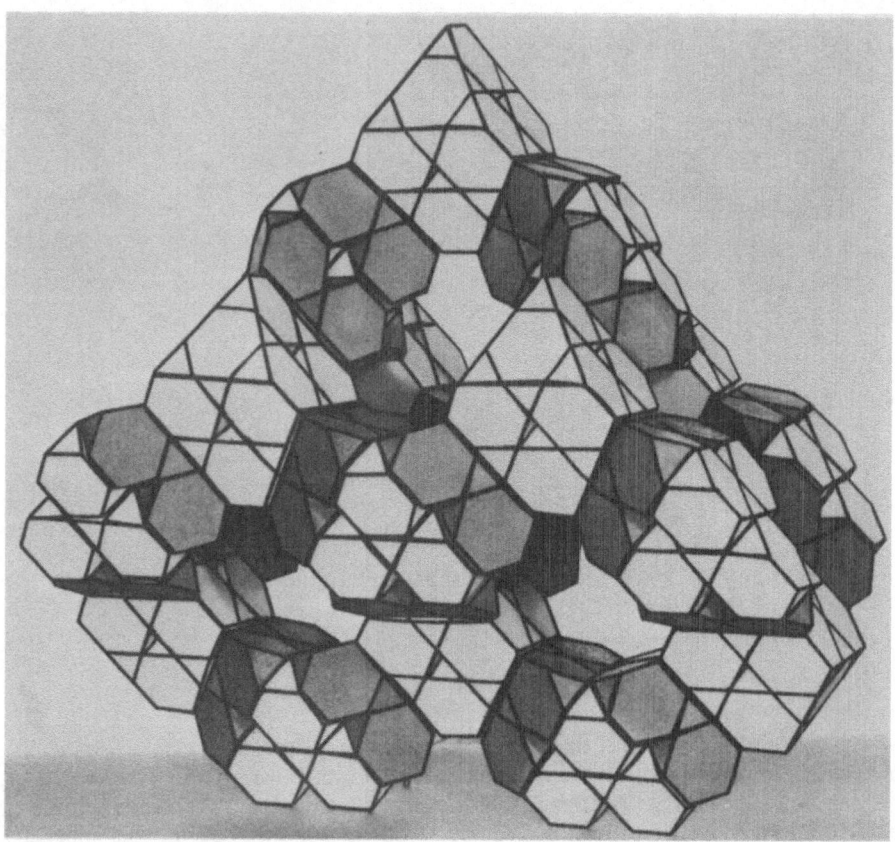

Figure 19.   The three complexes shown in Figs. 18a,b,c are
             arranged in a zigzag fashion to form an infinite,
             three-dimensional framework, part of which is shown
             here.

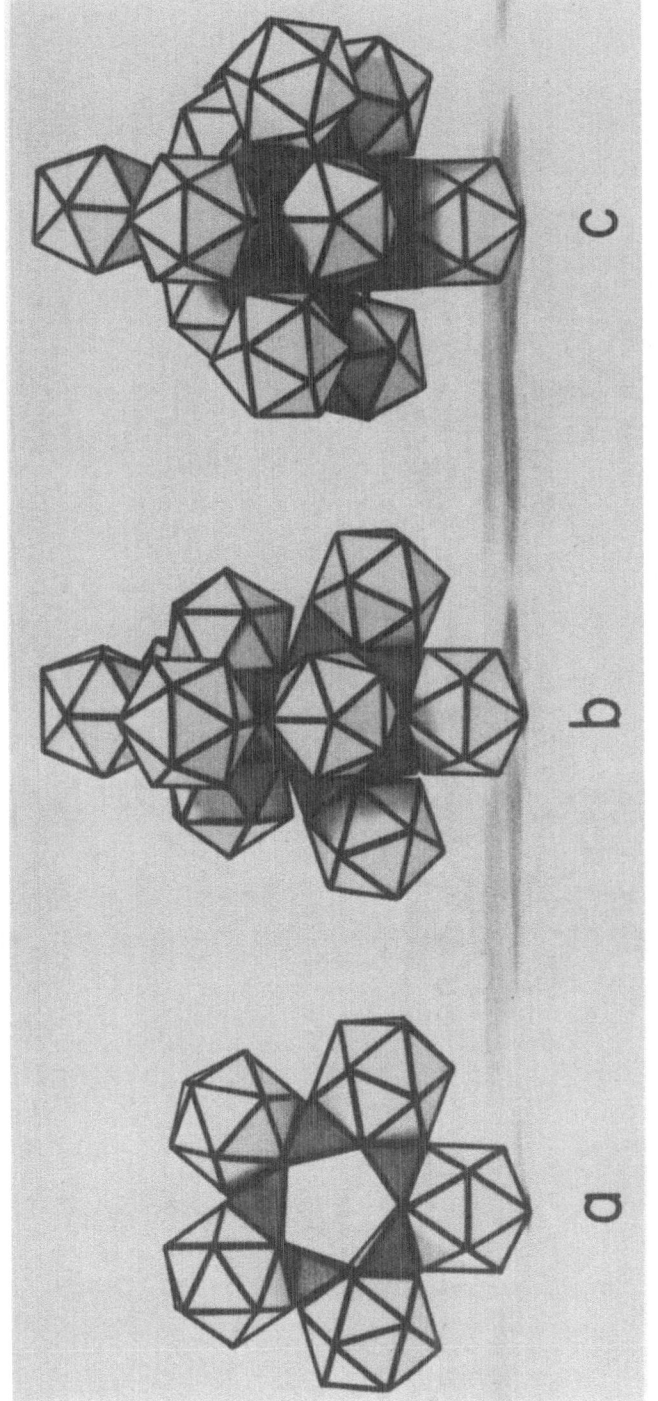

Figure 20. (a) Five icosahedra, sharing corners, arranged about an approximate fivefold axis of symmetry. (b) Two such fivefold rings interpenetrate at right angles. (c) Six interpenetrating fivefold rings form a complex of fourteen icosahedra and forty-two centered pentagonal prisms.

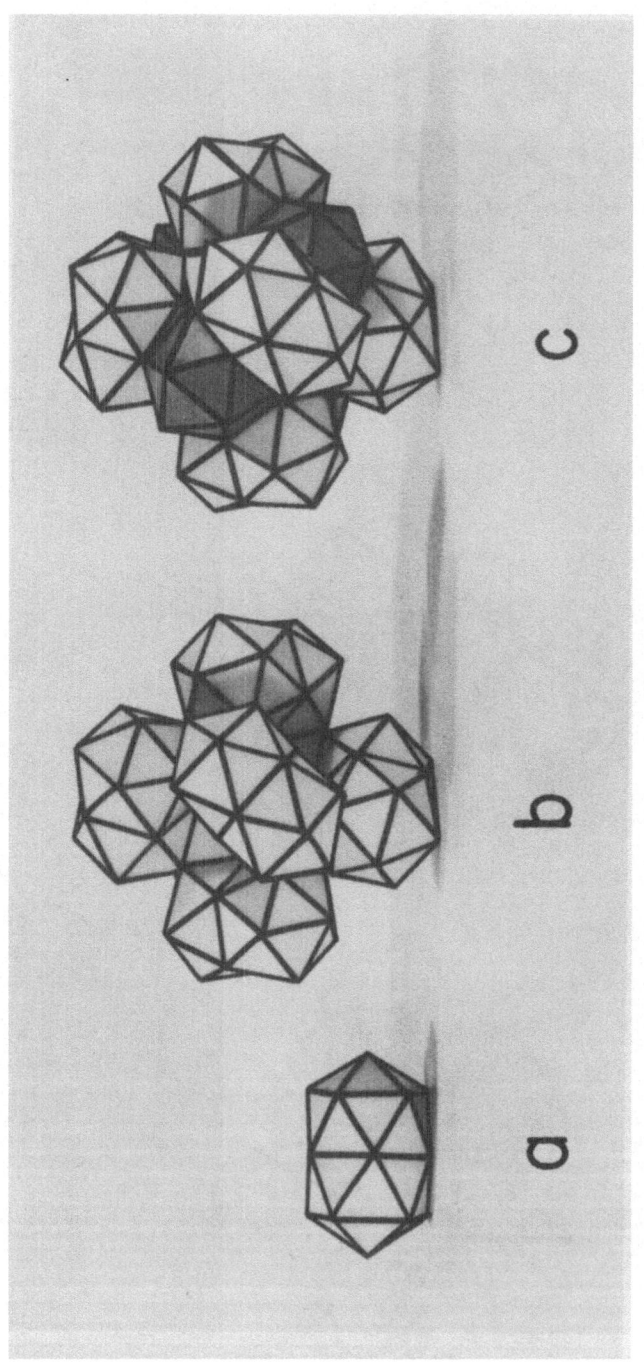

Figure 21. (a) Two interpenetrating icosahedra, referred to as a pair. The center of each of these icosahedra represents a vertex of the other. (b) A set of six such pairs arranged about the vertex of an octahedron of T_d symmetry. (c) Four more icosahedra, the dark ones, have been added to form a complex of sixteen icosahedra and eighteen pentagonal prisms.

other, as can be made out in Fig. 21a. The center of each pair
(that is, the center of each shared pentagon) is at the vertex of
an octahedron of T_d symmetry; accordingly, each one of two diametrically opposed pairs has its fivefold axis (long axis) at a right
angle to that of the other (Fig. 21b). Four more icosahedra have
to be inserted into this complex. Their centers are at the vertices
of a regular tetrahedron, and each of these icosahedra shares six
triangles with three "icosahedron pairs" that surround it, as is
shown in Fig. 21c. The four icosahedra are the dark ones in Fig.
21c. Each vertex that is shared between two icosahedron pairs is,
again, the center of a pentagonal prism, which has two atoms at the
extended poles and two additional atoms that penetrate two of the
prism faces (ligancy 14). Several of the pentagonal prisms can be
made out in Fig. 21b, especially at the upper left.

The aggregate shown in Fig. 21c thus represents sixteen icosahedra and eighteen pentagonal prisms.

The cubic unit contains four aggregates of the kind shown in
Fig. 20c and four of the kind shown in Fig. 21c. The former
aggregates are arranged about the points $\frac{3}{4},\frac{3}{4},\frac{3}{4}$, etc. (point set 4d)
and the latter about the points 0,0,0, etc. (point set 4a). Both
types of aggregates are connected with one another by shared vertices
in such a way that twelve more pentagonal prisms (plus two atoms at
the poles, ligancy 12) are formed between each two complexes. The
two types of aggregates (Figs. 20c and 21c) thus form the infinite,
three-dimensional framework shown in Fig. 22. The dark icosahedra
shown in Fig. 21c have been omitted in this large model, since they
are difficult to insert.

The framework of icosahedra (Fig. 22) fits into the cavities of
the Friauf-polyhedra framework shown in Fig. 19. Both frameworks
share vertices in such a way that additional coordination shells are
produced, most of them icosahedra, that penetrate the Friauf polyhedra as well as the icosahedra described above.

The unit cube of structure contains 124 Friauf polyhedra (L16);
144 μ-phase polyhedra (L15); 120 centered pentagonal prisms, each
one with two atoms at the extended poles, and two additional atoms
out from the centers of two prism faces (L14); 168 centered
pentagonal prisms, each one with two atoms at the extended poles (L12);
568 centered icosahedra (L12).

In terms of the unit-cell content the composition can be written
$Cu_{640}Cd_{484}$, and the 1124 atoms are distributed among 29 sets of
equivalent positions.

There appears to be a correlation between the two interpenetrating framework structures and the structures of the two phases
that form the metastable eutectic mixture that reacts to give this

Figure 22. The two complexes of icosahedra shown in Figs. 20
            and 21 share vertices with one another to form the
            diamondlike, infinite, three-dimensional framework
            shown here. This framework fits into the cavities
            formed by the Friauf polyhedra framework shown in
            Fig. 19.

phase (see introduction). One of these is CdCu$_2$ with the C36 type
of structure, the other is Cd$_8$Cu$_5$ with the D8$_2$ type of structure
(γ-brass), which is known to be essentially icosahedral. Parts of
these two structures are retained in the two frameworks. The zigzag
chain of Friauf polyhedra shown in Fig. 19 corresponds to a mixture
of the C14, C15, and C36 (MgZn$_2$, MgCu$_2$, MgNi$_2$) types of structure;
refer also to Fig. 24.

Some Notes on the Derivation of this Structure

In this case, we had to resort to six crystallographically
different Friauf polyhedra and five different icosahedra to
describe the model, whereas in the case of βMg$_2$Al$_3$ only five different Friauf polyhedra were needed. This implies that the derivation of the trial structure of Cu$_4$Cd$_3$ by means of packing maps
involves the fitting of more than twice as many different polyhedra
as does the derivation of the idealized ordered model of βMg$_2$Al$_3$.

The Friauf polyhedra framework shown in Fig. 19 was relatively
easy to derive. A perfect geometrical fit of the set of six polyhedra (F1 to F6; see original paper[6] and Fig. 23) on the packing
map enabled me to place reliance upon this part (about 50%) of the
structure. From here on I had to resort to a few other rules that
I formulated on the basis of observation of an extensive number of
other structures that incorporate Friauf polyhedra.[21] Two of these
rules are: (1) In infinite, three-dimensional frameworks of Friauf
polyhedra, each hexagon of a truncated tetrahedron, not shared by
another truncated tetrahedron, is usually shared by a hexagonal
antiprism that has two atoms at the extended poles (L14) or by a
μ-phase polyhedron (L15). These two polyhedra often terminate
groupings of Friauf polyhedra; see also Figs. 7 and 8. (2) In each
case where Friauf polyhedra form rows, layers, or three-dimensional
frameworks, each truncated tetrahedron is penetrated most often by
twelve icosahedra and sometimes by nine or more icosahedra and three
or less coordination shells of ligancy 13 or 14.

In accordance with these principles, I explored the possibility
of constructing an icosahedron around each vertex of a truncated
tetrahedron that was laid out on the packing map. The orientation
of each icosahedron is determined by the Friauf polyhedron. There
were, furthermore, only two kinds of hexagons of truncated tetrahedra (F2 and F4) that were not shared with other truncated tetrahedra, and these then became shared with μ-phase polyhedra. After
addition of a few more atoms, the structural motif represented by
the five icosahedra about the fivefold axis of symmetry unfolded;
see Fig. 20.

A packing map of Cu$_4$Cd$_3$ is shown in Fig. 23. Here are emphasized by shaded areas only those polyhedra that create the structural

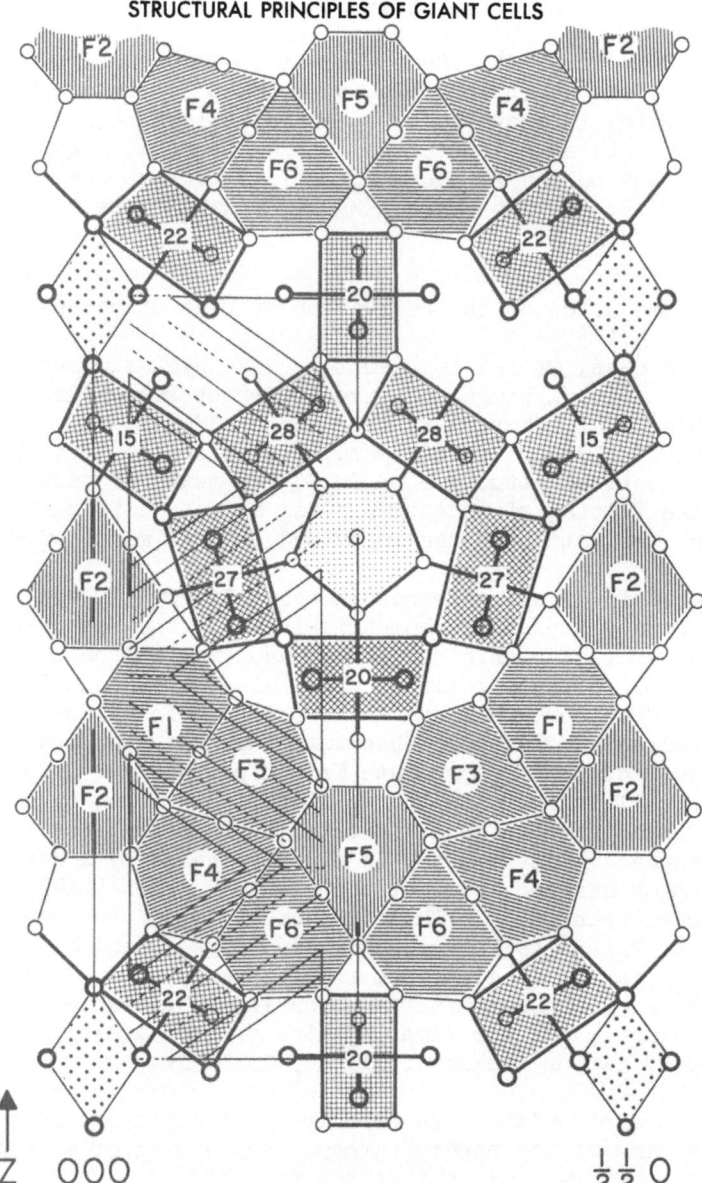

Figure 23.   The packing map of the structure of Cu_4Cd_3. Here are emphasized by shaded areas the polygonal sections of only those polyhedra that produce the structural motifs shown in Figs. 18 to 22. F1 to F6 represent sectioned Friauf polyhedra and the numbers 15, 20, 22 ⋯ are inside sectioned icosahedra. It is seen that the rectangles (compare with Fig. 9d) are slightly deformed.

motifs discussed in the preceding section. The rectangles marked
20, 27, and 28 are the sectioned icosahedra of the complex shown in
Fig. 20, the rectangles marked 15 and 22 form the complex shown in
Fig. 21, and F1 to F6 represent the framework shown in Fig. 19.
The interpenetrating icosahedra around the vertices of the truncated
tetrahedra that were used for the derivation, as discussed above,
are not emphasized here, but they can be traced out by the reader.
The μ-phase polyhedron that shares a hexagon with F2 can be traced
out also. In fact, every structural detail of Cu_4Cd_3 can be read
from this figure by a trained user of packing maps.

COMMON STRUCTURAL FEATURES IN INTERMETALLIC
COMPOUNDS CONTAINING FRIAUF POLYHEDRA

The different types of structures that contain Friauf polyhedra
are: (1) $MgCu_2$ and $MgZn_2$ [C15 and C14 types, Friauf phases[10,11]];
(2) $MgNi_2$ [C36 type, Laves phase[12]]; (3) MgCuAl [mixture of the two
Friauf phases and the Laves phase, Komura[22]]; (4) W_6Fe_7 [μ phases[13]];
(5a) P(Mo-Ni-Cr) [P phases[14]]; (5b) δ(Mo-Ni) [δ phases, closely
related to the P phase[23]]; (6a) $Mg_3Cr_2Al_{18}$ [E phase[19]]; (6b) $ZrZn_{22}$
and other AB_{22} compounds [closely related to (6a)[20,29]]; (6c) αVAl_{10}
[closely related to (6a)[24,32]]; (7a) αMn [X phases[25,26]];
(7b) $\gamma Mg_{17}Al_{12}$ [closely related to (7a)[16]]; (8a) R(Mo-Co-Cr)
[R phases[27]]; (8b) $\epsilon Mg_{23}Al_{30}$ [closely related to R phases[17]];
(9) $Mg_{32}(Zn,Al)_{49}$ [T phase[28]]; (10a) $NaCd_2$[15]; (10b) βMg_2Al_3
[closely related to (10a)[9]]; (11) Cu_4Cd_3.[6]

Figure 24 shows the basic simple modes in which truncated
tetrahedra are connected with each other and with other kinds of
coordination shells. Each of the structures referred to in (1), (2),
and (3) above can be described in terms of layers of close-packed
Friauf polyhedra (see also Fig. 2); they differ only in the way the
layers are superimposed on one another.

In $MgCu_2$ each Friauf polyhedron of the second layers is turned
60° with respect to that of the first, as is shown in Fig. 24a
(mode I), whereas in $MgZn_2$ the second Friauf polyhedron is a mirror
image of the first one (mode II, Fig. 24b). In $MgNi_2$ both modes
occur alternately in the hexagonal c direction. A variation in
sequence of the two modes (I and II) leads to an ever-increasing
length of the period (hexagonal c axis). Striking examples of this
are the two known modifications of MgCuAl, one with c_0 = 21.05Å, the
other with c_0 = 37.89Å.

The zigzag chain of the Friauf polyhedra framework of Cu_4Cd_3
(Fig. 19) exhibits the two modes in the sequence I-II-II-I.

Each hexagon of F2 and F4 in Cu_4Cd_3 that is not shared with
another truncated tetrahedron is shared with a μ-phase polyhedron,

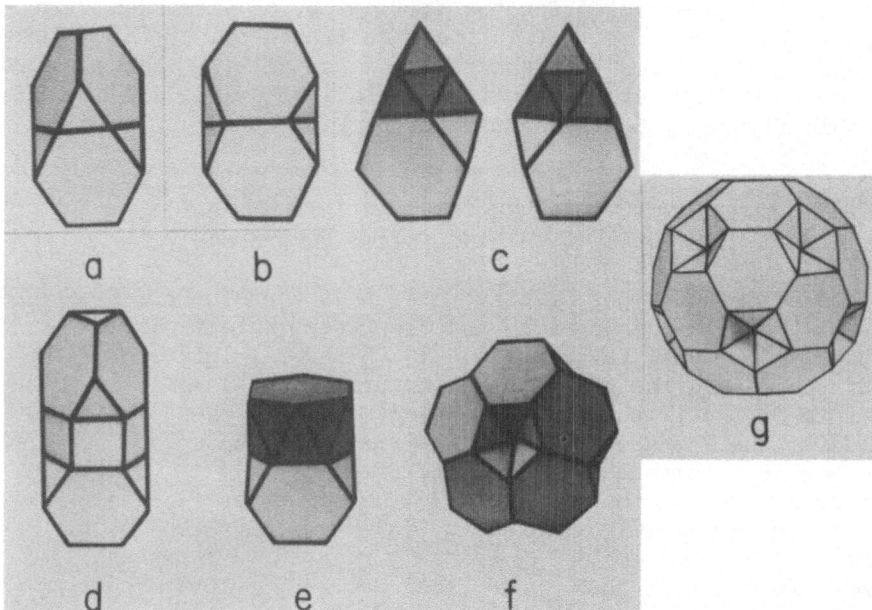

Figure 24. The basic, simple modes in which truncated tetrahedra
are connected with each other and with other polyhedra: (a) Two Friauf polyhedra forming mode I; C15
type, βMg_2Al_3, and Cu_4Cd_3.   (b) Two Friauf polyhedra
in mode II; C14 type, βMg_2Al_3, and Cu_4Cd_3.   (c) A
Friauf polyhedron (light) and a μ-phase polyhedron
(dark).   Both figures represent the same mode III
(see text);   P phase, $Mg_{32}(Zn,Al)_{49}$, and Cu_4Cd_3.   (d)
Two Friauf polyhedra connected via a hexagonal prism.
Two modes are possible but only this one (mode IV)
has as yet been observed; $Mg_3Cr_2Al_{18}$, AB_{22}, and αVAl_{10}.
(e) A Friauf polyhedron (light) and a hexagonal anti-
prism (dark) referred to as mode V; μ phase, P phase,
and βMg_2Al_3.   (f) Five Friauf polyhedra forming the VF
polyhedron (mode VI); βMg_2Al_3 and $Mg_{32}(Zn,Al)_{49}$.   (g)
Twenty Friauf polyhedra forming a truncated icosahedron;
$Mg_{32}(Zn,Al)_{49}$.   Note that the C36 type and MgCuAl represent a mixture of the modes I and II.

as shown in Fig. 24c. The geometrical nature of the two polyhedra is such that there exists only one mode (mode III); rotation of the μ-phase polyhedron (dark) by 60° with respect to the Friauf poly-hedron (light) results in a picture equivalent to that obtained by rotating the whole aggregate 180° so that its rear view can be seen. Both front and rear views are shown in Fig. 24c. This combination occurs also in the structures referred to in (5a,b), (8a,b), (9), and (10a,b) above.

The combination of Friauf polyhedra and hexagonal prisms (Fig. 24d) is extremely rare and occurs only in the structures (6a,b,c). It seems to eliminate the possibility that the atoms at the corners of the shared hexagons surround themselves with icosahedral coordi-nation shells. The structures (6a,b,c) exhibit a relatively large number of square configurations and octahedral interstices.

Association between a truncated tetrahedron and a hexagonal antiprism (Fig. 24e) is observed   in all cases except in (1), (2), (3), (6a,b,c), and Cu_4Cd_3 (11) above.

The VF aggregate (Fig. 24f) is represented not only by $NaCd_2$ and βMg_2Al_3 (10a,b) but also by $Mg_{32}(Zn,Al)_{49}$ (9). Here, each one of twenty truncated tetrahedra has its center at the vertex of a pentagonal dodecahedron. The resulting configuration is the truncated icosahedron shown in Fig. 24g. It is seen that here twelve interpenetrating VF aggregates are arranged about the vertices of an icosahedron. At the center of this complex is an icosahedron that is created by the triangles of the twenty contiguous truncated tetrahedra. This aggregate represents 113 atoms.

MISFIT OF POLYHEDRA

Manifestation of Disorder

The arrangements of Friauf polyhedra shown in Fig. 25 exhibit geometrical properties that deserve special attention. It is seen that in each of these the atom out from a certain hexagon (A atom), which normally is of the large kind, represents simultaneously the vertex of a truncated tetrahedron, at which a small atom (B atom) normally is expected. This vertex, accordingly, fills two disparate functions and therefore is called the bifunctional or hybrid vertex.[17,21] The coordination shell around this vertex is sometimes of the intermediate ligancy 14 (βMg_2Al_3, $\epsilon Mg_{23}Al_{30}$, and R phases) and sometimes of ligancy 13 ($\gamma Mg_{17}Al_{12}$ and X phases).

There are three different modes, in which hybrid vertices are created, as is shown in Fig. 25. If each truncated tetrahedron were regular and each of its 18 edges were of length a, then the dihedral angle between the hexagon and the equilateral triangle shown in

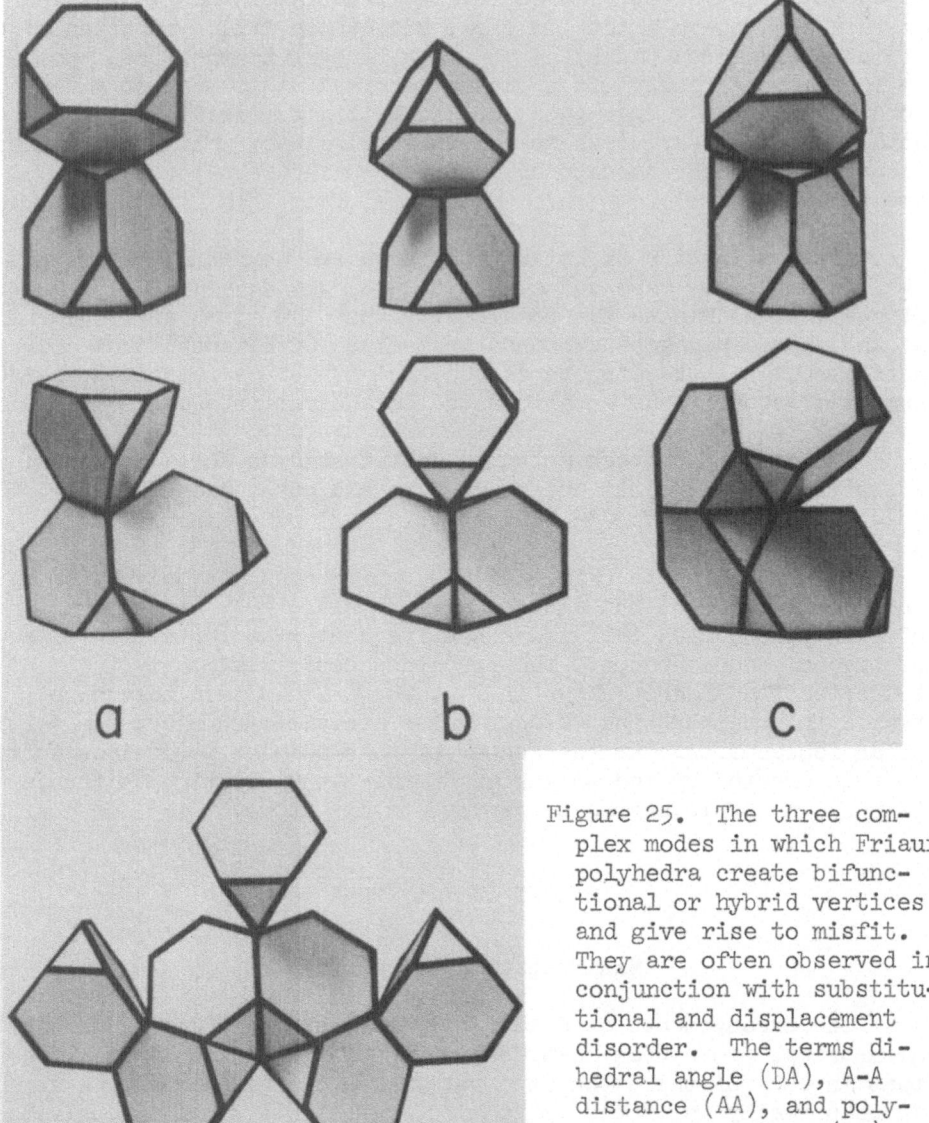

Figure 25. The three complex modes in which Friauf polyhedra create bifunctional or hybrid vertices and give rise to misfit. They are often observed in conjunction with substitutional and displacement disorder. The terms dihedral angle (DA), A-A distance (AA), and polyhedron distortion (PD) are explained in the text: (a) Mode VII. DA = 31°36', AA = 1.07 a, PD = 1.17 a; in $\gamma M_{17}Al_{12}$, χ phases, and βMg_2Al_3. (b) Mode VIII. DA = 35°16', AA = 1.12 a, PD = 1.22 a; in βMg_2Al_3 only. (c) Mode IX. DA = 38°56', AA = 1.17 a, PD = 1.29 a; in $\epsilon Mg_{23}Al_{30}$ and R phases. (d) Mixture of modes VII and VIII, as observed in βMg_2Al_3. The four Friauf polyhedra forming the arch belong to a VF polyhedron.

Fig. 25a (mode VII) would be 31°36', in Fig. 25b (mode VIII) 35°16', and in Fig. 25c (mode IX) 38°56'. In each case the hybrid vertex lies below the normal that passes through the center of the hexagon. In order that this vertex be equidistant from all six corners of the hexagon (with the dihedral angles unchanged) the equilateral triangle has to be converted into an isosceles triangle in which the two shanks are of length 1.17 a (mode VII), 1.22 a (mode VIII), or 1.29 a (mode IX), which involves a significant distortion. The distance between the hybrid vertex and the center of the adjacent Friauf polyhedron (A-A distance) is 1.07 a for mode VII, 1.12 a for mode VIII, and 1.17 a for mode IX, whereas the center-to-center distance between two truncated tetrahedra that share a hexagon [normal modes I and II (Fig. 24a,b)] is 1.23 a.

It is seen that at least one of the Friauf polyhedra must be significantly distorted, although all the truncated tetrahedra may be regular, as is the case in Fig. 25. The creation of the hybrid vertices, accordingly, is associated with a considerable degree of misfit. This feature, in addition to the bifunctional character, suggests that the hybrid vertices may favor substitutional and displacement disorder.

Mode VII is represented by $\gamma Mg_{17}Al_{12}$ in which slightly more than eight magnesium atoms (large) per structural unit can be replaced by aluminum atoms (small) to give $\gamma Mg_{13}Al_{16}$ (radius ratio Mg/Al ~ 1.14). The aluminum-rich compound is metastable at room temperature and transforms on annealing at or below 370°C to $\epsilon Mg_{23}Al_{30}$,[17] which represents mode IX (Fig. 25c). It seems likely that the substitutional disorder, which must be associated with the large homogeneity range, is confined to the bifunctional vertices. There are 24 such vertices in the unit of structure, and the composition $Mg_{17}Al_{12}$ is the one that corresponds to the 100 per cent occupancy of these vertices by large atoms (magnesium).

The well-known X phases[26] are isostructural with $\gamma Mg_{17}Al_{12}$ and the R phases,[27] which show a high degree of disorder, are nearly isostructural with $\epsilon Mg_{23}Al_{30}$.

In βMg_2Al_3 the two kinds of hybrid vertices shown in Figs. 25a and b occur simultaneously and affect the Friauf polyhedra called F3 and F5, as is shown in Fig. 25d. Here, the vertex of mode VIII was found to be displaced part of the time (see atom No. 19 in Fig. 11 of the original paper[9]). The Friauf polyhedron which is on top of the two contiguous polyhedra, as shown in Figs. 25b and d, is the one called F5 in βMg_2Al_3, and the partial disorder observed here is confined to this region of the structure, as has been pointed out earlier and also in the original paper.[9]

In view of the resulting misfit, the frequent occurrence of the hybrid vertices seems surprisingly high. As we have seen, the hybrid vertices seem to be associated with large homogeneity ranges, substitutional disorder, or displacement disorder, with the possible exception of $Mg_{23}Al_{30}$, which seems to have a narrow homogeneity range.[33]

INTERACTION BETWEEN Friauf POLYHEDRA AND ICOSAHEDRA

In almost each type of structure so far discussed, the Friauf polyhedra are by far outnumbered by icosahedra. This is caused by the fact that between ten and twelve vertices of each truncated tetrahedron are centers of icosahedral coordination shells. These penetrate the Friauf polyhedra in such a manner that in almost each case about one-half of their vertices are occupied by large atoms (A atoms) and the other half by small atoms (B atoms). Thus, the average effective size of the atoms at vertices of each icosahedron corresponds to a central atom that is approximately ten per cent smaller, the nominal radius ratio Mg/Al, corrected for the proper ligancies (L16 and L12, respectively), is about 1.14.

In the infinite, three-dimensional framework of Friauf poly- hedra shown in Fig. 19, each corner of a facet represents an icosa- hedron center. The basic building block thus appears to be the icosahedron, and the Friauf-polyhedra framework may be regarded as describing the modes, in which the icosahedra are arranged with respect to one another.

The two kinds of interpenetrating polyhedra are not commensu- rate, that is, both cannot be regular simultaneously; either one or both of them have to be deformed. The nature of deformation depends upon the mode of interpenetration, which, in turn, depends upon the orientation of the Friauf polyhedra with respect to one another in the various types of structures.

Deformation provides flexibility with regard not only to the effective size but also to the effective shape of the atom at the center of a coordination shell. The degree of distortion of the one shell relative to the other probably depends upon the power of each species of atoms (say, A atoms and B atoms in Friauf structures) to enforce its coordination requirements. It seems possible, therefore, that the diversity of environment in the extremely complex structures results from the fact that the atoms have assumed positions relative to one another which are characteristic of their individual sizes, shapes, and bond-forming powers, whereas a simple structure may involve adjustment of one atom to another in such a way that only an average value of atomic properties finds expression.

POSSIBLE FACTORS DETERMINING STABILITY

A striking feature of the most common coordination shells observed here is the frequent occurrence of triangular faces; the Friauf polyhedron is bounded by 28 triangles, the μ-phase polyhedron by 26, the hexagonal antiprism (with occupied poles) by 24, and the icosahedron by 20. In each of these, each corner is connected with five or six others that form a nearly planar pentagon or hexagon, respectively. It is possible (Pauling, private communication) that these configurations are more stable than squares because they reduce repulsions that may exist between atoms at nonadjacent corners. With atoms arranged at the distance a at the corners of a square, the next-nearest-neighbor distance is 1.41 a, while in a pentagon it is 1.62 a, and in a hexagon 1.73 a. In a triangular arrangement there are no unbonded neighbors to produce repulsion. In this context it may be of interest to direct attention to Fig. 20 and Figs. 24f and g. Here we observe a pronounced tendency toward the creation of large atom complexes with fivefold axes of symmetry.

A second important factor of triangular configurations is that they are a prerequisite for the formation of tetrahedral interstices; square configurations give rise to octahedral interstices. The volume of a central sphere that touches six contiguous spheres forming an octahedron is 6.3 times larger than a sphere that is at the center of a corresponding tetrahedron. Thus, tetrahedral configurations result in a minimum of interstitial space.

One of the most important factors stabilizing the icosahedral configuration is probably the shortened center-to-vertex distance. The most tangible experimental evidence for the tendency of atoms of unlike atomic radii to arrange themselves into these configurations has been brought forth in βMg_2Al_3. Here, the disorder, which provides the stability, brings about a substantial increase in the number of icosahedral coordination shells, and the apparent requirement is satisfied that about one-half of the vertices should be occupied by large atoms (magnesium) and the other half by small atoms (aluminum). It is possible that other disordered structures, which for a long time have defied detailed analyses, may be attributed to the same cause.

The icosahedron seems to be the one coordination shell that is most preferred in intermetallic compounds.

METALLIC RADII AND ATOMIC SIZES

The metrical properties of the polyhedra so far discussed are clearly inconsistent with sphere packing, as has been shown in the initial sections of this article. We have also seen that the atom

at the center of a Friauf polyhedron (A atom) appears to be tetrahedrally deformed, whereas the μ-phase polyhedron results in
trigonal deformation.

The ratio between the distances A–B and B–B = a (a = average
edge of the truncated tetrahedra) has the characteristic value of
1.17 that does not change significantly from one Friauf polyhedron
to another[6,9,17] or even from one structure type to another,
although the nominal radius ratios A/B of the atoms may differ
drastically. The most instructive example is αMn, which is isostructural with γMg$_{17}$Al$_{12}$; see also Fig. 25a. Here, the nominal
radius ratio is unity, but the manganese atoms behave as if they
were of drastically different sizes. Of the 58 atoms in the body-
centered cubic cell of edge length 8.91Å, ten are inside Friauf
polyhedra (large Mn atoms), 24 are at hybrid vertices of the kind
shown in Fig. 25a (mode VII, intermediate size), and 24 occupy
centers of icosahedra (small Mn atoms).

In αVAl$_{10}$, which is isostructural with Mg$_3$Cr$_2$Al$_{18}$,[19] there are
no large atoms (for example, magnesium) to occupy the centers of
the Friauf polyhedra, and these are, in fact, empty part of the
time.[32] The formula may be written (Al$_2$, hole)$_3$ V$_2$Al$_{18}$. It can be
seen[21] that this phenomenon is explicable on the basis of the VEC
rule. In the AB$_{22}$ compounds[20] (Fig. 24d) the B atoms seem to be of
drastically different sizes.

Rudman[30] observed that in many structures of the Friauf type
(C14 and C15), aluminum behaves as if its atomic size were
considerably smaller than in the pure metal, and that this behavior
is also reflected in the nature of substitutional solid solutions
with aluminum. The Al-Al distances observed in βMg$_2$Al$_3$ (and these
are extremely numerous) are clearly consistent with this observation, but a still more striking example is found in Mg$_2$Cu$_6$Al$_5$.[31]
Here, an aluminum atom is at the center of an icosahedron that has
twelve copper atoms at the vertices. This arrangement indicates
that the effective size of the aluminum atom is about ten per cent
smaller than that of the copper atoms, whereas in pure aluminum the
metallic "radius" is 12 per cent larger than in pure copper.

The apparently large size of the magnesium atoms inside the
Friauf polyhedra in Mg$_2$Al$_3$ may be attributed to the fact that two
metallic valences are engaged in sixteen bonds, and thus each
represents a weak bond that corresponds to a large interatomic
distance.

"Metallic radius" or "atomic size" does not seem to represent
a meaningful concept unless it is used in connection with terms that
describe the conditions under which it is observed, such as the
ligancy, bond number, and valence, or perhaps some other important
properties that have yet to be defined.

ACKNOWLEDGMENT

Acknowledgment is made to the National Science Foundation for its financial support (Grants GP-1701 and GP-4237), which enabled me to carry out the investigations described here.

REFERENCES

1. L. Pauling, J. Am. Chem. Soc. $\underline{45}$: 2777, 1923.

2. L. Pauling, Am. Sci. $\underline{43}$: 285, 1955.

3. H. Perlitz, Nature (London) $\underline{154}$: 607, 1944.

4. H. Perlitz, Chalmers Tekniska Högskolas Handlingar $\underline{50}$: 1, 1946.

5. F. Laves and K. Möller, Z. Metallk. $\underline{30}$: 232, 1938.

6. S. Samson, Acta Cryst. $\underline{23}$: 586, 1967.

7. R. Borg, AIME Trans. $\underline{221}$: 527, 1961.

8. E. A. Owen and L. Pickup, Proc. Roy. Soc. (London) $\underline{A139}$: 526, 1933.

9. S. Samson, Acta Cryst. $\underline{19}$: 401, 1965.

10. J. B. Friauf, J. Am. Chem. Soc. $\underline{49}$: 3107, 1927.

11. J. B. Friauf, Phys. Rev. $\underline{29}$: 34, 1927.

12. F. Laves and H. Witte, Metallwirt. $\underline{14}$: 645, 1935.

13. H. Arnfelt and A. Westgren, Jernkontor. Ann. $\underline{119}$: 185, 1935.

14. D. P. Shoemaker, C. B. Shoemaker, and F. C. Wilson, Acta Cryst. $\underline{10}$: 1, 1957.

15. S. Samson, Nature (London) $\underline{195}$: 259, 1962.

16. F. Laves, K. Löhberg, and P. Rahlfs, Nach. Ges. Wiss. Göttingen, Fachgr. IV $\underline{1}$: 67, 1934.

17. S. Samson and E. K. Gordon, Acta Cryst. $\underline{B24}$: 1004, 1968.

18. S. Samson, Acta Cryst. $\underline{17}$: 491, 1964.

19. S. Samson, Acta Cryst. $\underline{11}$: 851, 1958.

20. S. Samson, Acta Cryst. $\underline{14}$: 1229, 1961.

21. S. Samson, Chapter in book; "Structural Chemistry and Molecular Biology", W. H. Freeman and Company, San Francisco, 1968, pp. 687-717.

22. Y. Komura, Acta Cryst. $\underline{15}$: 770, 1962.

23. C. B. Shoemaker and D. P. Shoemaker, Acta Cryst. $\underline{16}$: 997, 1963.

24. P. J. Brown, Acta Cryst. $\underline{10}$: 133, 1957.

25. A. J. Bradley and J. Thewlis, Proc. Roy. Soc. (London) A115:
     456, 1927.

26. J. S. Kasper, Acta Met. 2: 456, 1954.

27. Y. Komura, W. G. Sly, and D. P. Shoemaker, Acta Cryst. 13:
     575, 1960.

28. G. Bergman, L. T. Waugh, and L. Pauling, Acta Cryst. 10: 254,
     1957.

29. E. D. Sands, Q. C. Johnson, A. Zalkin, O. H. Krikorian, and
     K. L. Kromholtz, Acta Cryst. 15: 832, 1962.

30. R. S. Rudman, AIME Trans. 233: 874, 1965.

31. S. Samson, Acta Chem. Scand. 3: 809, 1949.

32. A. E. Ray and J. F. Smith, Acta Cryst. 10: 604, 1957.

33. J. B. Clark and F. N. Rhines, AIME Trans. J. Metals 2: 6, 1957.

STRUCTURAL PROPERTIES OF SOME σ-PHASE RELATED PHASES

Clara B. Shoemaker and David P. Shoemaker

Massachusetts Institute of Technology

Cambridge, Massachusetts 02139

ABSTRACT

A survey is given of tetrahedrally close-packed structures which are dense packings of atoms of slightly different sizes with only tetrahedral interstices. The coordination numbers present are 12, 14, 15 and 16. The radii of the atoms depend on the chemical composition of the phase and increase with the coordination number. In the higher coordinations (14, 15, 16) the atoms are definitely non-spherical, each having in effect two different radii.

STRUCTURAL PRINCIPLES

The intermetallic compounds to be discussed here - mostly, but not exclusively, formed by transition metals - represent relatively dense packings of atoms of slightly different sizes (about 10% difference in CN12 radii). The name "σ-phase-related phases" is not a particularly good one, but was chosen since it was not until some years after the structure determination of the sigma phase in 1950[1] that it became clear that there is a large group of structures, some already known for a long time, with many common characteristics. In view of what follows a better name for these structures might be: <u>tetrahedrally close-packed structures</u>.

It is well known that in packing spheres of equal sizes the best space filling is achieved in the cubic or hexagonal close-packed structures and their variants. Each atom in these arrangements has twelve neighbors in the first coordination shell; in the cubic closest packing these are at the corners of a cuboctahedron, a polyhedron having 6 square and 8 triangular faces. There are tetrahedral and (half as many) octahedral interstices or "holes" in these packings. The local mean atom density is in effect higher at the tetrahedral holes than at the octahedral holes. A denser arrangement might be achieved if all interstices could be made tetrahedral. However, it appears impossible to fill space with regular tetrahedra throughout.

By introducing some variability in the sizes of the spheres one can achieve packings with only tetrahedral interstices. The tetrahedra are now no longer regular, although the ratio of the length of the longest tetrahedron edge in a given structure to that of the shortest need not exceed about 4/3. The crystal structure may be regarded as a space filling of these somewhat irregular tetrahedra, sharing faces, edges, and vertices. Within the above-mentioned limits of edge-length variability, a given tetrahedron edge -- i.e. an interatomic ligand -- is shared either among five tetrahedra or among six tetrahedra in structures now known.* Assuming that only five or six tetrahedra may share a given edge, the number that share a given vertex is limited to the four values 12, 14, 15 and 16. The tetrahedra sharing a given vertex form a coordination polyhedron with triangular faces and five-fold or six-fold vertices (i.e. five or six faces meeting at a point). The four possible coordination polyhedra are shown in Fig. 1 and their main properties are described in Table 1. The

---

* Sharing among four is possible, and sometimes encountered, but the four tetrahedra then combine to make a somewhat distorted octahedron, and the structure is then perhaps somewhat better described as having some octahedral interstices as well as tetrahedral ones. Sharing among seven would seem to be allowable within the above limit but seems not to have been encountered in structures of this type.

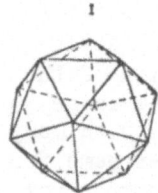 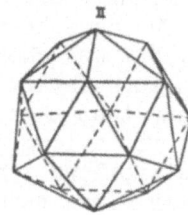 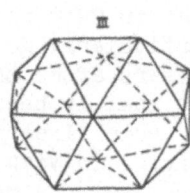 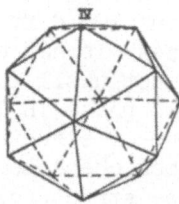

Figure 1. The four triangulated coordination polyhedra with
         5-fold and 6-fold vertices: I, CN12 (icosahedron);
         II, CN16; III, CN15; IV, CN14. (From Bergman, Waugh
         and Pauling[24]).

Table 1.   Triangulated coordination polyhedra with
5-fold and 6-fold vertices

Type	No. of vertices		No. of faces	Ideal point symmetry	Ideal sublevel degeneracies	
	5-fold	6-fold			p	d
CN12*	12	0	20	I_h - $5\bar{3}\frac{2}{m}$	3	5
CN14	12	2	24	D_{6h} - $\overline{12}\cdot 2\cdot m$	1,2	1,2,2
CN15	12	3	26	D_{3h} - $\bar{6}m2$	1,2	1,2,2
CN16	12	4	28	T_d - $\bar{4}3m$	3	2,3

* Regular, or approximately regular, icosahedron.

point symmetries given apply to the idealized polyhedra; in the actual structures all polyhedra are somewhat irregular. All structures in this family contain icosahedra (CN12) and at least one other coordination type. The importance of these polyhedra was pointed out by Kasper[2] in Cleveland in 1955, and independently by us after the structure determination of the P phase.[3] The polyhedra were also mentioned briefly by Bergman, Waugh and Pauling.[24]

The icosahedron may be described as a pentagonal antiprism with one atom outside each of the two opposing pentagonal faces. Each ligand from the center atom to a vertex atom is surrounded by five vertices describing an approximately planar pentagon; thus the icosahedron has twelve five-fold vertices, or the central atom forms twelve five-fold ligands ("minor ligands" according to Frank and Kasper[17]). The CN14 polyhedron (IV in Fig. 1) is a hexagonal antiprism with one atom outside each of the two opposing hexagonal faces. This polyhedron has twelve five-fold ligands around the waist, but the ligands to the other two atoms are surrounded by atoms defining planar hexagons, so there are two six-fold ligands on a straight line ("major ligands" according to Frank and Kasper). The CN15 polyhedron may be described as a truncated trigonal prism with extra atoms outside the three prism faces which have become hexagons as a result of the truncation. (In Fig. 1, III the trigonal axis is horizontal). The three ligands from the center atom to these last three atoms are six-fold and lie in a plane about 120° apart, the other twelve ligands are five-fold. This polyhedron is called by Samson[26] the "μ-phase polyhedron". The CN16 polyhedron may be described as a trucated tetrahedron with extra atoms outside the tetrahedral faces which have become hexagons as a result of the truncation. The ligands to these four atoms are six-fold and are in tetrahedral directions; the other twelve vertices are again five-fold. In Fig. 1, II a three-fold axis is almost perpendicular to the plane of the paper through the center of the central triangle. One six-fold vertex is in back of the figure, on the same three-fold axis, and the other three are surrounding the central triangle, at the upper left, upper right, and bottom of the illustration. This polyhedron is called by Samson[26] the "Friauf polyhedron".

These polyhedra interpenetrate each other, so that every vertex atom is again the center of a polyhedron. The six-fold vertices have to be occupied by atoms with CN larger than 12 and the orientation of their polyhedra has to be such that all six-fold ligands point in the required directions. As a result there is a continuous framework of six-fold ligands (the "major skeleton" of Frank and Kasper[18]) which may be linear, planar or three-dimensional. Sometimes two or more non-interconnected three-dimensional arrays of major bonds occur.

Most of the intermetallic structures in this family are formed by combinations of A and B transition elements (A elements are to the left of Mn in the periodic table and B elements are to the right of Mn; Mn itself is ambivalent). As will be discussed more fully later the B elements occupy preferentially the icosahedral positions, the A elements occupy the larger CN15 and CN16 positions, and A elements or mixtures of A and B elements occupy the CN14 positions. In the ternary alloys that contain Si or Al these elements share the CN12 positions with the B elements.

CLASSIFICATION OF STRUCTURES

Some tetrahedrally close-packed structures have been known for many years. In 1927 Friauf determined the structures of $MgCu_2$[4a] and $MgZn_2$[4b] and in 1934 Laves and Löhberg published the structure of $MgNi_2$[5a] which they showed to be related to the structures of $MgCu_2$ and $MgZn_2$[5b]. These Friauf-Laves phases are the first examples of structures with only tetrahedral interstices. The next representative to be found was the so-called β-W structure determined by Hartmann, Ebert and Bretschneider in 1931[6a]. The first compound found with all four kinds of coordination polyhedra was the μ-phase, the structure of which was determined by Arnfelt and Westgren in 1935[7].

In the fifties after the structure of the σ phase became known[1] Professor Paul A. Beck and his co-workers, first at Notre Dame University and later at the University of Illinois, discovered many new intermetallic phases in ternary transition-metal systems at elevated temperatures (about 1200°)[9]. These phases have many properties in common with the σ phase: they are hard and brittle, with low electrical conductivity. They have at most only small magnetic moments, even alloys containing Fe, Mn, Co, Ni. The X-ray powder diffraction patterns are extremely complex, with strong lines occurring in bunches and large gaps in between. Usually these phases have no equilibrium melting points and are formed by solid state reactions, so that as ordinarily prepared they are polycrystalline with very small crystallites. Professor Beck was kind enough to send us specimens of these new compounds and we were able to determine the structures of the P(Mo-Ni-Cr)[3], the R(Mo-Co-Cr)[10] and the δ(Mo-Ni)[11] phases, all variations on the structure of the σ phase, with CN12, 14 and 15 as in the σ phase and in addition CN16. The main handicap in the crystallographic work was the small size of the single-crystal fragments. Typically the intensity data were collected on fragments with largest dimension between 0.03 and 0.06 mm.

One of the simplest representatives of the tetrahedrally close-packed structures, the structure of Zr_4Al_3 (having CN12, 14 and 15) was -- curiously enough -- determined as late as 1960 by Wilson, Thomas and Spooner[12].

Recently many new intermetallic compounds have been discovered by Beck and co-workers[13] and also by Hladyshevskii and co-workers[14] in ternary systems of transition metals with non-transition elements, notably Si or Al. The structures of these compounds, many probably belonging to the group of σ-phase related structures, remain to be determined. We were able to determine the structure of the M phase (Nb-Ni-Al)[15a], discovered by Benjamin, Giessen and Grant[15b] and found it to be closely related to the μ phase and the P phase. The D phase (V-Fe-Si) appeared related to the δ phase, but contains some irregular coordination polyhedra also[16].

We will not give detailed descriptions of the crystal structures since many review articles on the subject have appeared (Kasper, in: "Theory of Alloy Phases", 1956[2]; Frank and Kasper, 1958[17], 59[18]; Nevitt, in: "Electronic Structure and Alloy Chemistry of the Transition Elements", 1963[19]; Wernick, in "Intermetallic Compounds", 1967[20]; Shoemaker and Shoemaker, in: "Structural Chemistry and Molecular Biology", 1968[21]). Table 2, taken from the last reference, lists some crystallographic properties of the intermetallic compounds under discussion. The most exhaustive treatment of the geometrical and topological features of structures of this kind has been given by Frank and Kasper[17,18]; these authors concluded that most of the possible structures with atoms only in "normal" coordinations (i.e., with only the four coordination polyhedra described above and hence tetrahedrally close-packed) are layered structures. They then showed how a whole family of structures may be derived by stacking similar layers in different ways and by introducing

Table 2. Crystal structures of sigma-phase-related phases with tetrahedral interstices

Phase	Example	Space group	Lattice constants (Å)			Atoms per cell	% sites with CN					
			a	b	c		11	12	13	14	15	16
A15	β-W (W$_3$O), Cr$_3$Si (75,25)	O_h^3 - Pm3n	5.036 4.564			8		25		75		
σ	Cr$_{46}$Fe$_{54}$	D_{4h}^{14} - P4$_2$/mnm	8.800		4.544	30		33		53	13	
Zr$_4$Al$_3$	Zr$_{57}$Al$_{43}$	C_{3h}^1 - P6̄	5.433		5.390	7		43		28	28	
P	Mo$_{42}$Cr$_{18}$Ni$_{40}$	D_{2h}^{16} - Pbnm	9.070	16.983	4.752	56		43		36	14	7
δ	Mo$_{50}$Ni$_{50}$	D_2^4 - P2$_1$2$_1$2$_1$	9.108	9.108	8.852	56		43		36	14	7
R	Mo$_{31}$Cr$_{18}$Co$_{51}$	C_{3i}^2 - R3̄	10.903		19.342	159*		51		23	11	15
μ	Mo$_{46}$Co$_{54}$	D_{3d}^5 - R3̄m	4.762		25.615	39**		54		15	15	15
M	Nb$_{48}$Ni$_{39}$Al$_{13}$	D_{2h}^{16} - Pnam	9.303	16.266	4.933	52		54		15	15	15
Mg$_{32}$(Al,Zn)$_{49}$	Mg$_{32}$(Al,Zn)$_{49}$	T_h^5 - Im3	14.16			162		61		7	7	25
C15	MgCu$_2$(33,67)	O_h^7 - Fd3m	7.080			24		67				33
C14	MgZn$_2$(33,67)	D_{6h}^4 - P6$_3$/mmc	5.16		8.50	12		67				33
C36	MgNi$_2$(33,67)	D_{6h}^4 - P6$_3$/mmc	5.27		13.3	24		67				33
A12	α-Mn X-Mo$_{17}$Cr$_{21}$Fe$_{62}$	T_d^3 - I4̄3m	8.912 8.920			58		41	41			17
D	V$_{26}$Fe$_{44}$Si$_{30}$	D_4^4 - P4$_1$2$_1$2	8.83?		8.646	56	14	29	14	14		29
β-Mg$_2$Al$_3$(NaCd$_2$)	Mg$_{40}$Al$_{60}$	Fd3m	28.24			1168		58				22
ϵ-Mg$_{23}$Al$_{30}$	Mg$_{43}$Al$_{57}$	R3̄	12.83		21.75	159*	11	45	4	25		15

* 53 atoms per primitive rhombohedral cell.   ** 13 atoms per primitive rhombohedral cell.

"tessellation faults" in the layers. Their catalogue of possible structures included many hypothetical ones, some of which were later found to exist (Zr_4Al_3 and the M phase). Their treatment did not describe those structures with "normal" coordinations that are not formed by stacking planar layers, except that of Mg_{32} $(Al,Zn)_{49}$,[24] which deviates from a strictly layered structure only by minor rumpling. Nevertheless some structures deviating importantly from planarity were found to exist (R phase,[10] δ phase[11]).

I. Layered Structures

Our treatment of these structures is related to that of Frank and Kasper, except for the Friauf-Laves and the μ phases. Frank and Kasper derived the structures of these phases and their variations by stacking hexagonal layers perpendicular to the three-fold axes in different ways. We will describe these structures by viewing them in the direction of [1$\bar{1}$0] for cubic structures or in the direction of [110] for hexagonal structures. In this way these phases appear to be more closely related to the other layer structures.

1. Main layers formed by tessellations of hexagons and triangles. All structures to be discussed here consist of two main layers formed by hexagons and triangles separated by one half lattice repeat in the perpendicular direction (solid-line net at z = o and broken-line net at z = 1/2) and two identical secondary layers (filled-in circles) half-way between the main layers (at z = 1/4 and 3/4) centering the hexagonal antiprisms. The coordination numbers of the atoms increase with the sizes of the circles. The structures are arranged in order of increasing complexity of the secondary networks.

1A. Secondary network of type 4^4, that is, four squares connecting at each atom (Fig. 2a). The two main layers are equivalent. Each hexagon consists of four CN14 atoms and two CN12 atoms. The atoms in the secondary layers are at the centers of CN14 polyhedra, which share triangular faces perpendicular to the main layers with four adjacent CN14 polyhedra. Perpendicular to the paper the CN14 polyhedra interpenetrate so that adjacent polyhedra share hexagons and the center atom of one polyhedra is a vertex atom of the next. This is structure type A-15 (A_3B, "β-tungsten", Cr_3O or Cr_3Si)[6a,b] projected down a cube axis. The distortion of the regular hexagons is such that the distances from an "in-between" or secondary-layer atom to the CN14 atoms in the hexagons is about 10% larger than the distance from that atom to a CN12 atom. The distance between the atoms in adjacent secondary layers (six-fold ligand) is 15% smaller than the distance between a secondary CN14 atom and a CN14 atom in a main layer (five-fold ligand). The framework of 6-coordinated ligands, or the "major skeleton",

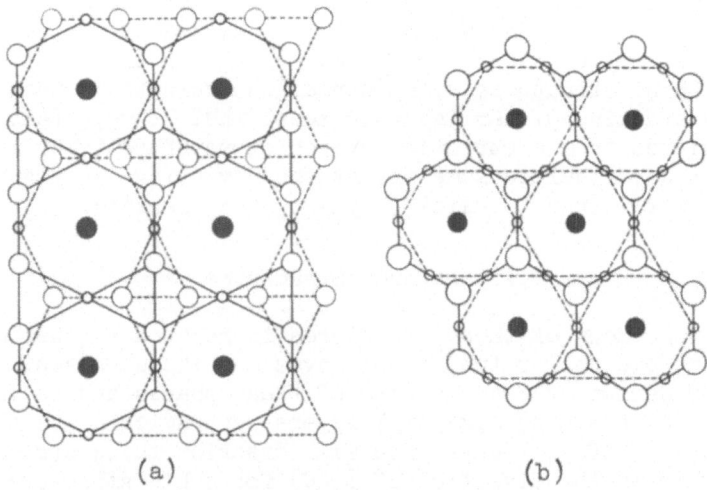

(a)   (b)

Figure 2. Two basic ways of joining rows of hexagons in layer
          structures.
          (a) Secondary network with tessellations of type 4^4.
          Projection of "β-W" or Cr_3Si down a cube axis.
          (b) Secondary network with tessellations of type 3^6.
          Projection of Zr_4Al_3 down the hexagonal axis.
          In these and following figures open circles represent atoms
          on planar "main layers" half the cell repeat apart. Filled
          circles represent atoms on "secondary layers" halfway
          between the main layers. Circles increase in size with
          coordination number.

consists of linear rows of CN14 atoms in the cube directions, that
is, in the z direction perpendicular to the paper and in the x and
y directions parallel to the paper.

    1B. <u>Secondary network of type 3^6</u>, that is six triangles
connecting in each atom (Fig. 2b). The two main layers are now
different. One main layer (solid-line) consists of hexagons
sharing sides with 6 neighboring hexagons without triangles in
between. The atoms in these hexagons are all 15-coordinated. The
next main layer (broken-line net) consists of smaller hexagons of
12-coordinated atoms sharing corners with neighboring hexagons such
that triangles are formed between the hexagons. The atoms in the
secondary layers (filled-in circles) are again at the centers of
CN14 polyhedra, which now share triangular faces perpendicular to
the main layers with six other polyhedra of this kind. The
interpenetration of the CN14 polyhedra perpendicular to the paper
is similar to that found in β-W. Fig. 2b is a projection down the

hexagonal axis of Zr_4Al_3[12], in which Zr occupies the CN14 and CN15
positions and Al the CN12 positions. Perpendicular to the
hexagonal axes there are three layers of Zr alternating with one
layer of Al. Perhaps this feature causes this structure to be less
stable, explaining why no other compounds with this crystal
structure have been found. The same arrangement does occur, however,
as a part of the μ-phase structure (see under 2C). The major
skeleton consists of rows of ligands connecting CN14 atoms and
planar arrangements of ligands connecting CN15 atoms (every other
main layer). The existence of this structure was predicted by
Frank and Kasper who called it the "split-calcium CaZn$_5$" structure[18].

The secondary networks described under 1A and 1B may be
combined in different ways leading to more possible structures.
Only one of these will be described here.

1C. Secondary network of type $4.3.4.3^2$, resulting from
alternating stacks of unit cells (with origin at a secondary-
layer atom) of the β-W structure (square pattern of secondary-
layer atoms) with stacks of unit cells of the Zr_4Al_3 structure
(triangular pattern of in-between atoms). This alternation occurs
both in horizontal and vertical directions causing the secondary-
layer atoms to form zig-zag lines in both directions. The
resulting structure is tetragonal and in fact is that of the σ
phase (Fig. 3a) with coordinations 12 and 14 (as in β-W) and 15
(as in Zr_4Al_3). The two main layers are equivalent and thus CN15

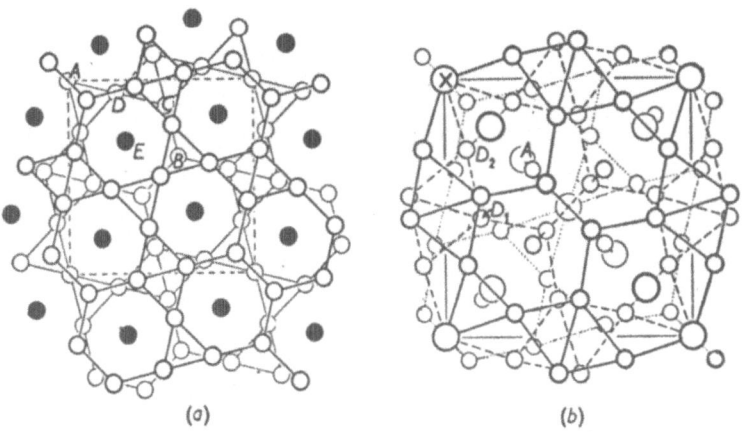

(a)   (b)

Figure 3. (a) Hexagon-layer structure with secondary network of
           type $4.3^2.4.3.$ that is a combination of the types shown in
           Fig. 2. Structure of the σ phase, projected down the
           tetragonal z axis. A, D have CN12; C, E: CN14; B: CN15.
           (b) Structure of α-Mn or χ phase; layers are rumpled, but
           note resemblance to σ-phase layers.

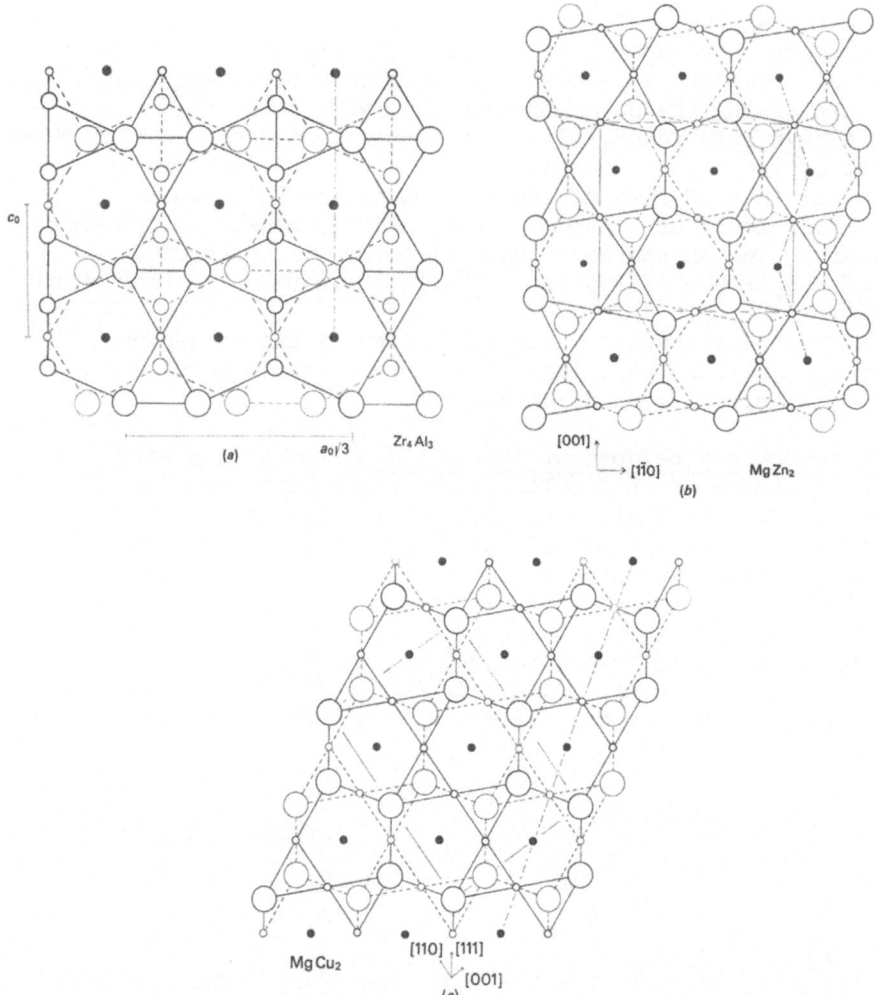

Figure 4.   Three basic ways of joining rows of pentagons in layer
        structures.
        (a) Secondary network of type 4^4. Structure of Zr_4Al_3,
        normal projection on (110).
        (b) Secondary network of type 3^6 with two different ways
        of joining rows of pentagons, as indicated by dash-dot
        line. Structure of $MgZn_2$, normal projection on (110).
        (c) Secondary network of type 3^6 with only one of the ways
        of joining rows of pentagons as displayed in Fig. 4b, repeat-
        ed throughout. Structure of $MgCu_2$, normal projection on
        ($1\bar{1}0$).

atoms occur in both main layers. Each CN14 polyhedron centered by
a secondary-layer atom shares triangular faces with five surrounding polyhedra of this kind. The irregularity of the hexagons is
such that the distances from the secondary layer atom to the atoms
in the hexagons increase in length with the CN of these atoms.
The major skeleton consists of lines connecting CN14 atoms, and
planar networks of ligands connecting CN14 and CN15 atoms in the
main layers ("Kagomé tiles").

2. <u>Main Layers Formed by Tessellations of Pentagons and
Triangles</u>.
    2A. <u>Secondary networks of type 4^4</u>. Fig. 4a shows the
combination of this secondary network with two equivalent pentagon-
triangle main layers. Each pentagon consists of two CN15 atoms,
two CN14 atoms and one CN12 atom. The atoms in the secondary
layers are at the centers of icosahedra which share a triangular
face with each of the two neighboring icosahedra in the left-right
direction, and an edge with each of the two neighboring icosahedra
in up-down direction (in the paper). This asymmetry causes the
secondary net to be rectangular, rather than square as in case 1A.
Perpendicular to the paper the icosahedra interpenetrate in such a
way that successive icosahedra share a five-membered ring and the
center atom of one icosahedron is a vertex atom of the next one.
This is again the structure of Zr_4Al_3 (1B) viewed now along the
short diagonal of the hexagonal cell.

    2B. <u>Secondary nets with 3^6 tessellations</u> combined with main
layers formed by pentagons and triangles (Fig. 4b, c). Each
icosahedron centered by a secondary-layer atom is surrounded by six
similar icosahedra, sharing a triangular face with four of them and
an edge with the remaining two. The triangles in the secondary net
have two short edges (icosahedra sharing triangles) and one long

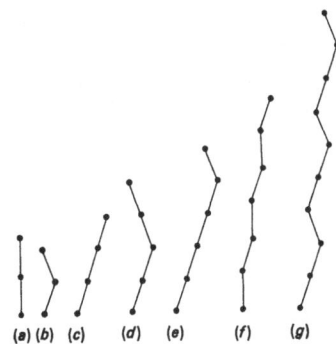

(a) (b) (c)   (d) (e)   (f)   (g)

Figure 5. Lines connecting secondary-layer atoms (see Fig. 4) in:
        (a) Zr_4Al_3, (b) $MgZn_2$, (c) $MgCu_2$, (d) $MgNi_2$,
        (e) 5-layer Komura phase, (f) μ phase, (g) 9-layer
        Komura phase.

edge (icosahedra sharing an edge). The dash-dot lines in Figs. 4b
and 4c trace the short edges of successive triangles. Fig. 4b with
a zig-zag dash-dot line shows the structure of $MgZn_2$ (C14)[4b] as
projected down the short diagonal of the hexagonal cell and Fig. 4c
with a straight dash-dot line shows the structure of $MgCu_2$ (C15)[4a]
viewed along a face diagonal of the cubic cell. Fig. 5 shows how
different combinations of these two ways of stacking rows of
icosahedra (or joining up rows of pentagons in projection) lead to
the $MgNi_2$[5a] structure and Komura's 5L and 9L phases[23]. For each
phase a pentagon consists of three CN16 and two CN12 atoms,
averaged over the structure. The major skeleton consists of the
ligands connecting the CN16 atoms. For $MgZn_2$ the major skeleton
has the wurtzite structure, for $MgCu_2$ the diamond structure, and
for the variations of these phases combinations of the two
structures occur.

2C. Secondary layers with tessellations $4^2.3^3$ are obtained
by alternating rows of pentagons as found in Zr_4Al_3 with rows of
pentagons as found in $MgCu_2$. The resulting structure is that of
the μ phase, projected down $[110]_{hex}$ in Fig. 6a, the first
structure to be determined (in 1935) that combined all four types
of coordination polyhedra[7,22]. The major skeleton consists of
planar networks joining CN15 atoms (graphite layers) and a three-
dimensional framework joining CN14 atoms and CN16 atoms. Many more
structures are possible by alternating in different ways the three
possible connections between the pentagons in the vertical
direction (Fig. 5). As we have pointed out above these structures
were derived by Frank and Kasper by stacking the hexagonal layers
perpendicular to the z axis in different ways and some possible
structures are listed in their Table 1.

2D. Secondary networks of type $4.3.4.3^2$ may be combined with
main layers consisting of pentagons and triangles to form a
structure analogous to the σ-phase (Fig. 6b). The pentagons in
one main layer consist of CN16 and CN14 atoms and in the other
main layer of CN15 and CN12 atoms. The secondary network shows
the same asymmetry as found in the Friauf-Laves phases and the
unit cell is orthorhombic. This hypothetical structure is entry
8 of Frank and Kasper's Table 3[18]. The ratios of atoms with CN12,
14, 15 and 16 are the same as found for the μ phase.

2E. By introducing a zig-zag line of points with tessellations
$4^2.3^3$ into the secondary network of the previous structure a
structure is obtained (shown in Fig. 6c) described as a hypothetical
structure in entry 9 of Frank and Kasper's Table 3 and later found
by us to be the structure of the M phase (Nb-Ni-Al)[15a]. The ratios
of atoms with coordinations CN12, 14, 15 and 16 is again the same
as for the μ phase (the symmetry of the structure is higher than
given by Frank and Kasper and the numbers of atoms with CN14 and
CN15 are in error). The two main layers are equivalent. The major

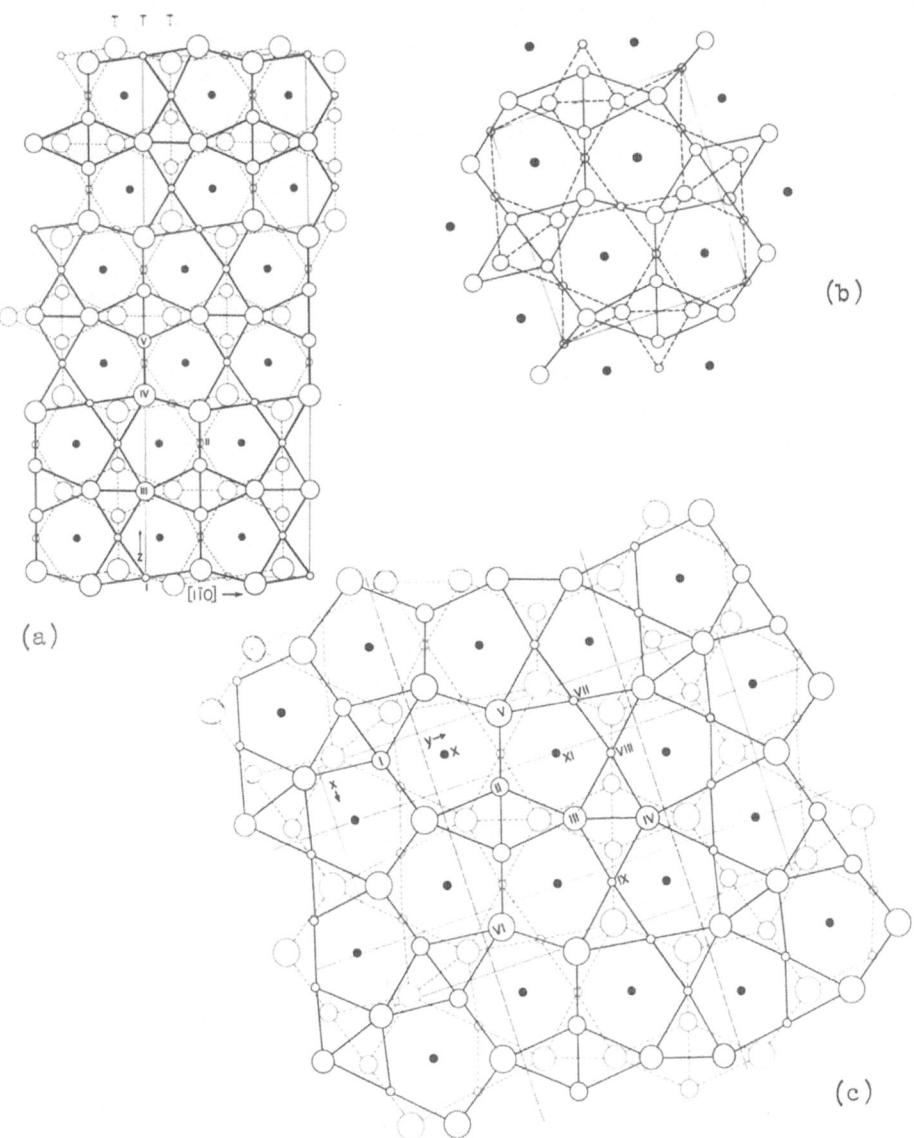

Figure 6.   Pentagon-layer structures with secondary networks that
are combinations of the types shown in Fig. 4.
(a) Secondary network of type $4^2.3^3$. Structure of the
μ phase, Mo_6Co_7, normal projection on (110).
(b) Secondary network of type $4.3^2.4.3$. Hypothetical
structure of pentagon-analogue of the σ phase.
(c) Secondary network with tessellations of both previous
types. Structure of the M phase (Nb-Ni-Al) projected down
the z axis.

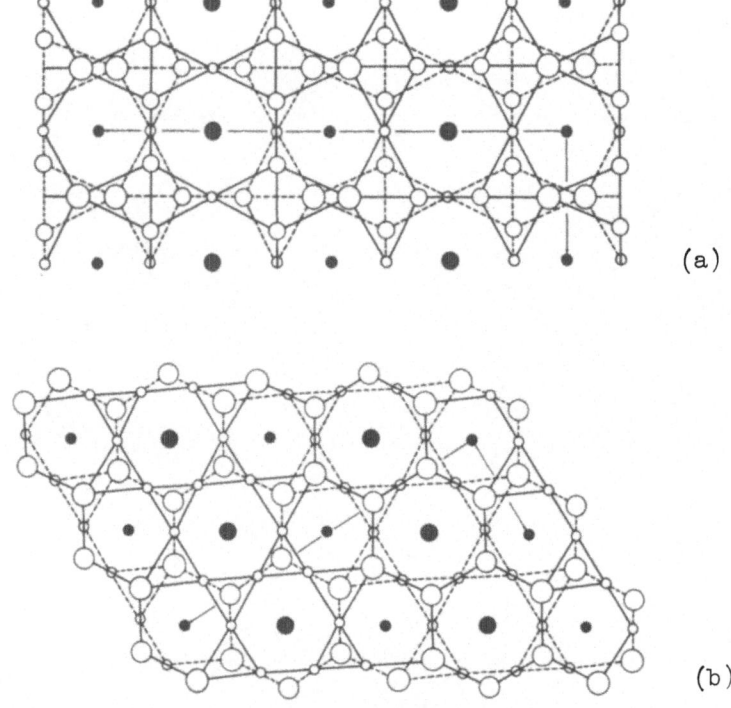

Figure 7.   Two basic ways of joining rows of alternating hexagons
            and pentagons in layer structures.
            (a)   Secondary network of type 4^4.   Hypothetical structure.
            (b)   Secondary network of type 3^6.   Hypothetical structure.

skeleton of the M phase consists of two non-interconnected three-dimensional frameworks.

3. <u>Main layers formed by alternating pentagons, hexagons, and triangles.</u>

3A. Fig. 7a shows a hypothetical structure with <u>secondary layers of the type 4^4</u> and alternating hexagons and pentagons in left-right direction. This is the projection of the all-hexagon analogue of the μ phase in a direction perpendicular to the long axis (entry 1 of Frank and Kasper's Table 3). The ratios of atoms with CN12, 14 and 15 are the same as for the σ phase.

3B. Fig. 7b shows a hypothetical structure with <u>secondary layers of the type 3^6</u> and alternating hexagons and pentagons in

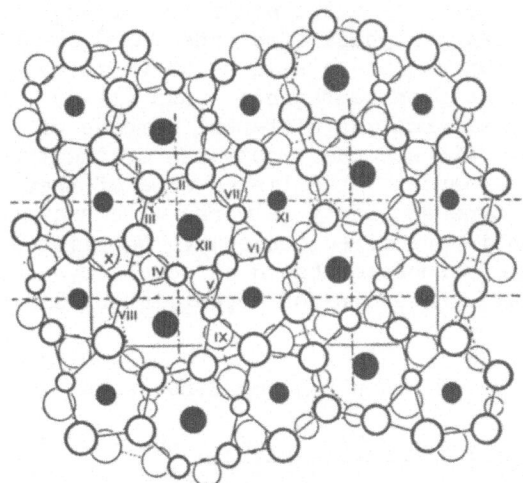

Figure 8.   Hexagon-pentagon layer structure with a secondary
            network that is a combination of the types shown in Fig. 7.
            Structure of the P phase (Mo-Ni-Cr) projected down the z
            axis.   The secondary network is topologically the same as
            that for the M phase (Fig. 6c).

left-right direction.   The ratios of atoms with all four coordination types are the same as for the μ phase.

    3C.   The first structure in which pentagons and hexagons were
found to alternate was that of the P phase[3], discovered by Beck
and coworkers in the Mo-Ni-Cr system[9].   Later the pattern of the
secondary-layer atoms in the M phase with all-pentagon layers (see
above) was found to be the same as that found previously in the P
phase (Fig. 8).   Table 2 shows the similarity of the two phases:
the cell dimensions are almost the same and the space groups are
identical, but the glide plane types are interchanged between the
a and b axes and there are more atoms per unit cell in the P phase.
The percentage of sites with CN12 increases from σ to P to M, and
the same is true for sites with CN16 (0% for σ).   The major
skeleton of the P phase contains two interlocking but mutually
non-interconnected infinite three-dimensional frameworks, similar
to those of the M phase, but the P phase major skeleton contains
in addition vertical rows of CN14 atoms.   So far the P phase
appears to be the only structure in which pentagon and hexagon
strips alternate throughout the structure.

    Structures with the same secondary networks as the μ phase
and the σ phase but with alternating hexagons and pentagons are
possible, but have not been found to exist.   However, there are

regions in the R phase where the atoms are arranged as expected
for a hexagon-pentagon μ phase (see under II-2) and similarly in
the δ phase there appear to be regions where the atoms are
arranged as for a hexagon-pentagon σ phase (II-3).[*] More structures
may be derived by alternating different numbers of pentagons and
hexagons combined with secondary networks of the types described
above. Frank and Kasper[18] have listed some of these structures
based on σ-phase type secondary networks with sequence faults of
that tessellation in one direction only. Not listed in their
Tables 2 and 3 are structures based on secondary networks which
do not have a sigma-phase type of repeat in any direction[43].

II.   Non-Layered Structures

1. The structure of $Mg_{32}(Al,Zn)_{49}$ has been discussed by
Frank and Kasper[24],[18]. It is a complicated cubic structure, with
layers formed of hexagons, pentagons and triangles, which are
only slightly puckered. All four coordination polyhedra occur
(see Table 2).

2. The R phase[10], which has been found in binary and ternary
transition-metal systems, including many ternary systems of two
transition metals with Si, has a complicated rhombohedral
structure in which all the atoms have the "normal" coordinations
and all interstices are tetrahedral, but in which the atoms are
not confined to four planes as in the layer structures. This
structure was determined by considering the packing of the
coordination polyhedra along the three-fold axes (Fig. 9). The
relationship with the layered structures shows up in the normal
projection on $(\bar{1}35)$, shown in Fig. 10 (a). There appear to be
two main layers of atoms (with deviations from the plane
generally less than ±0.2Å) arranged in pentagons, hexagons and
triangles as expected for a μ-phase analogue consisting of rows
of alternating pentagons and hexagons. There are however also
regions where the linear rows of pentagons and hexagons come to
an end and where the atoms are not confined to main and secondary
layers. The atoms with high CN seem to concentrate in the regions
where deviations from planarity occur. In fact the non-planarity
results from the unsymmetrical way in which one CN16 atom is
connected by major ligands to two CN16 atoms, one CN14 and one

---

[*] Among the layered structures most known structure types have
all-pentagon layers. All-hexagon layered structures of
structure types analogous to the μ phase and the M phase are
possible, but have not been found to exist, (Frank and Kasper's
Table 3, entry 1 and 3).

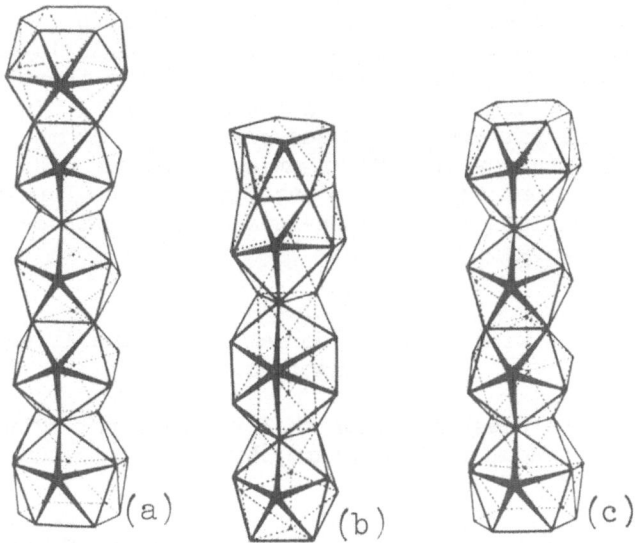

Figure 9. Strings of interpenetrating polyhedra along 3-fold
           axes in the R phase and some related structures.
           (a) R phase, entire c-repeat: CN16, 12, 12, 12, 16.
           (b) μ phase, one-half c-repeat: CN12, 15, 16, 14.
           (CN12 is shared with lower half, below).
           (c) MgNi$_2$, entire c-repeat: CN16, 12, 12, 16.

CN15 atom. The major skeleton is a complicated three-dimensional
framework.

    3. The δ phase[11] is another example of a structure in which
all atoms have triangulated coordination shells and CN12, 14, 15
and 16, but with regions where the planar layering of the atoms
has been disturbed. The structure is orthorhombic, space group
P2$_1$2$_1$2$_1$, very strongly pseudo-tetragonal. The space group does
not impose any symmetry on any of the coordination polyhedra,
which however approach very closely the more ideal geometry of
the polyhedra in the layered structures. The non-planarity
results from the unsymmetrical way that the CN16 is connected
by major ligands to two CN14 and two CN15 atoms. When projected

normally on to the $(04\bar{1})$ and $(40\bar{1})$ planes (Fig. 10(b), (c)) the
atoms are in almost planar pentagon-hexagon-triangle layers
(normal distance to the plane less than 0.2Å), but there are
again regions where the planarity is not maintained. In the
planar regions the pattern of the secondary-layer atoms is as in
the σ phase. The ratios of numbers of atoms in the different
types of coordinations are the same as found in the P phase. The
major skeleton consists of four non-interconnected three-
dimensional frameworks.

III.   Structures with Largely "Normal", but some Irregular
                      Coordinations

       In these structures the major skeleton does not extend
continuously throughout the structure, but stops at atoms with
irregular CN.

       1. α - Mn and χ phases. The structure of α - Mn was deter-
mined as early as 1927 by Bradley and Thewliss[25]. It has a large
cubic cell (a_0 = 8.912Å) and the coordination polyhedra present
are: CN12, CN16, but also CN13. Most of the interstices are
(slightly irregular) tetrahedra, but there are also some distorted
octahedral interstices and α - Mn is therefore not a true
representative of the group of structures we are concerned with,
even though it has striking similarities with the σ phase (Fig. 3).
The α - Mn structure might be considered as the forerunner of the
group of very complex intermetallic structures which have been
determined by Sten Samson[26]. In these structures tetrahedral
interstices predominate, but there are also more irregular regions.
The next representative of those structures is β-Mg_2Al_3 (Table 2)
with coordinations 12 and 16, but the remaining 20% of the
polyhedra are irregular with ligancies of 10 - 16[27].

       2. The ϵ-$Mg_{23}Al_{30}$ phase, determined by Samson and Gordon[30],
is closely related to the R phase. Both phases have space group
$R\bar{3}$ and 53 atoms per rhombohedral unit cell (Table 2). In fact
the atomic coordinates are very similar, the largest difference
being about 0.5Å. The atomic displacements cause some of the
coordinations to be different. Two large atoms (Mg) occupy
positions that are CN12 in the R phase (on the three-fold axes,
see Fig. 9(a)), and the polyhedra have been slightly enlarged so
that these CN's are now 13 for the top and bottom icosahedra and
14 for the central one. Another large atom is placed in a position
that is CN15 in the R phase and here has its CN reduced to 14.
Finally one small atom (Al) is placed in a (CN14) R position with
ligancy reduced to 11. Distorted octahedral interstices occur
where the ϵ phase deviates from the R phase. In the ϵ phase there
is not a continuous three-dimensional framework of major ligands

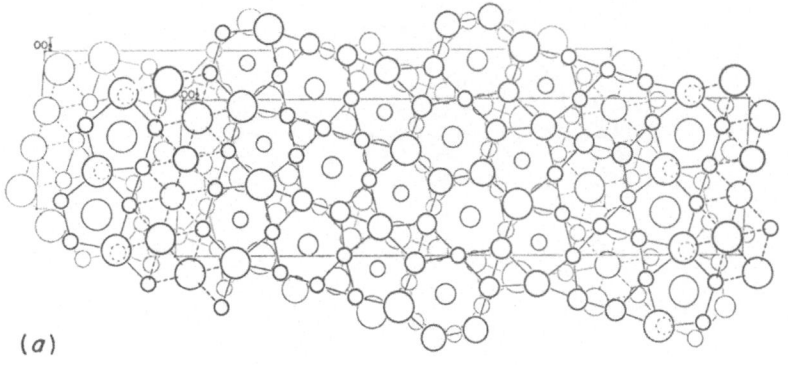

(a)

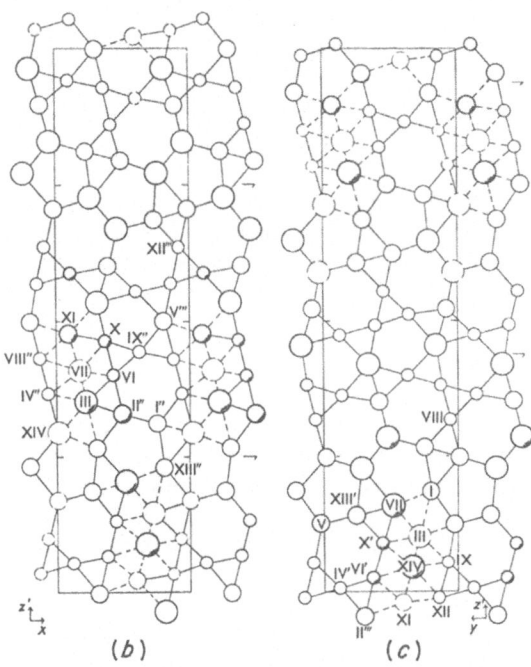

(b)   (c)

Figure 10.   Non-layer structures with almost planar regions.
   (a)   Structure of the R phase, Mo-Co-Cr, showing two main
layers (atoms joined by solid lines) parallel to $(\bar{1}\bar{3}5)_{hex}$.
Atoms to which no lines are drawn are in secondary layers,
approximately half-way between main layers.
   (b)   Structure of the δ phase, Mo-Ni, showing one main
layer parallel to the $(0\bar{4}\bar{1})$ plane. Atoms in secondary
layers are not shown.
   (c)   The δ phase structure, showing one layer parallel to
the $(40\bar{1})$ plane. Atoms in secondary layers are not shown.
Dashed lines denote ligands to atoms intermediate between
main and secondary layers. Atoms not on the mean planes of
the layers are projected normally onto them.

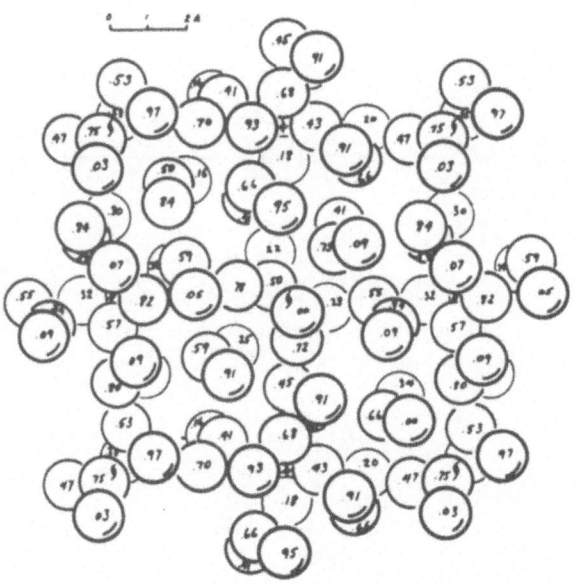

Figure 11. Proposed structure of the D phase (V-Fe-Si),
    projected down the tetragonal z axis. Numbers indicate
    fractional z coordinates. Note general resemblance with
    the σ phase, but severely rumpled layers.

as exists in the R phase. Apparently the tendency to form strong
bonds in particular directions is more pronounced for the
transition elements than for the non-transition elements Mg and Al.

    3. The D phase (V - Fe - Si) was described by Professor
Beck[13] and coworkers and also by Hladyshevskii and Shvets[29]. The
cell dimensions resemble closely those of the δ phase (Table 2)
and the Russian authors assumed that the two phases were isostructural. Our single-crystal study[16] revealed that the D phase
is truly tetragonal and that the σ-phase related layers are
approximately perpendicular to the c axis rather than to the a and
b axes as in the δ phase. The structure proposed for the D phase,
which has not yet been refined, has atoms in CN12, 14, and 16, but
some irregular CN's of 11 and 13 occur also (Fig. 11).

    4. The I phase (V - Ni - Si)[31], which is probably
isostructural with the S phase (Mn-Co-Si)[14] has a large monoclinic
cell, space group C2/c or Cc[32]. Its structure may be related to
that of the μ phase, but if so, it is considerably distorted.

INTERATOMIC DISTANCES AND ATOMIC RADII

It has been recognized for a long time that the bonding radius of an atom is not a well-defined constant. Pauling[28] has ascribed the variability of the bonding radius of an atom to variability of the bond number n according to the equation:

$$R(n) = R(1) - 0.300 \log n. \qquad (1)$$

The effective sizes of the atoms in the Cr_3Si and the Friauf-Laves phases have been discussed by many authors (see e.g. Nevitt[19]). It appears difficult in these phases to define atomic radii, since the concept of spherical atoms in contact in all directions does not apply. Usually one assumes contact in particular directions and then bases atomic radii on these distances. The radii thus defined depend not only on the R(CN12), the familiar CN12 radius of the element concerned, but also on the CN12 radius of the element it is alloyed with. It has been customary to plot the "contractions" of the atoms against the ratio $R(CN12)_A/R(CN12)_B$ in order to show a roughly linear relationship with this ratio[19].

In our studies of the more complex σ-phase-related phases we found a large number of independent interatomic distances varying from 20 in the σ phase to 94 in the δ phase. These distances cover a range generally within a limiting ratio 4:3; for the δ phase, for example, they range from 3.28 to 2.36Å. In interpreting these distances a number of observations may be made[34]:

1. The "size" of the atoms increases with the coordination number. A Mo atom, for instance, displays a larger radius in a CN16 position than in a CN15 position.

2. The atoms in the CN12 positions are essentially spherical and may be defined by a single radius (r_{12}) which applies to all CN12 positions in the particular compound.

3. The atoms with coordinations higher than 12 are not spherically symmetrical. In the direction of the major (or six-coordinated) ligands these atoms display a radius which is about 0.2Å shorter than in any other direction. The CN14 atom has two such ligands in opposite directions along a straight line, in which directions it displays a radius r_{14}^*. In the directions of the other 12 neighbors the CN14 atom displays a larger radius r_{14}. The CN15 atom has a radius r_{15}^* in the directions of the three major ligands (lying in a plane) and a larger radius r_{15} to the remaining 12 neighbors. The CN16 atom displays a radius r_{16}^* in the four tetrahedrally disposed major ligands and a larger radius r_{16} to the remaining 12 neighbors.

In the complex structures there may be many distances of a particular type that are not required to be equal by symmetry. We have expressed the experimentally determined distances in observational equations as the sums of the above defined characteristic radii. We have then determined the best least-squares values for these seven radii: r_{12}, r_{14}, r^{*}_{14}, r_{15}, r^{*}_{15}, r_{16}, r^{*}_{16}. Despite the range of experimental distances in a given phase (0.9Å) we have found that we can predict their values for a particular compound by the sums of these atomic radii with an average deviation of 0.06Å or less. By contrast, a least-squares calculation for the P phase in which each of the twelve crystallographically independent atoms was allowed one single radius gave an agreement between "oberved" and "calculated" distances which was very unimpressive, particularly for the atoms with higher coordinations.

The values for the characteristic radii obtained for the σ, R, P, δ and μ phases for which the "B" element is in the first long period are <u>roughly</u> given by the equation:

$$r = 0.1 \ (CN) \ - \ 0.2\Delta \ (\text{Å}) \tag{2}$$

where $\Delta = 1$ for major ligands and $\Delta = 0$ for minor ligands; however, the values of the radii are somewhat dependent on composition and structure type, as discussed later.

4. In the case of the phases of simplest structure (Cr_3Si phases and the cubic Friauf-Laves phases) the ligands of a particular type are constrained by symmetry to be equal, and the values of the radii can be obtained directly from the cell dimensions as shown in Table 3. Similarly, for Zr_4Al_3, r^{*}_{14}, and r^{*}_{15} follow from the cell dimensions ($r^{*}_{14} = c/4$, $r^{*}_{15} = a\sqrt{3}/6$) but the evaluation of r_{12}, r_{14}, and r_{15} depends upon the determination of four interatomic distances. While it might appear that in more complex layered structures (σ, P, M) r^{*}_{14} could be determined similarly ($r^{*}_{14} = c/4$), the $14^{*} - 14^{*}$ ligands perpendicular to the layers are not the only ones in the structure and the average r^{*}_{14} might be slightly different. The evaluation of other radius values depends for these structures and for the non-layered structures on the determination of several bond distances, and therefore on a quantitative refinement of the atomic positional parameters with diffraction data.

5. The values of the radii thus defined appear to be somewhat dependent on the identities of both A and B constituent atoms, the atom which is present in largest amount having the largest influence. For a given R(CN12) of atom A the apparent size of A increases with increase of R(CN12) of atom B, particularly in AB_2 compounds. Similarly for a given R(CN12) of atom B the apparent

Table 3. Relations Defining Major and Minor Ligand
Radii in terms of Lattice Parameters
$\beta W : A_3 B (A15)$

Atom A $r_{14}^* = \frac{1}{2} d_{A-A}^*$ (major) $= \frac{1}{4} a$ $= .25 \, a$

Atom A $r_{14} = \frac{1}{2} d_{A-A}$ (minor) $= \frac{\sqrt{6}}{8} a$ $= .3062 \, a$

Atom B $r_{12} = d_{A-B} - r_{14}$ $= (\frac{\sqrt{5}}{4} - \frac{\sqrt{6}}{8}) a = .2528 \, a$

Cubic Friauf-Laves phases: AB_2 (C15, or $MgCu_2$ type)

Atom A $r_{16}^* = \frac{1}{2} d_{A-A}^*$ (major) $= \frac{\sqrt{3}}{8}$ $= .2165 \, a$

Atom B $r_{12} = \frac{1}{2} d_{B-B}$ $= \frac{\sqrt{2}}{8} a$ $= .1768 \, a$

Atom A $r_{16} = d_{A-B} - r_{12}$ $= (\frac{\sqrt{11}}{8} - \frac{\sqrt{2}}{8}) a = .2378 \, a$

---

size of B increases with increase of R(CN12) of atom A, particularly in A_3B compounds. We have defined a quantity $\bar{R}$, which is the average of the R(CN12) radii (given for instance by Pauling[28]) for the atoms present in the alloy, weighted according to the atomic chemical composition[34]. We have then plotted the radii defined above against $\bar{R}$. It then appears that (for example) r_{12} displays a roughly linear dependence on $\bar{R}$ and that the points representing the Friauf-Laves phases and the Cr_3Si phases (for which the r_{12}'s are defined in Table 3) and those for the more complicated phases (for which r_{12} was obtained by a least-squares procedure) all scatter around the same straight line (Fig. 12).

A similar plot may be made for r_{14} and r_{14}^*, which would contain the Cr_3Si phases and the complex phases. For the Cr_3Si phases there exists a linear relationship between r_{12} and r_{14} or r_{14}^* (see Table 3) and therefore the same roughly linear dependence

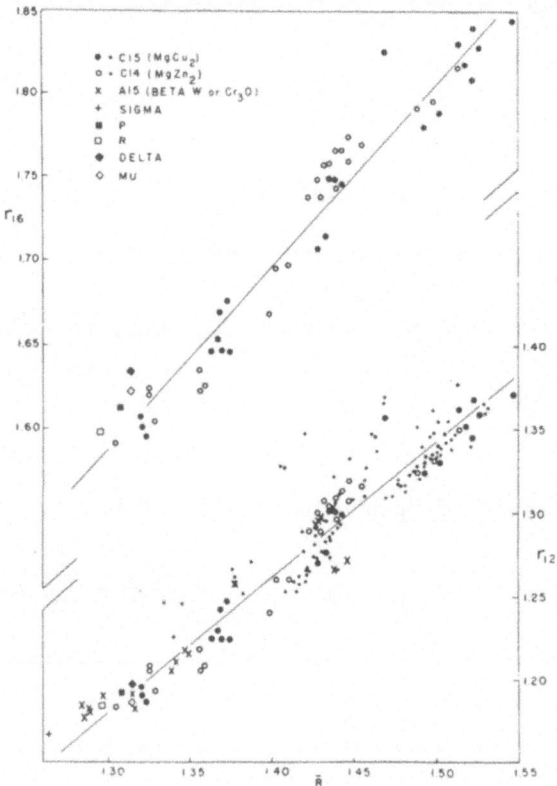

Figure 12. Plot of r_{12} (bottom) and r_{16} (top) versus $\bar{R}$ for
    tetrahedrally close-packed transition-element intermetallic
    compounds. The phases containing lanthanides or actinides
    are plotted with small circles for r_{12} and have been omitted
    for r_{16}.

Table 4. Equations for predicting Atomic Radii
in Phases with "Normal Coordinations"

For $M_m N_n \ldots$: $\bar{R} = (m \cdot R(CN12)_M + n \cdot R(CN12)_N + \ldots)/(m+n+ \ldots)$

$r_{12} = 1.200 + 0.810\ (\bar{R} - 1.325)\ =\ 0.127 + 0.810\ \bar{R}$

$\left. \begin{array}{l} r_{14}^{*} = \quad .989 \times r_{12} \\[6pt] r_{14} = 1.211 \times r_{12} \end{array} \right\}$ ratios from β-W phases

$\left. \begin{array}{l} r_{16}^{*} = 1.225 \times r_{12} \\[6pt] r_{16} = 1.345 \times r_{12} \end{array} \right\}$ ratios from cubic Laves phases

$\left. \begin{array}{l} r_{15}^{*} = 1.11 \ \ \times r_{12} \\[6pt] r_{15} = 1.28 \ \ \times r_{12} \end{array} \right\}$ ratios interpolated between those for CN14 and CN16

----

on $\bar{R}$ that exists for r_{12} must hold for r_{14} and r_{14}^{*}. The data points for the complex phases fit very well in the case of r_{14}, but in the case of r_{14}^{*} the experimental values are up to 0.05Å larger than expected.

The plots for r_{16} and r_{16}^{*} contain the Friauf-Laves phases and the complex phases. For the Friauf-Laves phases there is again a linear relationship between r_{12} and r_{16} or r_{16}^{*} (See Table 3) and therefore we find again the same roughly linear dependence on $\bar{R}$ for r_{16} and r_{16}^{*}. The complex phases fit well on the r_{16} plot (upper part of Fig. 12), but on the r_{16}^{*} plot their data points are as much as 0.05Å lower than expected.

The radii r_{15} and r_{15}^{*} occur only in the complex phases and in Zr_4Al_3, and therefore a relationship is harder to establish.

The straight line in the lower part of Fig. 12 represents the best fit to the plotted points and defines the linear relationship between r_{12} and $\bar{R}$ given in Table 4. The other equations in Table 4 define r_{14}^*, r_{14}, r_{16}^* and r_{16} as a function of r_{12}, based on the experimental ratios as found in the Cr_3Si and the cubic Friauf-Laves phase. The ratios for r_{15}^* and r_{15} are interpolated between those found for CN14 and CN16.

The validity of the equations in Table 4 may be tested in a number of ways.

A. Distances in Friauf-Laves and Cr_3Si Phases

The recognition that the radius of an atom depends on the type of ligand, the number of neighbors and the average radius of the constituent elements, resolves some controversies regarding atomic distances that have existed in the past. Pauling[33a] reported that the observed distances in the Cr_3Si phases agree well with the sum of the weighted CN12 radii ($2\bar{R}$). Using our equations we find that in the Cr_3Si phases the (minor) A - B distance is given by:

$$d_{A - B} = r_{14} + r_{12} = 0.28 + 1.80\bar{R} \qquad (3a)$$

which for the observed values of $\bar{R}$ is equal to $2\bar{R}$ within 1%.

It has been observed that for the Friauf-Laves phases the A - B distances are, on the other hand, better represented by the unweighted sum of the $R(CN12)$[33b]. In our treatment the A - B (minor) distance in the Friauf-Laves phases is given by:

$$d_{A - B} = r_{16} + r_{12} = 0.30 + 1.90\bar{R} \qquad (3b)$$

which exceeds $2\bar{R}$, and happens to be in fact quite close to the unweighted sum of the $R(CN12)$.

B. Distances in Complex Phases

In the derivation of the structure of the M phase[15a] in the Nb-Ni-Al system we arrived at a trial structure using the cell dimensions, the symmetry elements and the resemblance of the structure to the P phase in the Mo-Ni-Cr system. All the "normal" coordination polyhedra are represented in the trial structure and it was therefore expected that the interatomic distances could thus be expressed as sums of the characteristic radii, the values of which could be predicted from the $\bar{R}$ for this phase and the expressions given in Table 4. Accordingly, by a preliminary least-squares "refinement" without X-ray data we adjusted the trial positional parameters to fit the interatomic distances to the predicted values. The conventional agreement index R (later calculated from observed and calculated X-ray structure factors) for this structure was 24%, which was then reduced to 12% by refinement with the X-ray data. During this refinement there was no coordinate change larger than 0.13Å and no change in inter- atomic distances larger than 0.11Å; the average changes were much smaller. The minor radii determined by least-squares from the final parameters differed by no more than 0.01Å from the predicted radii, the largest difference for the major radii being 0.05Å. Thus it appears that the relationships in Table 4, which were derived from the simpler $MgCu_2$ and Cr_3Si phases, may also be applied for the more complex phases.

C.   Cell Dimensions of Sigma Phases

The dependence of the radii on the composition of a phase (and hence on $\bar{R}$) may further be checked by a comparison of the cell dimensions of different sigma phases. The values of the positional parameters have only been determined for a small number of sigma phases and in our treatment we will assume that the parameters derived for[36] (Mo_3Co_2) σ apply to all sigma phases since its $\bar{R}$ is near the middle of the range of $\bar{R}$'s for σ phases in the literature. With these parameters we may express the twenty crystallographically independent distances as functions of a and c. We may also express these distances as sums of the characteristic radii or with the aid of Table 4 as functions of r_{12}. We thus arrive at 20 observational equations relating a and c with r_{12} (or rather a^2 and c^2 with r_{12}^2). The best values of the cell dimensions as functions or $\bar{R}$ were then determined by the method of least squares:

$$a = 7.633\ r_{12} \quad \text{or} \quad a = .969 + 6.18\ \bar{R} \qquad (4a)$$

$$c = 3.938\ r_{12} \quad \text{or} \quad c = .500 + 3.19\ \bar{R} \qquad (4b)$$

$$^c/a = 0.516$$

In Fig. 13 we have drawn the lines representing these relationships and the observed values of a and c as functions of $\bar{R}$. The observed values were taken from references 41, 42 and 35.

The roughly linear dependence of a and c of sigma phases on $\bar{R}$ has been demonstrated before by Stüwe[35], who averaged all distances between atoms in the main layers (excluding distances between atoms in different planes) for the (Fe-Cr) σ phase, expressed this average distance as a fraction of the cell dimension a and set it equal to $2\bar{R}$. An empirical value was used for the $c/\bar{R}$ ratio because of the "contraction" of the CN14 atoms in the

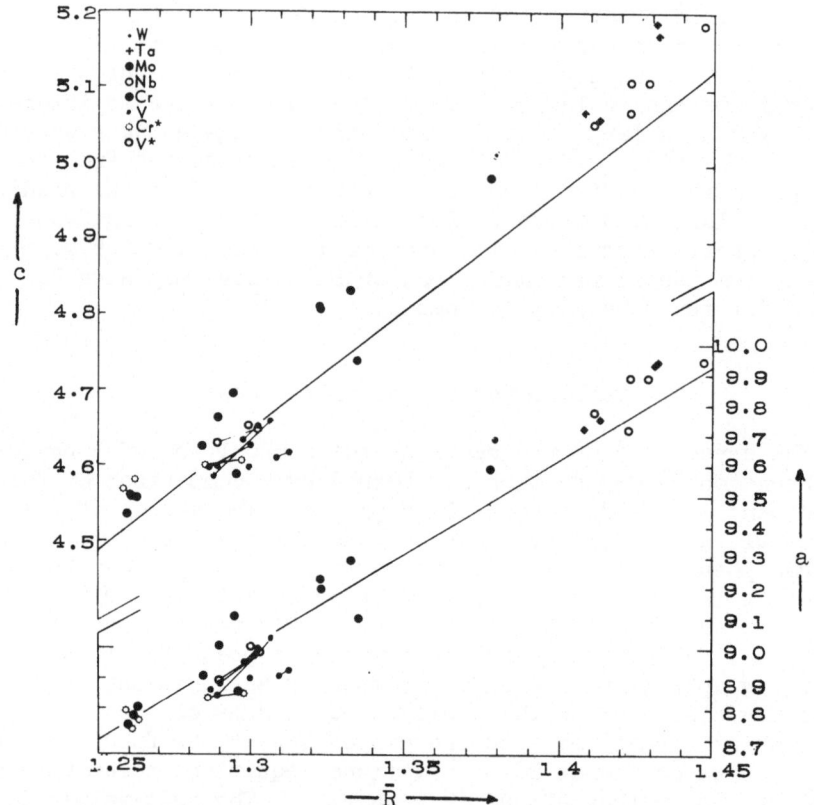

Figure 13. Plot of a (bottom) and c (top) versus R̄ for σ phases.
The lines indicate the predicted values on the basis of Eq.
(4a) and (4b). The different type of points identify only
the A component. Cr* and V* indicate ternary phases
containing Si.

c direction. In our treatment a is determined by 18 near-neighbor
distances between the atoms, including those between atoms in
different main layers or subsidiary layers. In the same way c is
determined by 13 near-neighbor distances between atoms in
different layers.

Some of the scatter of the data points around the line may be
caused by the assumption of the same positional parameters for all
the sigma phases. We also have not taken into account a possible
variation of interatomic distances with degree of order. The
degree of occupancy of a particular site by a particular atom may
differ between different alloys and even within the same alloy at
different atomic compositions. The predicted values for c seem

to be systematically lower than the observed values, particularly for high values of $\bar{R}$. The largest deviation in $\underline{c}$ is found for TaOs and WOs and amounts to about 0.10Å; the largest deviation in $\underline{a}$ is found for WOs and amounts to 0.20Å. (In the calculation of the average radii the Si $R(CN12)$ has been taken as 1.26Å).

ORDERING OF THE ATOMS

The distribution of the atoms over the available atomic sites has been determined for some binary alloys with X-rays in cases where the difference in scattering power for the two components is large enough. Examples are the (Mo_6Co_7) μ phase[22], the $(MoNi)$ δ phase[11] and the (Mo_3Co_2) σ phase[36]. Ordering studies with neutron diffraction of powder samples have been carried out for example for $(V-Ni)$ σ phases[37]. Some ternary phases have been studied by a combination of X-ray and neutron methods (P and R phases)[38]. The results indicate that the elements to the left of the Mn column in the periodic table (A elements) prefer the CN15 and CN16 positions. The elements to the right of Mn (B elements) prefer the CN12 positions, and either kind or mixtures occupy the CN14 positions. Mn and Re behave as A or B elements depending on the elements they are alloyed with. Si and Al are found to occupy CN12 positions[15a, 39,40].

Assuming that only A elements occupy CN15 and CN16 positions and only B elements CN12 positions and a mixture of A and B elements CN14 (or CN13) positions one might predict possible composition limits for these phases (Table 5). Usually the composition is observed to be restricted to a narrow field that generally lies within the broad limits predicted by the table. The numbers of available sites with particular coordination type are apparently not of overriding importance in the choice of a particular structure. In some cases these numbers are the same for different structure types: P and δ, M and μ.

The fact that A elements prefer the higher coordinated positions may be attributed to their larger size, but may also be a consequence of their electronic structure. As has been pointed out the atoms with coordinations larger than twelve appear to have a smaller radius in specified directions. For CN14 these directions are linear diagonal, for CN15 trigonal planar and for CN16 tetrahedral. This suggests that some covalent bonding might exist in these directions.

Table 1, taken from reference[21], shows how the atomic p and d sublevel-degeneracies are affected by perturbations of the ideal symmetries of the coordination polyhedra. Perturbations conforming to the point group of the icosahedron do not affect the degeneracy of the p and d levels, but for all the other coordination symmetries

Table 5. Predicted Composition Limits, Assuming Only
A Elements in CN16 and CN15, Only B in CN12

| Phase | Percentages of sites with | | | | Predicted limits, at.% A |
	CN15 and CN16	CN14	CN13	CN12	
χ	17		41	41	17 - 58
M	30	15		54	30 - 45
μ	30	15		54	30 - 45
R	26	23		51	26 - 49
δ	21	36		43	21 - 57
P	21	36		43	21 - 57
σ	13	53		33	13 - 66

at least the d-level degeneracy is partly broken. This may
explain why the icosahedral positions are preferred by the B
elements which have filled or nearly filled d shells, and by Si
and Al with empty d shells. The A elements with d shells nearer
to half-filled might prefer the polyhedra of higher coordination
where the d orbitals may hybridize with s and p orbitals and form
directed bonds. Determination of magnetic moments of atoms on
different sites by neutron diffraction studies might test this
hypothesis.

ACKNOWLEDGEMENTS

    We are pleased to acknowledge the financial support of this
work by the Army Research Office (Durham). We are indebted to
Professor Paul A. Beck and Dr. Bill C. Giessen for specimens of
these phases with which we did our diffraction work, and for
helpful discussions. We thank Mr. Philip Manor for drawing
several of the illustrations. Computations were done in part at
the M.I.T. Computation Center.

REFERENCES

1. D. P. Shoemaker and B. G. Bergman, J. Am. Chem. Soc., 72, 5793 (1950).

2. J. S. Kasper, in Theory of Alloy Phases, Amer. Soc. of Metals Symposium (1956), p. 264.

3. D. P. Shoemaker, C. B. Shoemaker and F. C. Wilson, Acta Cryst., 10, 1 (1957).

4a. J. B. Friauf, J. Am. Chem. Soc., 49, 3107 (1927).

4b. J. B. Friauf, Phys. Rev., 29, 34 (1927).

5a. F. Laves and K. Löhberg, Nachr. Gött. Akad. d. Wiss, math. phys. Kl. IV, Neue Folge 1, Nr. 6, p. 59 (1934),

5b. F. Laves and H. Witte, Metallwirtschaft, 14, 645 (1935).

6a. H. Hartmann, F. Ebert, and O. Bretschneider, Z. anorg. allg. Chem., 198, 116 (1931).

6b. G. Hägg and N. Schönberg, Acta Cryst., 7, 351 (1954).

7. H. Arnfelt and A. Westgren, Jernkontor. Ann., 119, 185 (1935).

8. B. G. Bergman and D. P. Shoemaker, Acta Cryst., 7, 857 (1954).

9. S. Rideout, W. D. Manly, E. L. Kamen, B. S. Lement and P. A. Beck, Trans. TMS-AIME, 191, 872 (1951).

10. Y. Komura, W. G. Sly and D. P. Shoemaker, Acta Cryst., 13, 575 (1960).

11. C. B. Shoemaker and D. P. Shoemaker, Acta Cryst., 16, 997 (1963).

12. C. G. Wilson, D. K. Thomas and F. J. Spooner, Acta Cryst., 13, 56 (1960).

13. D. I. Bardos and P. A. Beck, Trans. TMS-AIME, 236, 64 (1966).

14. Yu. B. Kuzma and E. I. Hladyshevskii, Zh. Neorgan. Khim, 9, 674 (1964).

15a. C. B. Shoemaker and D. P. Shoemaker, Acta Cryst., 23, 231 (1967).

15b. J. S. Benjamin, B. C. Giessen and N. J. Grant, Trans. TMS-AIME, 236, 224 (1966).

16. C. B. Shoemaker and D. P. Shoemaker, to be published.

17. F. C. Frank and J. S. Kasper, Acta Cryst., 11, 184 (1958).

18. F. C. Frank and J. S. Kasper, Acta Cryst., 12, 483 (1959).

19. M. V. Nevitt, in Electronic Structure and Alloy Chemistry of the Transition Elements, P. A. Beck (ed)., Interscience, N. Y. (1963), pp. 101 ff.

20. J. H. Wernick, in Intermetallic Compounds, J. H. Westbrook (ed)., John Wiley and Sons, N. Y. (1967), p. 197.

21. D. P. Shoemaker and C. B. Shoemaker, in Structural Chemistry and Molecular Biology, San Francisco: W. H. Freeman and Company (1968).

22. J. B. Forsyth and L. M. d'Alte da Veiga, Acta Cryst., 15, 543 (1962).

23. Y. Komura, Acta Cryst., 15, 770 (1962).

24. G. Bergman, J. L. T. Waugh and L. Pauling, Acta Cryst., 10, 254 (1957).

25. A. J. Bradley and J. Thewlis, Proc. Roy. Soc. A, 115, 456 (1927).

26. Sten Samson, in Structural Chemistry and Molecular Biology, San Fransicso: W. H. Freeman and Company (1968).

27. Sten Samson, Acta Cryst., 19, 401 (1965).

28. L. Pauling, J. Am. Chem. Soc., 69, 542 (1947).

29. E. I. Hladyshevskii and G. N. Shvets, Izvest. Akad. Nauk S.S.S.R., Metally, 120 (1965).

30. Sten Samson and E. K. Gordon, Acta Cryst., B24, 1004 (1968).

31. D. I. Bardos, R. K. Malik, F. X. Spiegel and P. A. Beck, Trans. TMS-AIME, 236, 40 (1966).

32. C. B. Shoemaker and D. P. Shoemaker, Trans. TMS-AIME, 239, 937 (1967).

33a. L. Pauling, Acta Cryst., 10, 374 (1957).

33b. S. Geller, Acta Cryst., 10, 380 (1957).

34. C. B. Shoemaker and D. P. Shoemaker, Trans. TMS-AIME, 230, 486 (1964).

35. H. P. Stüwe, Trans. TMS-AIME, 215, 408 (1959).

36. J. B. Forsyth and L. M. d'Alte da Veiga, Acta Cryst., 16, 509 (1963).

37. J. S. Kasper and R. M. Waterstrat, Acta Cryst., 9, 289 (1956).

38. D. P. Shoemaker, C. B. Shoemaker and J. Mellor, Acta Cryst., 18, 37 (1965).

39. P. J. Brown and J. B. Forsyth, Acta Cryst., 14, 362 (1961).

40. D. I. Bardos, K. P. Gupta and P. A. Beck, Trans. TMS-AIME, 221, 1087 (1961).

41. W. B. Pearson, Lattice Spacings and Structure of Metals and Alloys, Pergamon Press, N. Y. 1958.

42. K. P. Gupta, N. S. Rajan and P. A. Beck, Trans. TMS-AIME, 218, 617 (1960).

43. W. B. Pearson and C. B. Shoemaker, Acta Cryst., in press (1969).

METAL-RICH METAL-METALLOID PHASES

H. H. Stadelmaier

North Carolina State University

Raleigh, North Carolina

INTRODUCTION

This is a review of borides, carbides, nitrides, and oxides of transition elements, often in combination with non-transitional elements. In the metal rich compounds the metalloid atom is isolated and surrounded by six or eight transition metal atoms in close contact with each other. The metalloid atom is therefore interstitial, sometimes in the strict sense that the phase fields include metalloid-free ("unfilled") compositions. Five structure types are singled out because they occur extensively and because the author is familiar with them through his own work. There may be some initial objection to having borides, carbides, nitrides, and oxides lumped together in one class. It is indeed rare, but not impossible, to find an individual structure type that spans the entire range from borides to oxides. On the other hand, there is so much overlap between types that a common classification does seem justified.

PHASES WITH THE PEROVSKITE OR FILLED $AuCu_3$ STRUCTURE TYPE

The metallic perovskite ($E2_1$ or $L'1_2$) type has the ideal composition T_3MX, where T is a transition metal, M is usually a non-transitional metal and X is a metalloid. When the compositions of the known phases are combined in a single diagram the composite phase field stretches from T_4X ($L'1$ or Fe_4N type) to T_3M ($L1_2$ or $AuCu_3$ type) as shown in Fig. 1. Also shown are the unit cells of the cubic structures. The actual compositions of a number of phases are seen in Fig. 2 with references indicated by superscripts [1-16]. A compilation of the carbides and

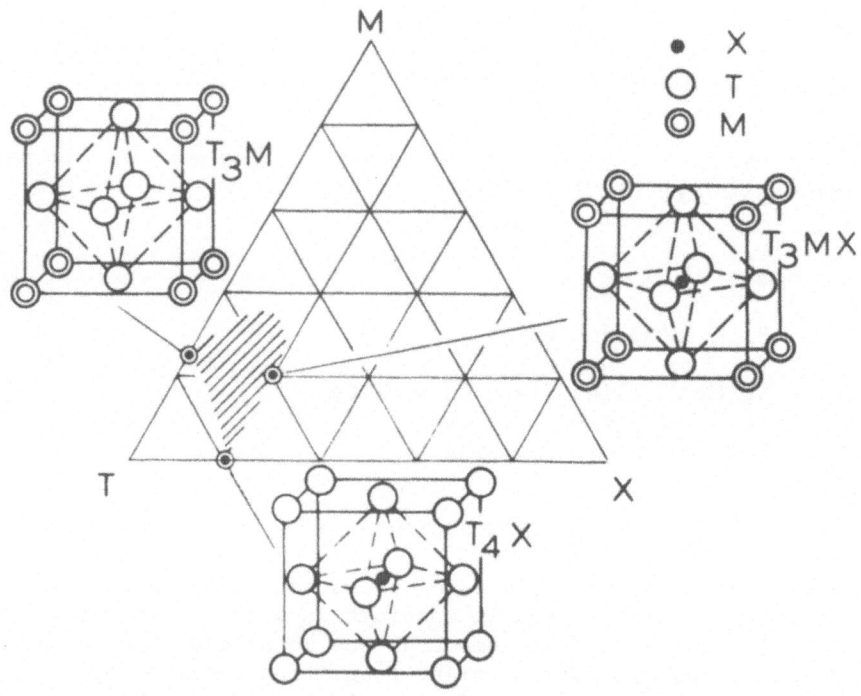

Fig. 1. Structure and composition of metallic perovskites and related binary phases. Shaded area: composite phase field.

nitrides known in 1961 is found in a review by the author [17]. Since then, a number of metallic perovskites based on early transition metals or rare earth metals has been found [18-22]. Nitrogen and oxygen stabilized $L1_2$ types based on Ti_3Au, V_3Au and V_3Pt [23] also belong to this group of phases. Two chromium phases $Cr_{2.4}Pt_{1.7}C$[24] and Cr_3GaN[25] are also known, and new manganese nitrides have been reported [25]. These more recent findings are compiled in Table I. From Fig. 2 it is obvious that neither the M nor the X content is necessarily stoichiometric. When the composition is T_3MX, perfect ordering according to Fig. 1 is the rule. This is verified by structure factor calculations for Mn_3ZnC[4], Mn_3AlC[26], Y_3AlC and Sc_3AlC[18] as well as Pt_3PbC[27]. At non-stoichiometric compositions the diffracted X-ray intensities are explained by the assumption that excess M occupies T positions statistically and vice versa. Substitution of vacancies for X at random explains a deficiency of the metalloid X [15].

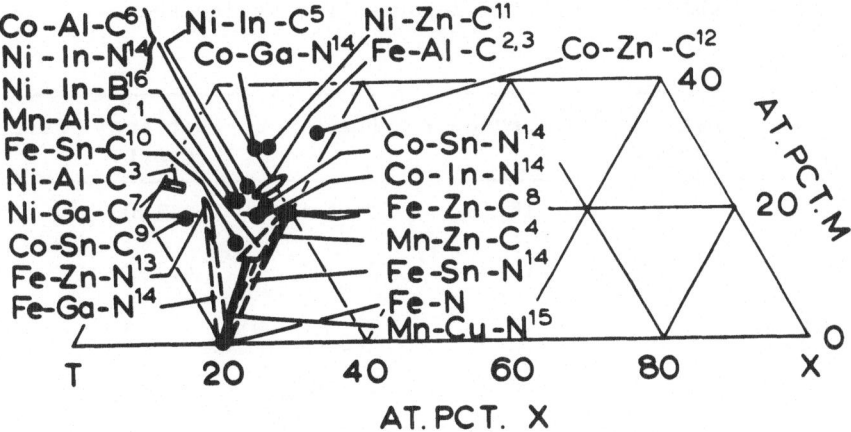

Fig. 2. Typical perovskite phase compositions.

Random substitution of M into T positions is supported by structure factor calculations in $Cr_{2.4}Pt_{1.7}C$ [24]; random substitution of T into M positions has been verified in $Fe_{13}Sn_3C_3$ [27].

It has been pointed out [17] that the metal M is frequently the alloying element found in the electron compounds of Hume-Rothery[28]. In many instances the M content of the phases shown in Fig. 2 varies with the group number of M in the manner suggested by this analogy. (In checking this, one should not mix carbides and nitrides.) For example the M(II) content in Ni-Zn-C is appreciably higher than the M(III) content in Ni-Al-C, Ni-Ga-C or Ni-In-C. The series Co-Zn-C, Co-Al-C, Co-Sn-C also follows this pattern. There is no obvious dependence of the M:T ratio on the group number of T in the series Fe-Al-C, Co-Al-C, Ni-Al-C, perhaps suggesting that iron, cobalt, and nickel have a common valency (Ekman rule[29]). In others (Fe-Zn-C, Co-Zn-C; Fe-Sn-C, Co-Sn-C) the M content clearly depends on the group number of T. The above mentioned series Co-Zn-C, Co-Al-C, Co-Sn-C also shows a systematic decrease in metalloid content with increasing group number of M which is less prominent in the series Fe-Zn-C, Fe-Al-C, Fe-Sn-C or Mn-Zn-C, Mn-Al-C and is even reversed for the nitrides. Nitrogen is expected to contribute three electrons where carbon contributes four, therefore the metalloid content of a nitride should be higher than that of a carbide containing the same components T and M. This is verified for pairs Fe-Sn-N, Fe-Sn-C and Co-Sn-N, Co-Sn-C but not in Fe-Zn-N, Fe-Zn-C. The group number dependence of the perovskite compositions is shown schematically in Table II. Any thought that M and X maintain an electronic balance is immediately dispelled by comparing isoelectronic compositions such as $Fe_4[M(III),N]$ with actual phase compositions which are invariably far removed from them.

Table I. Recently Reported Perovskite Phases

Phase	Lattice const.	Reference	Phase	Lattice const.	Reference
Sc_3AlC	4.48	18,19	Ti_3InN	4.190	22
Y_3AlC	4.89	18	Ti_3TlN	4.191	22
$Nd_3AlC_{0.9}$	4.96	18	$Ti_3Au(O,N)_x$	4.096	23
$Gd_3AlC_{0.9}$	4.90	18	Ti_3AuC_x	4.147	23
$Dy_3AlC_{0.7}$	4.85	18	$V_3Au(O,N)_x$	3.964	23
$Ho_3AlC_{0.7}$	4.81	18	V_3PtO_x	3.918	23
Y_3TlC_x	4.889	20	$Cr_{2.4}Pt_{1.7}C$	3.820	24
Ce_3InC_x	5.103	20	Cr_3GaN	3.8755	25
Ce_3TlC_x	5.096	20	Mn_3CuN	3.906	15,25
Ce_3SnC_x	5.101	20	Mn_3AgN	4.0195	25
Ce_3PbC_x	5.112	20	Mn_3GaN	3.898	25
Ti_3AlC	4.156	21	$Ni_3InB_{0.5}$	3.773-3.798	16
Ti_3InC	4.199	22			

Table II. Group Number Dependence of M and X Content in Perovskites

Component		T		M
Group No.		VII VIIIa VIIIb VIIIc		II III IV
Direction of Increasing Content of	M	$\longrightarrow$		$\longleftarrow$
	X	$\longleftarrow$		Carbides $\longleftarrow$ Nitrides $\longrightarrow$

A model that serves as a useful guide assumes that some fraction of the bond between T atoms in the face centered cubic structure is based on d orbitals. When a metalloid atom enters into an octahedral hole some of its sp or p orbitals overlap d orbitals of the surrounding T atoms. This eliminates some fraction of the d bonds otherwise present between T atoms. The corner atoms of the face centered cubic unit cell are not in direct contact with X and are left with what might be termed "unattached d bond links." Therefore perovskite formation is enhanced when the corner positions are occupied by M atoms with filled d shells. Direct evidence for the involvement of 3d electrons is found in ferromagnetic Fe_4N [30] and ferrimagnetic Mn_4N [15,31,32,33]. The T atom in the corner position has a higher magnetic moment than the face centering atoms, which is consistent with the assumption that the corner atom has an excess of unpaired d electrons. If the bond between M and T is then based mainly on sp electrons, the M content should be governed by electron compound rules, as suggested. If the bond between corner and face centering atoms is seriously weakened (when the d shell is nearly filled so that the bond between T and X takes up more sp states of the face centering atoms), the corner atom may be missing altogether as in Cu_3N (DO_9 type). A quantitative interpretation of the model is not suggested at this point, thus the number of carbon electrons spin pairing with d electrons on T is not necessarily four.

The magnetic properties of the manganese compounds and the iron nitrides have been studied extensively. Their interpretation is supported by neutron diffraction analysis to establish the magnetic ordering [30-32]. A discussion of the magnetic behavior of the manganese perovskites due to Bouchaud and Fruchart[33] follows, with some modifications by the present author. It is simpler than the detailed band model of Goodenough[34]. The moments in Mn_4N are

$$Mn_c \text{ (corner atom)} \qquad +3.9 \ \mu_B$$

$$\underline{3 \ Mn_f \text{ (face centering atoms)} \qquad -3 \times 0.9 \ \mu_B}$$

$$\text{Unit cell of } Mn_4N \qquad +1.2 \ \mu_B$$

Mn_c is unaffected when the nitrogen atoms are replaced with carbon or vacancies[33]. The assumption about carbon can be verified experimentally, the one concerning vacancies can only be valid as long as the vacancies remain isolated. The data of Takei, Heikes and Shirane [32] indicate that this assumption is reasonable below a vacancy concentration of 0.1. Each Mn_f loses one electron per added nitrogen vacancy for a total increase of $3\mu_B$ at these sites, and the moment at the 3 Mn_f sites is increased by a total of about 1 μ_B for each carbon atom substituted for nitrogen[32]. The assumption about carbon substitution appears valid over the whole range between Mn_4N and extrapolated Mn_4C [33]. Therefore nitrogen contributes three and carbon two electrons toward the face centered moment.

Mn$_4$N can be related to antiferromagnetic γ-Mn (atomic moment 2.4 μ$_B$ [35]) as follows. First, let γ-Mn be transformed to a hypothetical ferrimagnetic modification whose corner atom Mn$_c$ has the new moment +3.9 μ$_B$. Then fill the interstitial site with a nitrogen atom that spin pairs with the three Mn$_f$ atoms. For the first step 3n electrons are transferred from one Mn$_c$ to three Mn$_f$ sites. Hence,

(New moment on Mn$_c$) = (old moment on Mn$_c$) + (no. electrons trans-
                                                     ferred)

$$3.9 = 2.4 + 3n$$

$$n = 0.5$$

For ferrimagnetic γ the moment on Mn$_f$ is (2.4 - n), or 1.9. With the nitrogen atom one electron per Mn$_f$ is added, bringing the total moment on Mn$_f$ to 0.9, as observed. The net moment of "antiferromagnetic γ" is 3.9 - 3×1.9 = -1.8. This compares with an observed value of -2.36 extrapolated from the series Mn$_4$N$_{1-x}$□$_x$ where □ are vacancies[33]. The discrepancy is due to the questionable validity of the extrapolation that includes compositions far outside the range of isolated vacancies. Electron transfer must also be assumed between Mn$_f$ and non-transitional elements M occupying the corner position. The moment on one Mn$_f$ in the ferromagnetic perovskites is (in μ$_B$) 1.03 for Mn$_3$InN, 0.66 for Mn$_3$SnN, 1.58 for Mn$_3$ZnC, 1.26 for Mn$_3$AlC and 0.7 for Mn$_3$SnC [4,26,36,37]. The difference in moments between carbides whose components M have different group numbers, or between the indium and the tin nitride, is close to an integral or half-integral number of electrons per M atom. In a related structure, the unfilled L1$_2$ type Pd$_3$Sn, the effective number of 5s electrons for Sn is 0.5, as determined by the Mössbauer isomer shift[38]. Obviously the number of electrons that M contributes to the magnetic sublattice increases with the group number of M. The ability of manganese to accept varying numbers of electrons from M can be interpreted as a "multiple valence" of manganese and explains why the valence of M seems to have no influence on the composition of the manganese perovskites, as noted by Nowotny [39]. With the assumption of a constant moment on Mn$_c$ one can calculate the ferrimagnetic moments of Mn$_4$N as non-transitional M atoms are substituted into the corner position. In Fig. 3 they are compared with data taken from the paper of Juza and Puff [15]. The vacancies are a by-product of the experimental method in which x units of elemental M are added to (1-x) units of stoichiometric Mn$_4$N to yield compositions Mn$_{4-x}$M$_{x/4}$N$_{1-x/4}$. The calculation is based on isolated vacancies, each of which contributes 3 μ$_B$ to three Mn$_f$ atoms. The solid line assumes only

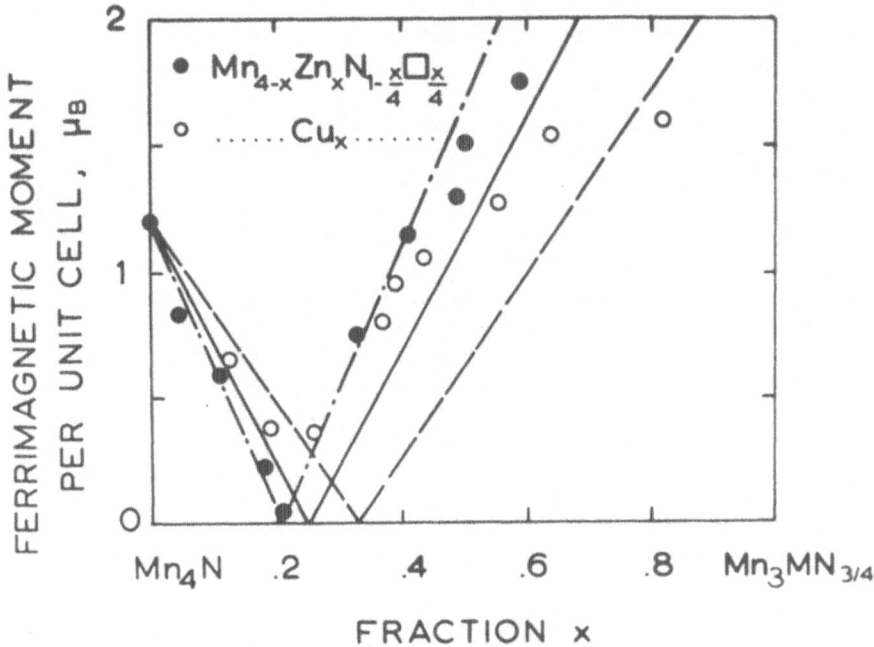

Fig. 3. Calculated and experimental ferrimagnetic moments of Mn_4N
with added zinc or copper. Solid line, no electron transfer; broken line, transfer to Mn_f; dot-dashed line, transfer from Mn_f.

removal of one corner moment for each M atom that is substituted,
the broken line transfer of one electron from M to 3 Mn_f, the
dot-dashed line transfer of one electron from 3 Mn_f to M. The
data points at low x (low vacancy concentration) and the location
of the ferrimagnetic compensation point indicate that the 3d band
associated with the Mn_f sublattice is being emptied. When nickel
is dissolved in Mn_4N, practically the same curve and compensation
point on the x-axis is found as for zinc in Fig. 3 [40]. When indium
and tin are substituted without nitrogen vacancies, the ferrimagnetic compensation points are at x = 0.26 and 0.38, respectively[41].
Calculated values for transfer of zero and one electron to 3 Mn_f
are x = 0.31 and 0.42, respectively, when no nitrogen vacancies
are assumed. Thus electron transfer, as it affects the magnetic
properties, is from Mn_f to M for nickel, copper, and zinc, the
same but approaching zero for indium, and from M to Mn_f for tin.
The same information is obtained from the ferromagnetic moments at
the ideal perovskite compositions. The moment of Mn in Mn_3InN
(1.03) is close to that of Mn_f in Mn_4N (0.9) but with the 3d band
slightly depleted, and the moment of Mn_f in Mn_3SnN (0.66) is
smaller than 0.9. That zero electron transfer is not approached

until group III is also suggested by the carbides. Substitution
of carbon for nitrogen in Mn_4N adds $\mu_B/3$ to the moment on each
Mn_f. The new moment on Mn_f $(0.9 + 0.33 = 1.23)$ agrees with the
moment per manganese atom in Mn_3AlC (1.26). The moment per manganese atom in Mn_3ZnC (1.58) is much higher, again suggesting electron transfer from Mn to Zn. For a fixed component M, the variation of composition with the group number of T shows two opposite
trends in Fig. 2 that can be understood as follows. In the series
Mn_3AlC, Fe_3AlC_x, Co_3AlC_x, Ni_3AlC_x there is probably no electron
transfer between T and Al. The carbon content is fixed by the
structure of the 3d band and decreases from manganese to nickel.
The zinc carbides with manganese, iron, and cobalt have no such
limitation on carbon content because electron transfer to zinc de-
pletes the 3d band. Consequently the carbon content can remain
stoichiometric while the zinc content varies. The latter increases
from manganese to cobalt because an increasing degree of d-band
emptying is required in that direction. Magnetic measurements of
the iron, cobalt, and nickel carbides would be useful to test this
speculation, but little is known beyond the fact that most iron
and cobalt phases are ferromagnetic at room temperature whereas
the nickel phases are not [3,42].

The atomic size of the metallic perovskites has been
examined often, most recently by Rosen and Sprang[18] who,
following Dwight and Beck[43] introduced fractional lattice
contractions into the discussion. In this way they were able to
remove an anomaly in the carbides whose lattice constants
increase in the order of component M as Al, Zn, Sn, Mg although
the corresponding CN 12 radii increase in the order Zn, Al, Sn,
Mg[3]. The fractional contractions follow the sequence of the
CN 12 radii[18]. For $R_M > R_T$, where R is the CN 12 radius, the
lattice contracts, but for $R_M < R_T$ it expands. From this Rosen
and Sprang[18] concluded correctly that it is mainly M which
contracts or expands and attributed this to efficient space filling.
The lattice expansion can be understood if T-T contacts are the
overriding geometrical factor, as suggested by the author[17] in a
different context*. For the perovskite family this leads to a
predicted lattice contraction or expansion that is too large.
A closer fit can be obtained with the hypothesis that the T-M
contact is adjusted through volume changes of spherical atoms.
This hypothesis is justified where electron transfer is suspected.
Let an $AuCu_3$ type T_3M with atomic radii equal to CN 12 radii t_{12}

---

*The size analysis of Rosen and Sprang is related to the near
  neighbor diagrams of Pearson (see article in this volume), spec-
  ifically Fig. 10, line B-B when referenced on line A-B as the
  zero level of compression.

and m_{12} change to the same arrangement with the final radii Δt and Δm through small size increments

$$t = t_{12} + \Delta t \text{ and } m = m_{12} + \Delta m$$

Next, invoke a space filling principle in which the total atomic volume is conserved. For Ll_2 and related types optimum space filling is obtained when the final radii are equal: $t = m$. It follows that

$$t_{12} - m_{12} = \Delta m - \Delta t \qquad (1)$$

The sum of the volume changes between three incompressible spheres of T and one of M is zero, or

$$12\pi t^2 \Delta t + 4\pi m^2 \Delta m = 0$$

With $t = m$ it follows that

$$\Delta t = -\Delta m/3 \qquad (2)$$

Combining (1) and (2),

$$\Delta t = (m_{12} - t_{12})/4 \qquad (3a)$$

$$\Delta m = -3(m_{12} - t_{12})/4 \qquad (3b)$$

In the Ll_2 structure type the lattice constant changes proportionally with

$$\Delta m + \Delta t = -(m_{12} - t_{12})/2 \qquad (4)$$

For $m_{12} < t_{12}$ it is obvious that $\Delta m + \Delta t > 0$, or the lattice expands. The reason is that the radius at which $t = m$, while being smaller than t_{12}, is greater than the arithmetic mean of t_{12} and m_{12} upon which the calculated lattice constant is based.* Equation (4) normalized to $(t_{12} + m_{12})$ is compared with lattice expansions in Fig. 4. The data points are the Ll_2 and $L'l_2$ types shown by Rosen and Sprang[18] with the addition of Ti_3AlC[21] and carbides with gallium[7], germanium, and indium[44]. No correction is made for interstitial carbon. Rosen and Sprang eliminate the carbon expansion by subtracting from the observed lattice constants a fraction proportional to the carbon content [18].

---

*The use of "contraction" or "expansion" requires some discretion. When $m_{12} < t_{12}$ the T sublattice contracts, equ. (3a), the M sublattice expands, equ. (3b), and the total lattice expands, equ. (4). Any one or several of these equations could be used with equal justification to analyze experimental data.

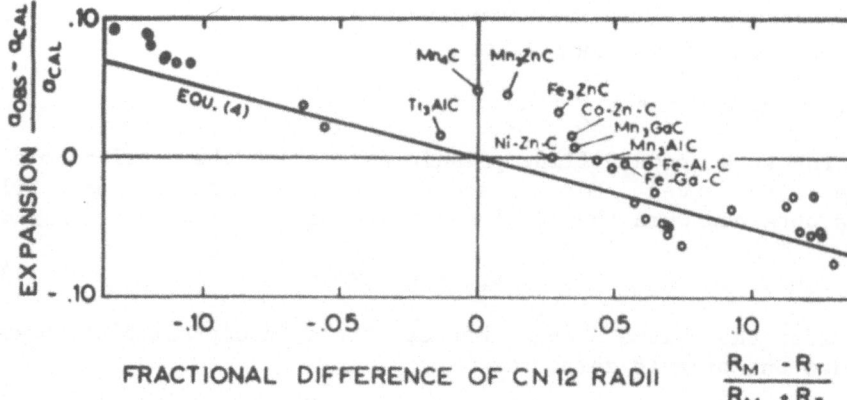

Fig. 4. Fractional lattice expansions of perovskite carbides.

Without this correction ($a_{obs} - a_{cal}$) goes through zero where the
magnetic data suggest no electron transfer between T and M. The
reasonable agreement between equ. (4) and the data points support
the model. It suggests that when the radius ratio is not ideal
for close packing, the atoms will readjust their size but maintain
a constant total volume in the process. The conservation of atomic
volume was introduced in a somewhat different context by Rudman [45].
It is not suggested that the atomic size actually varies in the
continuous fashion assumed in the model. Both the data scatter
and the integral jumps in electron transfer indicated by the magnetic moments discourage such an oversimplification.

Although most of the known perovskites are carbides and
nitrides, boron and oxygen stabilized phases can be found. The
amount of interstitial oxygen required to stabilized V_3Pt is not
known[23], but in the boride $Ni_3InB_{0.5}$ exactly one half of the interstitial positions are filled with boron atoms. Because of its
larger atomic size boron usually prefers the square antiprismatic
coordination (CN 8) to the octahedral site. Compounds TiB, HfB,
and ZrB with the NaCl structure also show that CN 6 is not impossible for boron. The lattice constant and composition of the nickel-indium boride are quite close to those in the nickel-indium
carbide [16]. The lattice constants of the monoborides are likewise
close to those of their monocarbide couterparts. If one has to
strain a point to bring boron into the perovskite family, no such
problem presents itself with the $Cr_{23}C_6$ type, as will become
evident.

Tetragonal variants of the Ll_2 and $L'l_2$ types are quite
common, but they shall not be discussed here. Another peculiarity
that shall only be mentioned is the coexistence of two perovskite

phases in the same system. This is found in Mn-Sn-C [1] and
Ni-Zn-N [13]. Judging from their melting behavior, the perovskites
do not possess high stability. $Ni_3InB_{0.5}$ decomposes into two
solid phases above 800°C, and the carbides that we know in this
respect melt incongruently.

H-PHASES

The perovskite phase field in Fig. 1 stays below 20 at.%
X at which one octahedral hole per unit cell is filled. This
enables the metalloid X to avoid contact with the M-atom, except
when the atomic ratio M:T exceeds 1:3. The ternary carbides and
nitrides that are based on hexagonal rather than cubic close
packing apparently follow the same principle of X atoms avoiding
contact with M atoms. They have been labeled H-phases by Jeitschko,
Nowotny, and Benesovsky [46-52]. The theoretical composition is
T_2MX, where the symbols have the meaning given in the perovskites.
The structure is based on a close packed hexagonal cell with three
times the usual axial length and an axial ratio c/a = $\sqrt{24}$ or
4.90. Like atoms are arranged in close packed atomic layers that
are familiar from ordered CuPt [53]. In the H-phase two layers of
T atoms alternate with one layer of M, as shown in Fig. 5. The
T layers have one layer of octahedral holes located half way
between them, and these holes are completely filled with interstitial X-atoms in a simple hexagonal arrangement. The remaining
octahedral holes are not filled because half of their neighbors

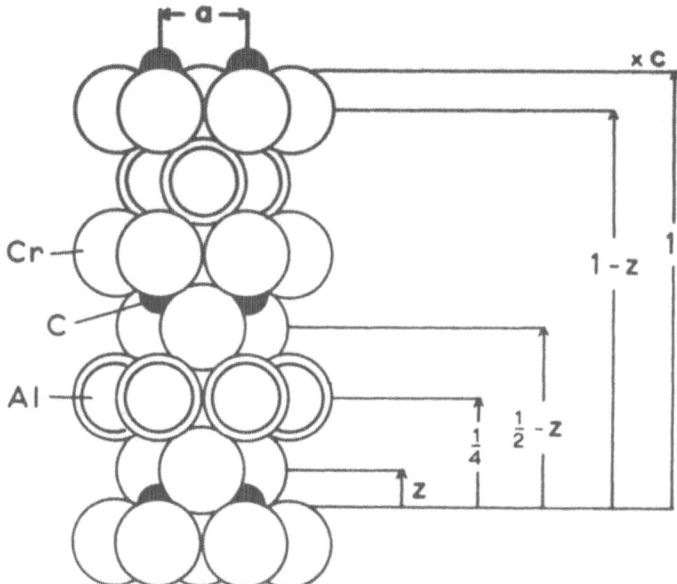

Fig. 5. Cr$_2$AlC Structure

are M atoms. The spacing between two T-layers is not expected to equal that between T and M layers, and therefore the coordinate of T along the $\underline{c}$-axis is not a simple fraction. The observed value of the parameter $\underline{z}$ that locates T is 0.086, that is somewhat larger than 1/12, the value for close packing of spheres of equal size. The coordinates of M are fixed. The octahedra surrounding X are contracted along the $\underline{c}$-axis because the overall contraction from $\underline{c}/\underline{a} < \sqrt{24}$ outweighs the expansion from $\underline{z} > 1/12$. Two interatomic distances result from the distortion of the octahedron, one equal to the lattice parameter $\underline{a}$ and another one a few per cent shorter. The atomic distance T-M is shorter than predicted by the CN 12 radius sum, see Table III. The axial contraction varies in an irregular fashion with the T-component but increases with the group number of M as shown in Fig. 6. The trend is similar to that of the perovskite contractions but, because of the uniaxiality, not amenable to the same simple analysis. The $\underline{c}/\underline{a}$ ratio of the nitrides is consistently larger than that of the carbides, due to a smaller lattice parameter $\underline{a}$ in the nitrides. This is analogous to the distance of nearest approach of the metal atoms in the mono carbides and nitrides, which is smaller for the nitrides. The relation to the compounds with the NaCl structure is twofold: the preference for T-components from group IV and V, and the existence of the structural element of the T-metal double layer containing interstitial X-atoms which is equivalent to two close packed titanium layers in TiC or TiN. Table III

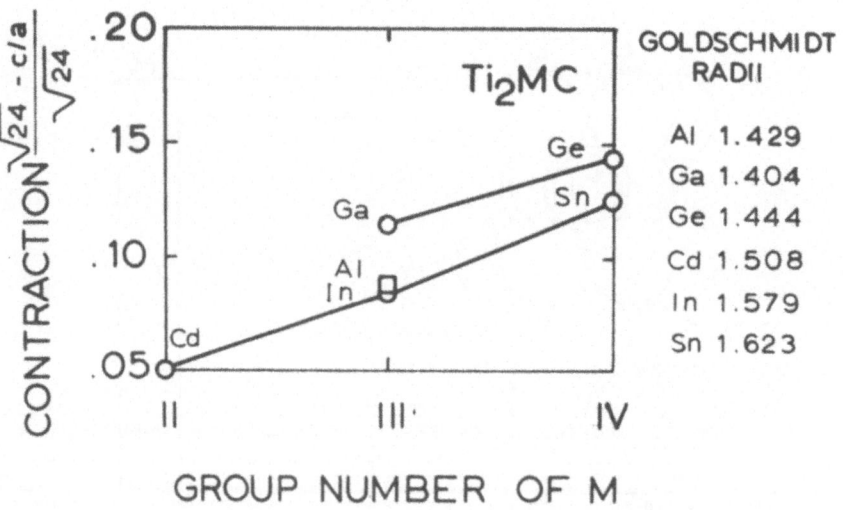

Fig. 6. Axial contraction of H-phases

Table III. Interatomic Distances in Ti-Al-C Phases, in Å [52]

Atom Pair	Ti_3AlC (Perovskite)	Ti_2AlC (H-Phase)	TiC (NaCl Type)	Sum of CN 12 Radii
Ti-C	2.08	2.12	2.16	-
Ti-Al	2.94	2.85	-	2.896
Ti-Ti	2.94	2.92/3.04		2.934

summarizes the interatomic distances of the three phases with
octahedrally coordinated carbon atoms in the system titanium-
aluminum-carbon. The distance Ti-C increases with the carbon
content of the phases, and the increasing average Ti-Ti distance shows that the metal lattice has to make room for the
carbon.

A list of the 36 known H-phases is recorded in the handbook
by Pearson [54]. All compositions are treated as stoichiometric [46-52]
although it is not known whether there might not be compositional
variations similar to those observed in the perovskites. The
T-component of the H-phases is not restricted to the first long
series elements titanium, vanadium, and chromium but also includes
the second and third series elements zirconium, niobium, and
hafnium. This is analogous to the formation of perovskite carbides
by palladium and platinum [55] or the rare earth elements [18].

PHASES WITH THE ETA CARBIDE, OR FILLED Ti_2Ni STRUCTURE

The first description of an eta carbide structure is due to
Westgren [56] and is still valid. Since there was no compelling
reason to introduce the many alternative parameter sets in
current use, all atomic parameters are converted to those of
Westgren in this review. The Westgren arrangement is originally
based on a composition W_2Fe_4C, and there is reason to doubt that
this composition really exists since no other known eta types
have it. Furthermore, the experimental information leading to
this assumed formula is somewhat sketchy because it is derived
from the investigation of commercial steel compositions [56-59].
From what is now known, one should distinguish between the "unfilled" compositions of ternary $T_3M_2^IM^{II}$, binary $T_3M_2T = T_2M$,
and the filled compositions T_6M_6X, T_4M_2X to T_3M_3X, as well as
$T_8M_4X_3$ to $T_6M_6X_3$, where T is a transition metal with a low
group number, M is a transition metal with a higher group
number or a non-transitional metal, and X is interstitial carbon,

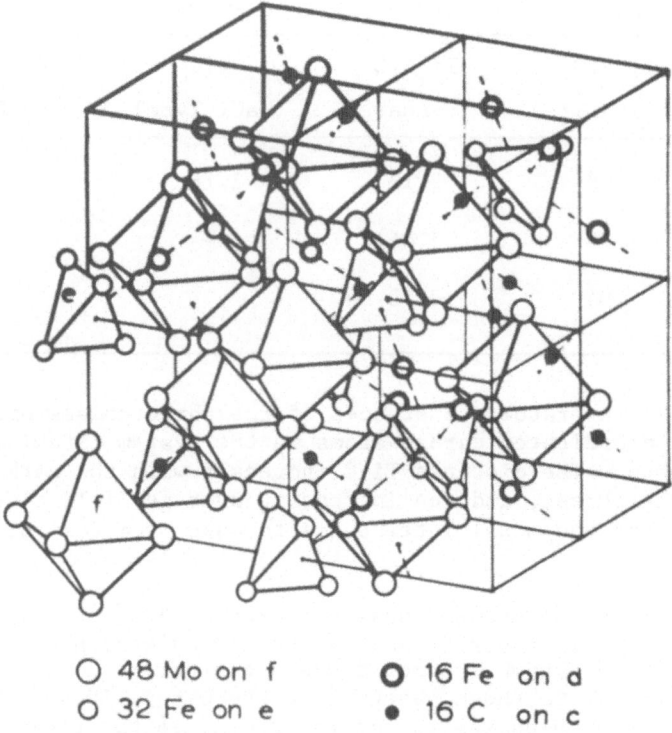

○ 48 Mo on f   ◉ 16 Fe on d
○ 32 Fe on e   ● 16 C   on c

Fig. 7.   Eta carbide structure

nitrogen, or oxygen. Quaternary compositions based on filling of
$T_3M_2^IM^{II}$ are also found, in which mutual replacement of the elements M^I and M^{II} over a wide range is possible. The listed compositions correspond to an ordered occupancy of the lattice by the two
or three atomic species, and only $T_8M_4X_3$ to $T_6M_6X_3$ is a completely
filled Ti_2Ni type, whereas the others should be considered partially filled. As a rule, there is no binary T_3M_3, presumably
because other structure types are much more stable at this
composition.

    The Westgren structure is illustrated in Fig. 7. The view
is selected so that it would show all of the atoms in a complete
unit cell if the hidden atoms were visible. There are eight
regular octahedra (position f) on a diamond cubic lattice and
eight regular tetrahedra (position e) on another diamond lattice
that interpenetrates the first one. Sixteen metal atoms on point
position d are tetrahedrally coordinated around the e tetrahedra,
and 16 metalloid atoms on c surround the octahedra in tetrahedral
coordination. The metalloid atoms are located in a trigonal antiprism, or distorted octahedron, between adjacent sides of a pair

of regular octahedra $\underline{f}$. For T_3M_3X, the occupancy is 48 T on $\underline{f}$,
16 M on $\underline{d}$, 32 M on $\underline{e}$, 16 X on $\underline{c}$. For T_4M_2X it is 48 T on $\underline{f}$, 16 T
on $\underline{d}$, 32 M on $\underline{e}$, 16 X on $\underline{c}$, as first proposed by Kislyakova [60].
The arrangement in Ti_2Ni determined by Yurko, Barton, and Parr [61]
is the same, except that the $\underline{c}$ position (based on Westgren param-
eters) is vacant. In unfilled $T_3M^I M^{II}$, M^I is on $\underline{e}$ and M^{II} on $\underline{d}$.
A unit cell containing only eight metalloid atoms on position
$\underline{a}$ in the center of the octahedra $\underline{f}$ has been described by
Leciejewicz [62]. The other 16 metalloid positions $\underline{c}$ stay vacant.
This accounts for a composition T_6M_6X as in W_6Fe_6C [58,62] and
W_6Co_6C [60]. If $\underline{a}$ and $\underline{c}$ are both occupied by a total of 24 atoms,
the metal to metalloid ratio is 4:1 as in $T_8M_4X_3$. Examples re-
ported by Jeitschko, Nowotny, and Benesovsky (ref. 63) are
Zr_2ZnN_x and Nb_2ZnC_x, where x = 3/4. That this composition is not
so abundant is probably due to its proximity to the perovskite
composition. The location of carbon on position $\underline{c}$ has been confirmed for W_3Fe_3C by Bojarski and Leciejewicz [64] using neutron
diffraction, and Leciejewicz [62] established that carbon is on $\underline{a}$
in W_6Fe_6C with the same technique.

The theoretical vaues of the adjustable parameters are
$x(\underline{e}) = 1/2 - \sqrt{231/512} = -0.172$ and $x(\underline{f}) = 3/16 = 0.1875$ for
spherical packing of the two species of metal atoms T and M.
The values determined from the X-ray intensities are usually quite
close to these. Thus Westgren [56] finds $x(\underline{e}) = 0.825 = -0.175$ and
$x(\underline{f}) = 0.195$. Some confusion about eta structures arose in connection with the parameters used by Mueller and Knott [65]. Because
the occupancy of $\underline{c}$ and $\underline{d}$ was the reverse of that found by
Westgren [56] it was surmised that the atomic arrangement in the eta
oxides is somehow different from that in the carbides [66]. That
this difference is spurious was shown by Stadelmaier and Meussner [67]
as well as Parthé et al. [68] who demonstrate that the oxide and
carbide structures are identical. The atomic arrangements on $\underline{c}$
and $\underline{d}$ happen to be the same, except for a translation of 1/2, 0,
0 [67] (or 1/2, 1/2, 1/2 [68]), and the adjustable parameters $x(\underline{e})$ and
$x(\underline{f})$ can be accomodated to make either $\underline{c}$ or $\underline{d}$ alternative sites of
the interstitial atoms. Details of the atomic arrangement in eta
phases are reviewed in a table by Parthé et al. [68]

The total list of unfilled and filled Ti_2Ni types is extensive. There are 64 entries in Vol. II of the handbook by
Pearson [54] in addition to over 20 not picked up from Vol. I [69].
Phases not tabulated in these compilations or by Nevitt [66,70]
are listed in Table IV [71-80]. In all tabulations the compositions
are best regarded as approximate. Nor does reference to different
compositions in the same system ensure the existence of separate
phases. The distribution of simple eta phases in the periodic
system is seen in Fig. 8. For convenience, only first long
series elements M are included. The T component is one of the
eight metals from titanium to tungsten, excluding chromium.

Table **IV**. Recently Reported Eta Phases

Phase	Lattice Constant	Ref.	Phase	Lattice Constant	Ref.
			CARBIDES		
V_3Fe_3C	10.877	71	Ta_3CrAl_2C	11.60	73
Zr_3Fe_3C	11.90	72	$Nb_{44}Co_6Al_{36}C_{14}$	11.70	
Zr_4Os_2C	12.41	72	Nb_3CoAl_2C	11.62	73
Nb_3Ni_3C	11.698	71	$Nb_3Co_{1.5}Al_{1.5}C$	11.60	
Nb_4Ni_2C	11.62	72	$Ta_{44}Co_6Al_{36}C_{14}$	11.56	
Nb_3Co_3C	11.633	71	Ta_3CoAl_2C	11.56	73
Nb_4Co_2C	11.63	72	$Ta_3Co_{1.5}Al_{1.5}C$	11.53	
Ta_3Co_3C	11.618	71	$Nb_{44}Ni_6Al_{36}C_{14}$	11.73	
Ta_4Co_2C	11.59	72	Nb_3NiAl_2C	11.64/11.50[*]	
Ta_4Ni_2C	11.61	72	$Nb_3Ni_{1.5}Al_{1.5}C$	11.54	73
Nb_3VAl_2C	11.77	73	$Nb_{44}Ni_{30}Al_{12}C_{14}$	11.55	
Ta_3VAl_2C	11.67	73	$Ta_{44}Ni_6Al_{36}C_{14}$	11.54	
$Nb_{44}Cr_6Al_{36}C_{14}$	11.71		Ta_3NiAl_2C	11.59/11.45[*]	
Nb_3CrAl_2C	11.71		$Ta_3Ni_{1.5}Al_{1.5}C$	11.51	73
$Nb_3Cr_{1.5}Al_{1.5}C$	11.69	73	$Ta_{44}Ni_{30}Al_{12}C_{14}$	11.49	
$Nb_{44}Cr_{30}Al_{12}C_{14}$	11.66		$Nb_{44}Cu_6Al_{36}C_{14}$	11.75	
$Nb_{44}Mn_6Al_{36}C_{14}$	11.73		Nb_3CuAl_2C	11.69/11.55[*]	
Nb_3MnAl_2C	11.69		$Nb_3Cu_{1.5}Al_{1.5}C$	11.66/11.51[*]	73
$Nb_3Mn_{1.5}Al_{1.5}C$	11.66	73	$Nb_{44}Cu_{30}Al_{12}C_{14}$	11.63	
$Nb_{44}Mn_{30}Al_{12}C_{14}$	11.65		Ta_3CuAl_2C	11.62	73
$Ta_{44}Mn_6Al_{36}C_{14}$	11.64		Ta_3ZnAl_2C	11.63	73
Ta_3MnAl_2C	11.61		W_6Fe_6C	10.934	58, 62
$Ta_3Mn_{1.5}Al_{1.5}C$	11.60	73	Mo_6Co_6C	10.897	74
$Ta_{44}Mn_{30}Al_{12}C_{14}$	11.58		Mo_6Ni_6C	10.893•	74

[*] Two eta phases   (Continued)

Table IV – Continued

Phase	Lattice Constant	Ref.	Phase	Lattice Constant	Ref.
			NITRIDES		
$Zr_{3.5}V_{2.5}N$	12.15	72	V_4Ni_2N	10.87	72
Zr_4Fe_2N	12.20	72	Nb_3Cr_3N	11.51	72
Zr_4Co_2N	12.16	72	$Nb_{3.5}Mn_{3.5}N$	11.42	72
Zr_4Ni_2N	12.17	72	Nb_4Fe_2N	11.33	72
Zr_4Ru_2N	12.30	72	Nb_4Co_2N	11.61	72
Zr_4Rh_2N	12.34	72	Nb_4Ni_2N	11.61	72
Zr_4Pd_2N	12.40	72	Ta_3Cr_3N	11.43	72
Zr_4Re_2N	12.32	72	Ta_3Mn_3N	11.35	72
Zr_4Os_2N	12.37	72	Ta_4Fe_2N	11.30	72
Zr_4Ir_2N	12.35	72	Ta_4Co_2N	11.53	72
Zr_4Pt_2N	12.43	72	Ta_4Ni_2N	11.52	72
V_4Co_2N	10.85	72			
			OXIDES		
Zr_4Co_2O	12.18	72	$Nb_{3.6}Fe_{2.4}O$	11.23	72
Zr_4Re_2O	12.35	72	Ta_4Fe_2O	11.28	72
Nb_4Ni_2O	11.58	72	Ta_4Ni_2O	11.57	72
Nb_3Ni_3O	11.20	72	Nb_4Co_2O	11.60	72
Ta_4Co_2O	11.54	72	$Nb_3Zn_3O_{0.4}$	11.545	67, 75
			UNFILLED TYPES		
Ni_2Zn_3Si	10.718	76	Zr_3Cu_2Al	11.98	80
Mn_3Ni_2Si	10.757	77	Zr_3Ag_2Zn	12.15–12.16	80
Nb_3Co_2Si	11.196	78	Zr_3Au_2Al	12.22–12.27/12.28[*]	80
Zr_2Fe	12.14	79	Hf_3Cu_2Zn	11.84	80
Zr_3Cu_2Zn	11.96–12.07	80	Hf_3Au_2Al	12.09–12.10	80

[*] Two eta phases

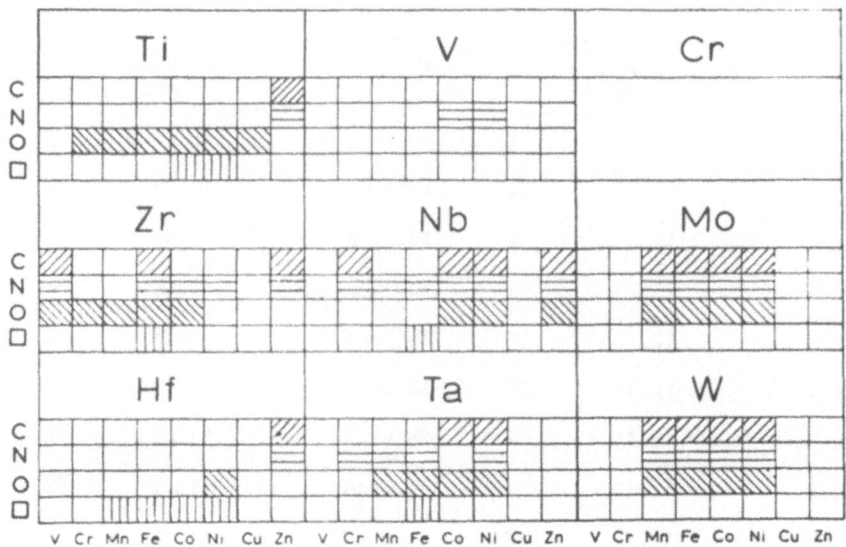

Fig. 8.   Filled (C, N, O) and unfilled (□) eta phases with M in
          the first long series of the periodic system.

The atomic radius of chromium is probably too small; where chromium is found in eta it has the function of an M component. A
very unusual element T is found in Zn_3Ni_2Si [76], in view of the
distribution of T found in Fig. 8. The component M includes a
wide range of elements from vanadium to zinc. When unfilled ternary or filled quaternary phases are considered (Table IV), it is
seen that copper, zinc, aluminum, and silicon also function as M
components. By analogy to the perovskites or tau phases (see below), it is expected that other non-transition metals will stabi-
lize eta.

     Phase compositions are best known for the oxides. The systems
Ti-Mn-O, Ti-Fe-O, Ti-Co-O, and Ti-Ni-O studied by Nevitt [81] and reviewed by the same author [66] show varying oxygen contents at a
constant atomic ratio T:M. Phase fields are also well established
for Zr-Pt-O, Zr-Fe-O [82], Ti-Cr-O [83], and Nb-Zn-O [75]. In spite of
much work reported in the literature, the constitution of carbide
systems was poorly known until quite recently. The eta phase in
Mo-Fe-C is described by extrapolation from steels [84]. For W-Co-C
the work of Rautala and Norton [85] shows conventionally determined
phase regions of two separate eta phases, corresponding roughly to
the W_6Co_6C and W_4Co_2C reported by Kislyakova [60]. The former show
them with large homogeneity ranges around W_3Co_3C and $W_6Co_3C_2$
(= $W_8Co_4C_{2.7}$). It needs to be decided whether $W_6Co_3C_2$ is almost
filled $T_8M_4X_3$, or whether the carbon content is assumed too high
in both phases and the compositions of Kislyakova [60] are the

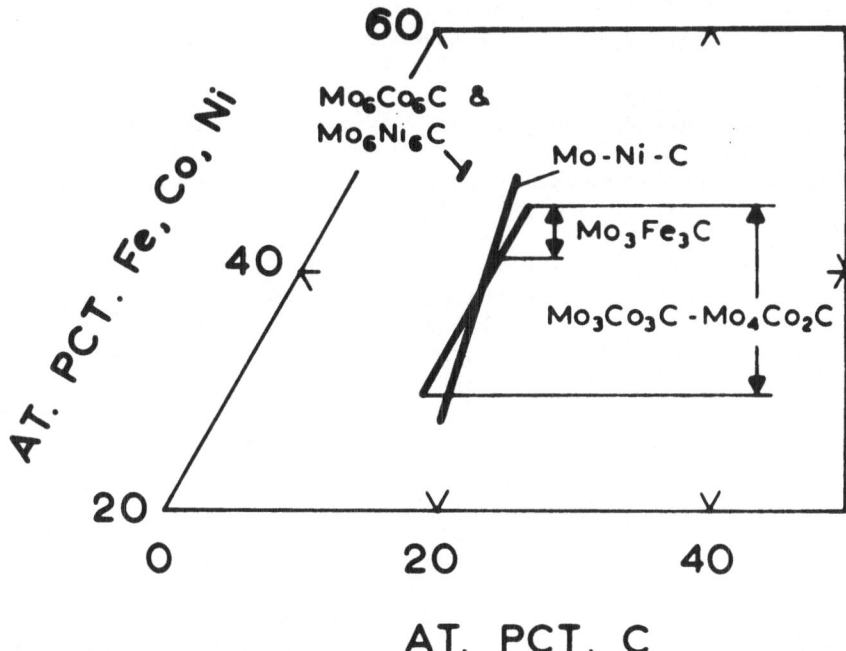

Fig. 9. Homogeneity ranges at 1000°C of eta phases in Mo-Fe-C,
         Mo-Co-C, and Mo-Ni-C

correct ones after all. Accurate work on analogous Mo-Co-C performed in the author's laboratory [74] supports the second alternative. In Mo-Co-C two eta phases coexist: stoichiometric Mo_6Co_6C
and a phase extending continuously from Mo_3Co_3C to Mo_4Co_2C in which
the carbon content remains exactly stoichiometric at 14.3 at.%.
The related system Mo-Ni-C is similar [74], the main difference is
that the carbon content of the second phase runs from a slight
carbon deficiency at Mo_3Ni_3C to a carbon excess at Mo_4Ni_2C, confirming the possibility of having carbon on lattice position $\underline{c}$ and
$\underline{a}$ simultaneously. In Mo-Fe-C only one eta phase, Mo_3Fe_3C, is observed [74], contrary to earlier information by Kuo [86]. The eta
phase regions of the three molybdenum systems are shown in Fig. 9.
In this connection, the designations η_1 and η_2, originally introduced by Kislyakova [60] to represent W_4Co_2C and W_6Co_6C, respectively, then adopted by Kuo[86] with the different meaning T_3M_3X and
T_4M_2X, respectively, have outlived their usefulness and should be
avoided. The disparity between the phase fields of the oxides
and carbides is striking. Whereas the oxides run along lines of
constant atomic ratio T:M at varying oxygen contents, the carbides
maintain a nearly stoichiometric carbon content while the ratio
T:M varies widely, as shown in Fig. 9. A dependence of the atomic
ratio T:M upon group number is not clearly established. A typical
series with increasing group number of T is Zr_6Fe_3O, Nb_3Fe_2,

Mo_3Fe_3O, and two series in which M progresses systematically are Ti_3Mn_3O, $Ti_4Fe_2O_x$, $Ti_4Co_2O_x$ and Ta_3Mn_3N, Ta_4Fe_2N, Ta_4Co_2N. They can be summarized by saying that T_3M_3X is favored by T and M from groups that are closer together and T_4M_2X is preferred when T and M are further apart in the periodic system. Nothing seems to be known about magnetic properties of the eta phases. The molybdenum carbides with iron, cobalt, and nickel have no permanent magnetic moment at room temperature. [74]

An analysis of atomic size in eta requires a knowledge of parameters $x(\underline{e})$ and $x(\underline{f})$ for which only limited information is available. Size effects can also be discussed on the basis of the lattice constant alone because the octahedra $\underline{f}$ in Fig. 7 form a three dimensional connection of T atoms. The lattice constant calculated from this sublattice is $\sqrt{2}$ 16/3 R_T. If the M atoms on $\underline{e}$ are included, all-round contact of spherical atoms on $\underline{e}$ and $\underline{f}$ is achieved when the radius ratio R_T/R_M equals $(8 - \sqrt{154/3})^{-1}$ or 1.19. The observed expansion could be discussed on the basis of this ratio. It is more correct to look at the size adjustments between the sublattice $\underline{f}$ and the continuous sublattice formed of atoms on positions $\underline{d}$ and $\underline{e}$. For the $\underline{d} + \underline{e}$ sublattice we define an average radius $\bar{R} = [2R_M(\underline{e}) + R_M(\underline{d})]/3$ for T_3M_3X, or $\bar{R} = [2R_M(\underline{e}) + R_T(\underline{d})]/3$ for T_3M_2TX and T_2M. All metal atoms fit together when, in units of one lattice constant, $R_T(\underline{f}) = 3/16\sqrt{2}$, $R_M(\underline{e}) = 3(4\sqrt{2} - \sqrt{77/3})/16$ and $R_M(\underline{d}) = R_T(\underline{d}) = \sqrt{1485/2 - 34\sqrt{462}}/16 - R_M(\underline{e})$. With these values the ideal radius ratio for close packing is $R_T/\bar{R} = 1.22$, or almost that of the Laves phases. If, in analogy to equ. (3a), only the change of the dimensions of the T sublattice is analyzed, we obtain a similar relation for constant volume space filling,

$$\Delta t/\bar{R}_{12} = -[(1.22)^3 + 1]^{-1} (t_{12}/\bar{R}_{12} - 1.22)$$

$$= -0.35(t_{12}/\bar{R}_{12} - 1.22) \tag{5}$$

It is compared with experimental data in Fig. 10. As a rule, the T metal sublattice is expanded. This can also be verified from the observed atomic radii in the few cases where they are derived from known $x(\underline{f})$. The agreement with the radius ratio and the slope predicted by equ. (5) is reasonably good. The radius ratio was already recognized by Kuo [86] and Nevitt et al. [82] to be significant and reported as having an average of 1.17 [82]. The parameter $x(\underline{f})$ is usually greater than 3/16 so that the octahedra surrounding the X atom are contracted toward X along a cube diagonal, i. e. with the same distortion as in the H-phases. In W_6Fe_6C [62] and Mo_6Ni_6C [87] $x(\underline{f})$ is still greater than 3/16 although the distorted octahedra are now vacant. In the two unfilled phases Ti_2Ni [61] and Ni_2Zn_3Si [87] whose parameters are known accurately, $x(\underline{f})$ is quite close to 3/16 so that the whole T-metal sublattice consists of undistorted

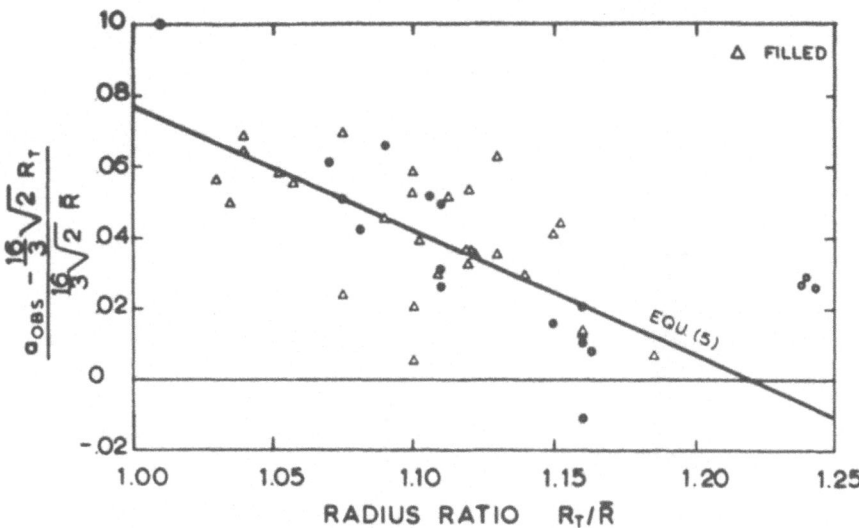

Fig. 10.   Expansion of T-metal sublattice vs. radius ratio in
eta phases

octahedra.   For unfilled Mn_3Ni_2Si, V_3Ni_2Si, and Nb_3Ni_2Si [88] parameters are also available but their refinement is probably too crude
to permit any useful conclusions.

TAU PHASES WITH THE $Cr_{23}C_6$ (OR $Cr_{21}W_2C_6$) STRUCTURE TYPE

The $D8_4$ structure type of $Cr_{23}C_6$ and $Cr_{21}W_2C_6$ has been known
since it was first described by Westgren [89] in 1933. Yet it was
not until 1961 that borides with this structure were reported [17].
The first known borides were probably those of Lavendel [90]. Their
compositions were assumed as Co_5TaB_2 and Ni_5TaB_2 and, while the
Bravais lattice and lattice constants were given correctly, the
structure type was not recognized. An extended report on borides
with non-transition metal stabilizers was written by the author and
Yun [91], followed by a paper concerned with the atomic ordering in
these borides [92]. A first review of these phases was presented in
1964 [93], and since then numerous phase diagrams involving borides
of this type have been determined in our laboratory. Significant
contributions have been added by workers in Vienna and Lvov.

The structure of tau is shown in Fig. 11 in a view that accounts for all of the atoms in a unit cell [94]. Four cubo-octahedra
h occupy the points of a face centered cubic lattice, and four cubes
f are on a second face centered cubic lattice filling the octahedral
interstices of the first one. The center of each cubo-octahedron
h contains one atom on position a for a total of four atoms, and

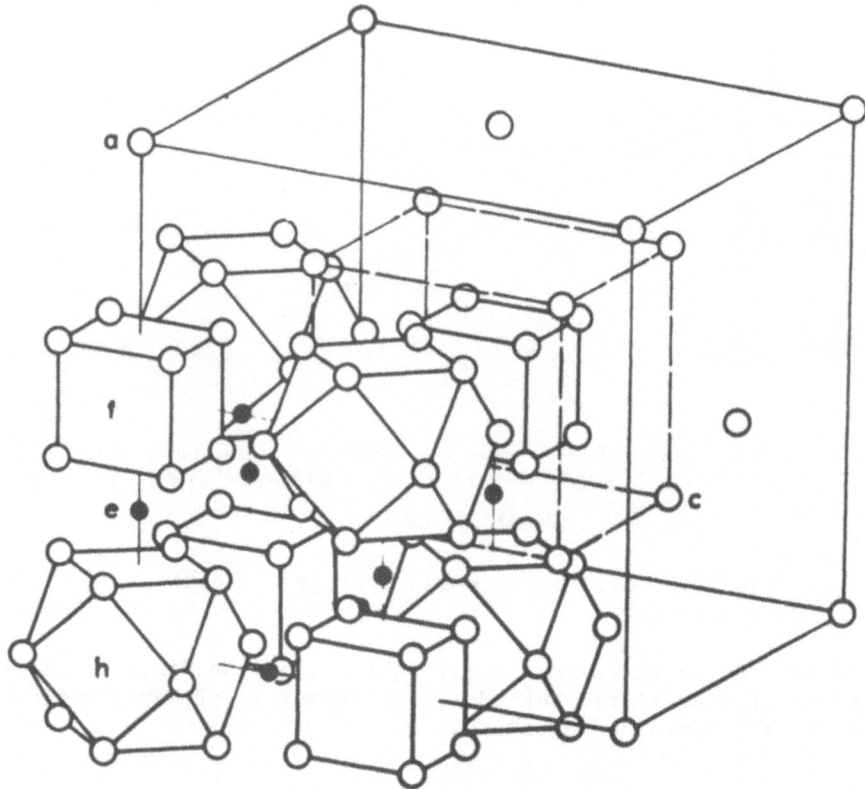

Fig. 11. $Cr_{23}C_6$ structure

there are eight additional atoms on position $\underline{c}$ at the corners of a cube whose edge length is half the cell edge. The metalloid atoms carbon or boron are located on $\underline{e}$ surrounded by eight neighbors (4 $\underline{h}$ and 4 $\underline{f}$) that form a square antiprism. For $Cr_{21}W_2C_6$ the occupancy is 48 Cr on $\underline{h}$, 32 Cr on $\underline{f}$, 4 Cr on $\underline{a}$, and 8 W on $\underline{c}$. Using the general formula $T_{21}M_2X_6$, one identifies the M atom on position $\underline{c}$ as the tau stabilizer. Without it only $Cr_{23}C_6$ and $Mn_{23}C_{5.4}$ would exist. The atoms on $\underline{c}$ are located in the center of a CN 16 Friauf polyhedron, as pointed out by Samson [95], but they are not comparable in size to the large atoms in the Laves phases. If there is excess M, the additional M atoms occupy position $\underline{a}$, and for $T_{<20}M_{>3}X_6$ they go into $\underline{h}$ and $\underline{f}$ at random [92]. The known tau borides are listed in Table V [96-126], and Fig. 12 summarizes much of this information in condensed form. In analogy to the decreasing carbon content in the perovskites progressing from manganese to nickel, a decreasing electron content of the metalloid in tau is observed in the sense that chromium and manganese form carbides, iron forms a carbo-boride $Fe_{23}C_3B_3$ [127] and cobalt and nickel form borides. The dependence of the variable M content on group

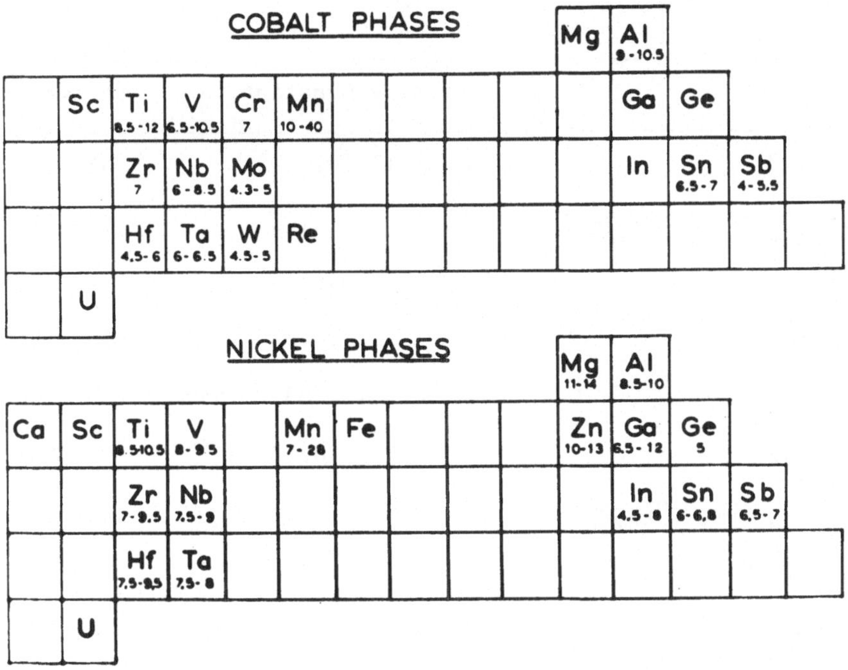

COBALT PHASES

											Mg	Al 9-10.5
	Sc	Ti 8.5-12	V 6.5-10.5	Cr 7	Mn 10-40						Ga	Ge
		Zr 7	Nb 6-8.5	Mo 4.3-5							In	Sn 6.5-7	Sb 4-5.5
		Hf 4.5-6	Ta 6-6.5	W 4.5-5	Re							
	U											

NICKEL PHASES

											Mg 11-14	Al 8.5-10
Ca	Sc	Ti 8.5-10.5	V 8-9.5		Mn 7-28	Fe					Zn 10-13	Ga 6.5-12	Ge 5
		Zr 7-9.5	Nb 7.5-9								In 4.5-8	Sn 6-6.8	Sb 6.5-7
		Hf 7.5-9.5	Ta 7.5-8									
	U											

Fig. 12. Distribution of tau stabilizers M in cobalt and nickel tau phases. Numbers indicate limits of M content in at.% at 800°C.

number has been reported for non-transition metals [91,93]. The known ranges of M content are shown in Fig. 12. The minimum values are all for tau phases in equilibrium with the binary system Co-B or Ni-B and therefore form a logical basis for comparison. If there is a trend toward decreasing M content with increasing group number of M as assumed before all the data were in, it is not without numerous exceptions. Comparing transition metal with non-transition metal tau stabilizers M, one observes that M elements in phases of like composition tend to be shifted by one group number. Examples are the pairs of cobalt compounds Ti/Al, Nb/Sn, Mo/Sb. That this behavior is not unique in tau is seen in Table VI where the same shift is observed in cobalt and nickel compounds with the CsCl structure. If titanium and aluminum are comparable tau stabilizers, they should be capable of replacing each other continuously. Alloys from $Ni_{71}Ti_8B_{21}$ to $Ni_{71}Al_8B_{21}$ show an uninterrupted series of tau phases (ref. 128). Other completely miscible pairs are $Ni_{72}Zr_7B_{21}$ to $Ni_{72}Sn_7B_{21}$, $Ni_{71}V_8B_{21}$ to $Ni_{71}Nb_8B_{21}$, and $Ni_{71}V_8B_{21}$ to $Ni_{71}Ta_8B_{21}$. Continuous replacement of cobalt by nickel can be expected in most cases where cobalt and nickel tau phases exist separately. This was demonstrated for $Co_{70}V_9B_{21}$ to $Ni_{70}V_9B_{21}$ [128].

Table V. Tau Borides

Composition	Lattice Constant, Å	Reference
$Mn_{11}Re_{12}B_6$	11.193	96
$Fe_{16}Re_7B_6$ to $Re_{20}Fe_3B_6$	11.0 - 11.3	96, 97
$Co_{21}Sc_2B_6$	10.546	96
$Co_{21}U_2B_6$	10.575, 10.620	98, 99
$Co_{20}Ti_3B_6$	10.542, 10.508–10.552	99, 100
$Co_{21}Zr_2B_6$	10.597, 10.582	98, 99
$Co_{21}Hf_2B_6$, $Co_{>21}Hf_{<2}B_6$	10.574	99, 101
$Co_{21}V_2B_6$, $Co_{20.5}V_{2.5}B_6$	10.486, 10.476–10.497	99, 102
$Co_{21}Nb_2B_6$	10.520, 10.540–10.574	99, 103
Co_5TaB_2, $Co_{21}Ta_2B_6$	10.565, 10.548–10.549	90, 104
$Co_{21}Cr_2B_6$	10.471	105, 106
$Co_{21}Mo_2B_6$; $Co_{21.7}Mo_{1.3}B_6$	10.505; 10.500–10.504	99, 107; 108
$Co_{21}W_2B_6$; $Co_{21.6}W_{1.4}B_6$	10.506; 10.500–10.503	99, 107; 109
$Co_{20}Mn_3B_6$ to $Co_{11.5}Mn_{11.5}B_6$	10.518–10.641	106
$Co_{21}Re_2B_6$	10.516, 10.494	105, 94
$Co_{23-m}Mg_mB_6$	10.541	27
$Co_{20}Al_3B_6$	10.48 - 10.52	110
$Co_{23-m}Ga_mB_6$	10.52	91
$Co_{21}In_2B_6$	10.580	99
$Co_{21}Ge_2B_6$	10.499	111
$Co_{21}Sn_2B_6$	10.601	112
$Co_{21}Sb_2B_6$; $Co_{21.6}Sb_{1.4}B_6$	10.594, 10.595; 10.540–10.596	98, 99; 113

(Continued)

Table V - Continued

Composition	Lattice Constant, $\overset{\circ}{A}$	Reference
$Ni_{23-m}Ca_mB_6$	10.472	27
$Ni_{21}Sc_2B_6$	10.56/10.60*	96
$Ni_{21}U_2B_6$	10.647, 10.652	98, 99
$Ni_{20}Ti_3B_6$, $Ni_{20.3}Ti_{2.7}B_6$	10.577, 10.507–10.538	99, 114
$Ni_{20}Zr_3B_6$, $Ni_{21}Zr_2B_6$	10.594, 10.561–10.681	99, 115
$Ni_{20}Hf_3B_6$, $Ni_{20.5}Hf_{2.5}B_6$	10.649, 10.585–10.669	99, 116
$Ni_{21}V_2B_6$, $Ni_{20.4}V_{2.6}B_6$	10.485, 10.473–10.492	99, 117
$Ni_{20.5}Nb_{2.5}B_6$	10.539–10.593	118
Ni_5TaB_2, $Ni_{20.7}Ta_{2.3}B_6$	10.565, 10.534–10.576	90, 119
$Ni_{21}Mn_2B_6$	10.495	99
$Ni_{21}Mn_2B_6$ to $Ni_{15}Mn_8B_6$	10.508–10.678	120
$Ni_{11}Re_{12}B_6$	11.25	96
$Ni_{20}Fe_3B_6$ (metastable)	10.501	125a
$Ni_{19.5}Mg_{3.5}B_6$	10.549–10.585	121
$Ni_{19.7}Zn_{3.3}B_6$	10.498–10.555	122
$Ni_{20.3}Al_{2.7}B_6$	10.495–10.552	123
$Ni_{23-m}Ga_mB_6$, $Ni_{21}Ga_2B_6$ to $Ni_{19.5}Ga_{3.5}B_6$	10.481–10.525	91, 125b
$Ni_{21}In_2B_6$	10.581–10.609	16
$Ni_{21.6}Ge_{1.4}B_6$ (metastable)	10.489	124
$Ni_{21}Sn_2B_6$	10.584–10.598	125
$Ni_{21}Sb_2B_6$	10.585, 10.598, 10.596	98, 99, 126

*Two Phases

Table VI. Group Number of M in CsCl Types CoM and NiM

Compound	Group No.	Compound	Group No.
CoSc	3	CoZn	2
CoTi	4	CoAl	3
CoZr	4	CoGa	3
CoHf	4	NiZn	2
NiSc	3	NiGa	3
NiTi	4	NiIn	3

Limited substitution of manganese for cobalt and nickel is found
in the two systems in which manganese also serves as tau stabilizer.
The composition ranges are $Co_{20}Mn_3B_6$ to $Co_{11.5}Mn_{11.5}B_6$ [106] and
$Ni_{21}Mn_2B_6$ to $Ni_{15}Mn_8B_6$ [120]. A similar series is $Fe_{16}Re_7B_6$ to
$Re_{20}Fe_3B_6$ [129]. It was suggested to have random distribution of
iron and rhenium atoms over the whole composition range [129], and
this has been largely verified, except at $Fe_{16}Re_7B_6$ where the
rhenium atoms occupy the c position preferentially. [130] In any
case, there is no ordering between positions f and h. The solubility of M in $Cr_{23}C_6$ is also of interest. It is, in at.%, 15 for
molybdenum [131], 12 for vanadium [132], 1 for niobium and tantalum [133],
4 for titanium and 1 for hafnium [134]. The group number dependence
of the maximum titanium and vanadium content in $Cr_{23}C_6$ is therefore
the reverse of that in the isotypic borides and follows that of the
solubility of titanium and vanadium in face centered cubic iron,
cobalt, or nickel. The main difference between borides and carbides
is that in the borides titanium and vanadium act as stabilizers,
whereas they are not required for stability in $Cr_{23}C_6$. The direction of the phase fields of tau is usually parallel to the side
T-M of the composition triangle, indicating that M is being substituted for T while the boron content remains fixed. When M is a
non-transition metal, the boron content is apparently more variable.
Reliable information is lacking to show where the excess boron atoms
go when the boron content is greater than the theoretical 20.7 at.%.
One conceivable position is inside the cube formed by atoms on
position f. When all boron positions are so occupied, the theoretical formula is $T_{21}M_2B_7$ with 23.3 at.% boron. This question
remains unsettled at the present time, and many tau phases do have
a boron content that is close to 20.7 at.%. The minimum metalloid

content is about 19 at.%, and no "unfilled" representatives of this structure type exist.

A poorly understood aspect of tau is the metastability of some phases. In one instance tau crystallized from the melt at $Ni_{75}Ge_5B_{20}$. This experiment was never repeated successfully, even with some variations in technique, and when the original sample with tau was remelted, only the equilibrium phases $Ni_3Ge + Ni_3B$ were obtained [124]. Likewise, $Ni_{20}Fe_3B_6$ is metastable with respect to (Ni, Fe) + (Ni, Fe)$_3$B [135]. The tau borides usually crystallize congruently from the melt at temperatures below 1200°C. Exceptions are the incongruently melting tau phases in Co-Mo-B, Co-W-B, and Co-Sb-B, all of which have the lowest known M contents. $Cr_{23}C_6$ solidifies peritectically, whereas $Mn_{23}C_6$ forms peritectoidally. Adding boron to the manganese carbide changes the equilibrium sufficiently to make the tau carboboride coexist with the melt [128]. The saturation magnetic moments extrapolated to 0°K and infinite field have been determined for tau in $Co_{20}Al_3B_6$ (57.0 e. m. u., T_c = 136°C), $Co_{21}Sn_2B_6$ (54.0 e. m. u., T_c = 152°C) [136], and $Co_{21}Ge_2B_6$ (45.0 e. m. u., T_c = 238°C)[111]. These correspond to moments in Bohr magnetons per cobalt atom of 0.67, 0.71, and 0.56, respectively. Inasmuch as the temperature dependence of the reciprocal susceptibility above the Curie point suggests ferrimagnetic behavior [136], any attempt to interpret these numbers would be speculative without supporting information on the magnetic order.

The atomic parameters associated with the point positions $\underline{e}$, $\underline{f}$, and $\underline{h}$ [92] are very close to those for incompressible spherical metal atoms of equal size on $\underline{f}$ and $\underline{h}$. The theoretical values are $x(\underline{e}) = 1/3 - \sqrt{2}/24 = 0.275$, $x(\underline{f}) = 1/2 - \sqrt{2}/12 = 0.382$, and $x(\underline{h}) = 1/6 = 0.167$. Under these conditions the $\underline{f}$ and $\underline{h}$ atoms are not in contact across the square antiprism, and the radius ratio of metalloid to T metal atom is 0.68 instead of the 0.65 required for an ideal antiprism in which all eight metal atoms touch. For the borides the radius ratios X:T are not much greater than 0.68. A size analysis of the metal atoms in tau can also be handled with a constant volume space filling model. The metal atoms on $\underline{a}$ and $\underline{h}$ form a fully connected sublattice. If it is assumed that the atoms on $\underline{a}$ and $\underline{h}$ are incompressible spheres of equal size, the lattice constant can be calculated and is $6\sqrt{2}\,R_T$. For $\underline{f}$ atoms of the same size, the cubes $\underline{f}$ "rattle" inside the $\underline{h} + \underline{a}$ sublattice, but this is avoided because each cube is also suspended between six boron and eight M atoms. The M atom on $\underline{c}$ just fits between its four closest neighbors on $\underline{f}$ to form a continuous sublattice $\underline{c} + \underline{f}$ when $R_T/R_M = (3\sqrt{6}/2 - \sqrt{3} - 1)^{-1} = 1.06$. The equation for the expansion of the $\underline{h} + \underline{a}$ sublattice analogous to equ. (3a) and (5) is now

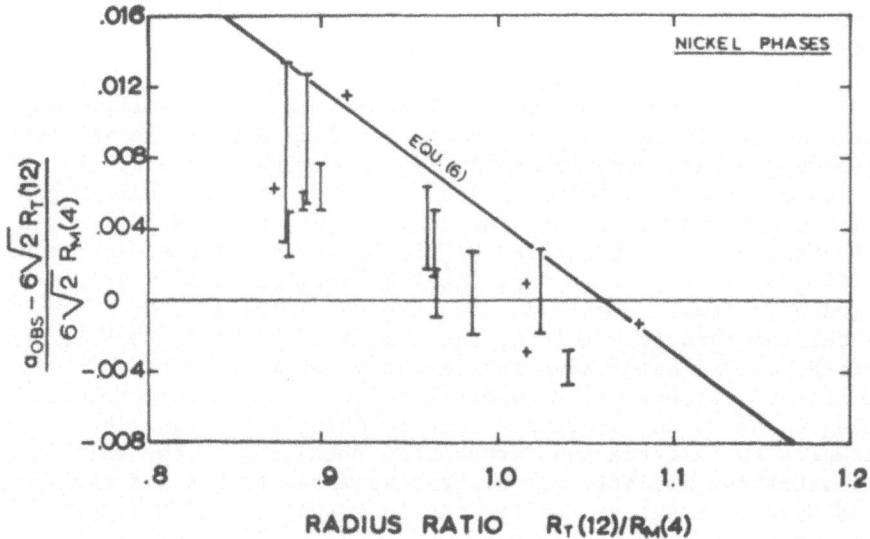

Fig. 13. Expansion of the $\underline{h}$ + $\underline{a}$ sublattice in tau as a function
of the radius ratio.

$$\Delta t/m_4 = [21\,(1.06)^3/2 + 1]^{-1}\,(1.06 - t_{12}/m_4)$$

$$= -0.074(t_{12}/m_4 - 1.06) \tag{6}$$

It is based on the geometry of the $\underline{f}$ + $\underline{c}$ sublattice for its radius
ratio and assumes that all atoms on $\underline{a}$, $\underline{f}$, and $\underline{h}$ must achieve the
same size at the expense of the M atoms on $\underline{c}$. The excess volume
from the eight atoms on $\underline{c}$ is therefore assumed to be distributed
equally over all 84 atoms on $\underline{a}$, $\underline{f}$, and $\underline{h}$. The subscript 4 on m_4
indicates that the tetrahedral radius of M, assumed to be 12%
smaller than the CN 12 radius, is used. It is introduced because
the coordination number for position $\underline{c}$ suggests it and because it
provides a better fit with the data.* The experimental data for
nickel phases are compared with equ. (6) in Fig. 13. Their scatter
below the theoretical line indicates a contraction that is not explained by the model. The data points for cobalt (not shown) are
parallel to those of nickel but shifted in the direction of greater contraction. The additional contraction is attributed to the
stronger attractive forces between cobalt and boron than between
nickel and boron. This trend is unmistakable in the C16 type
compounds that have the same coordination polyhedron around the

*Tetrahedral bond contraction of M, explored tentatively by
 Stadelmaier et al. [92], was reintroduced into the discussion by
 Samson [137].

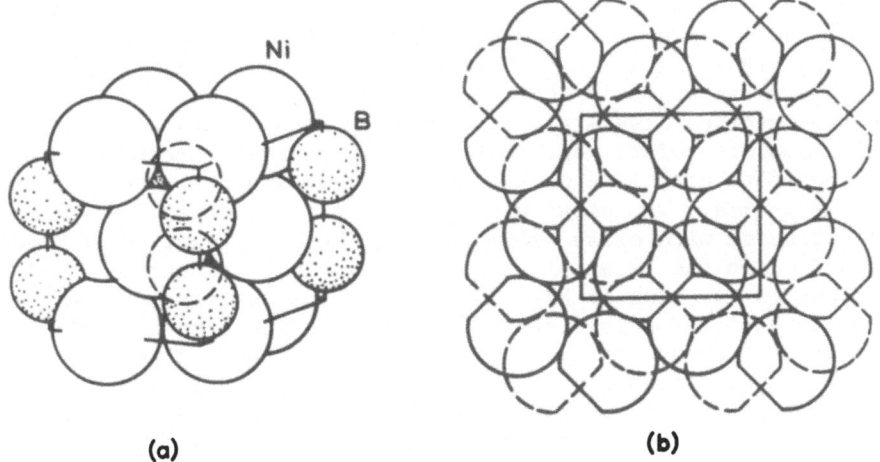

Ni

(a)   **(b)**

Fig. 14. Ni$_2$B structure. (a) Unit cell. (b) Close packing of
metal atoms, projection on basal plane.

boron atom as tau (see next section). Where the atomic parameters
are known [92], they confirm the often substantial size reduction required to make a large M atom fit into position c. An example of
an atom on c that is expanded, is found in the tau series in iron-
rhenium-boron [130]. The controlling influence exerted on the lattice constant by the atom on position c is also evident when it is
replaced continuously by an atom of different size. The slopes of
the lattice constant vs. composition curves show an unambiguous
correspondence with the size of the substituted M atom [128]. One
will conclude that the final size of M on c is controlled by tetrahedral bond contraction and a size adjustment dictated by space
filling. The effect of boron is apparently to contract the lattice,
contrary to carbon in the perovskites which expands it.

BORIDES WITH THE CuAl$_2$ STRUCTURE TYPE

The borides represented by the series Cr$_2$B to Ni$_2$B are related to tau because the coordination around the essentially isolated boron atoms is also square antiprismatic. The unit cell of
the tetragonal C16 structure type is seen in Fig. 14 (a). Beside
eight metal neighbors, each boron atom has two neighbors at only a
slightly greater distance. Kiessling interpreted this proximity
of the boron atoms as a tendency toward boron-boron bonding [138].
It is not possible to pack regular square antiprisms of spherical
atoms into a three-dimensional lattice without distortion. Therefore Laves, in his discussion of the geometry of this structure,
was compelled to introduce two limiting arrangements between which

all observed ones are found [139]. The first alternative is to make
up the basal plane with spheres joined into rigid squares which,
in turn, are assembled into a close packed layer. All nearest
neighbor metal atoms in the basal plane are then equidistant. When
such undistorted layers are stacked upon each other, the resulting
antiprisms are twisted out of the 45° symmetrical arrangement and
are therefore not close packed in the direction of the crystallographic c-axis. The other alternative is to maintain the symmetrical antiprism with close packing along the c-axis and admit some
distortion in the basal plane. The resulting arrangement of the
metal atoms is shown in Fig. 14 (b). If the nearest neighbor
pairs are sufficiently distorted that the other atoms in the
basal plane touch, one has optimum space filling. With this arrangement the radius ratio boron:metal is $R_B(8)/R_T(11) =$
$(2 + \sqrt{2}/2)^{1/2} - 1 = 0.645$ for a boron atom that just fits the
CN 8 hole. The axial ratio for this arrangement is c/a =
$[2/(\sqrt{2} + 1)]^{1/2} = 0.910$. The observed atomic coordinates place
the real structure halfway between these limiting cases [139]. Consequently the squares in the antiprism are not rotated exactly 45°
against each other but, on the other hand, the contracted atom
pairs in the basal plane are distorted less. In the place of a
simple space filling argument, we have a more complex explanation
of the geometry of this structure: 1. Space filling favors a 45°
symmetrical antiprism and c/a = 0.91. 2. Repulsion between closest
metal atom pairs introduces some twist, making the antiprism asym-
metrical. 3. Attractive forces between boron atoms and between
boron layers and metal layers reduce c/a below the value for close
packing to 0.82-0.85. A tacit assumption of the space filling argument is the size adjustment between boron and metal atoms required to fit the boron atom into a hole that is otherwise too
small. Support for point 3 comes from the observation of an increasing axial contraction going from Ni_2B to Mn_2B, that is, in
the direction of increasing stability of the borides indicated by
their increasing melting points [140] in the same direction. Fig. 15
shows the axial ratio plotted against the number of electrons per
unit cell [102,141]. The axial ratio appears to be a smooth function
of the electron concentration and relatively insensitive to the
source of electrons, as long as the alloying element is from the
same period. The extensive miscibility among these borides was
first demonstrated by Hägg and Kiessling [142]. It represents simple
substitution of the T-metal atoms for each other and applies even
when the solubility is limited, as 30 at.% Cr in Co_2B [106], or 8
at.% V in Co_2B [102]. On the other hand, the solution of non-transition metals is accompanied by vacancy formation indicated in some
of the compositions of Fig. 15. That the deviation from the stoichiometric boron content is due to boron vacancies was confirmed
by density measurements in cobalt-zinc-boron [143]. In nickel-mag-
nesium-boron [121] and iron-aluminum-boron [141] the boron vacancies
are suggested by the composition ranges of the boride. A similar
boron deficiency is also observed in binary tantalum-boron, where

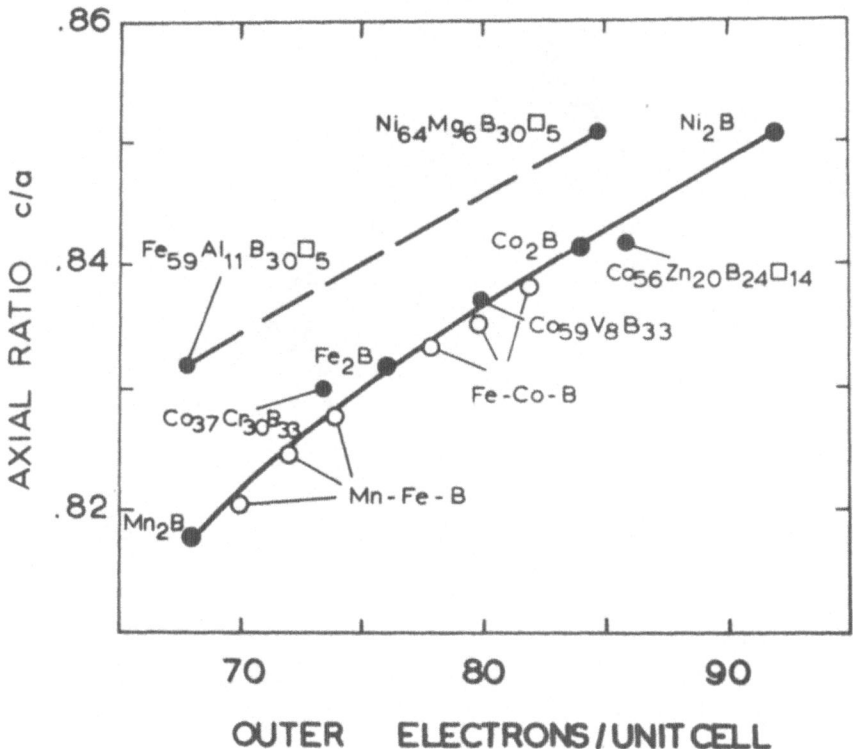

Fig. 15. Axial ratio <u>c/a</u> of C16 type borides of the first long
         period

the composition of the high temperature phase is reported as
Ta$_{2.4}$B [144]. The existence of two borides with the C16 structure
type does not ensure their miscibility: Mo$_2$B and Fe$_2$B show virtually no mutual solubility [145], nor do Mn$_2$B and Ni$_2$B [120]. The only
second and third long period elements that form this boride are tantalum, molybdenum, and tungsten. Although nothing appears to be
known about the miscibility of these borides, they do dissolve
some metallic elements, e. g. 15 at.% Nb in W$_2$B [146], or 16.7 at.%
Ti in Mo$_2$B [147]. Substitution of rhenium in Fe$_2$B up to
(Fe$_{0.33}$Re$_{0.67}$)$_2$B [97] indicates that hypothetical Re$_2$B is approaching
stability.

The magnetic properties of these borides have been studied
thoroughly by Cadeville and Meyer [148] and Cadeville and Daniel [149]
in investigations that also include the monoborides of the same
elements. Mn$_2$B and Ni$_2$B are paramagnetic, the susceptibility of
the latter is temperature independent [150]. The other borides T$_2$B
and their solid solutions are ferromagnetic and for Fe$_2$B

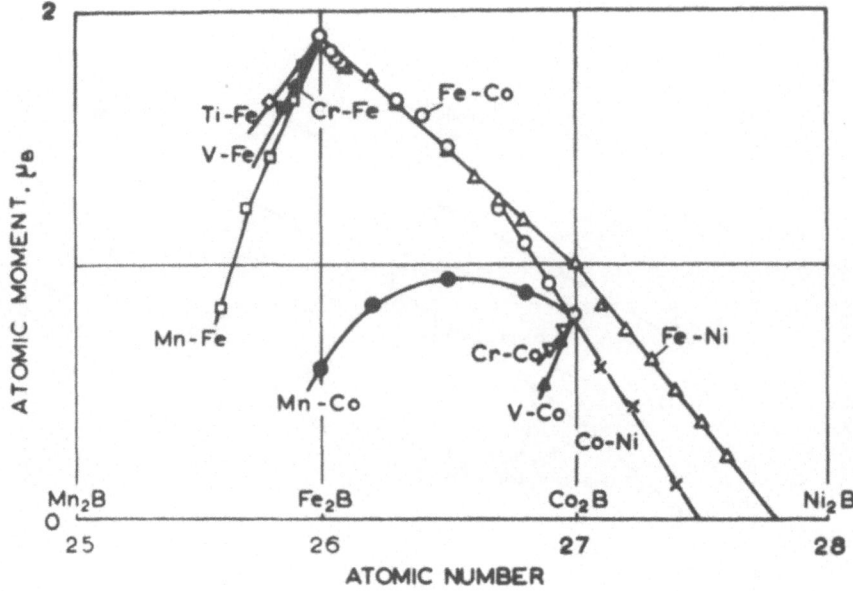

Fig. 16.   Atomic moment of borides T_2B from saturation magnetization at 20°K [149]

Mössbauer measurements by Shinjo et al. [151] verify that there is
only one type of magnetic atom [152]. The saturation magnetization
of the mixed borides has a group number dependence that is fully
analogous to that of the face centered cubic binary alloys [149].
The curves are shown in Fig. 16. Compared with the binary alloys,
they are shifted to the left, suggesting transfer of 1.7 s, p [149]
states from a boron atom to the 3d band of the transition metals.
The electron transfer of this magnitude appears to be a property
of the boron "solute" because it is constant and independent of the
crystallographic structure and of the identity of the T-component
for phases ranging from solid solutions with boron, through T_3B
and T_2B, to TB [152]. Dissolving small percentages of iron or manganese in Ni_2B produces a Curie-Weiss type of paramagnetism, and
moments of 2.6 and 3.6 μ_B are associated with the iron and manganese atoms, respectively [153]. Therefore Ni_2B is behaving as a
noble metal solvent with a transition metal solute that introduces
the localized moment.

CONCLUSIONS

The phases reviewed here are characterized by a continuous metal lattice containing more or less isolated metalloid atoms. Besides being small enough to fit into the CN 6 or 8 interstices without excessive distortion or size adjustment, the metalloid atoms are bonded partly through interaction of their s, p states with d states of the transition metals. Magnetic evidence indicates that typical numbers of electrons spin pairing with d electrons are 1.7 for boron, 2 for carbon, and 3 for nitrogen. Interstitial oxygen is a reality in the filled Ti_2Ni type. The metal lattice is clearly controlled by space filling, as seen by analysis of the size adjustments between two different species of metal atoms when they are present in an ordered arrangement. These size adjustments satisfy a space filling model in which atomic volume is conserved. Variable atomic parameters in cubic structures, or variable axial ratios in hexagonal or tetragonal structures relax the rigid space filling requirements when specific affinities between the atoms demand it. The group number behavior of the metallic components is complex but often parallels that in other metallic phases. In the metallic perovskites and borides manganese appears to act as a multivalent element. Zinc is also peculiar in that it occasionally fills the rôle of a transition element, as in the unusual composition Zn_3Ni_2Si, comparable to regular silicon stabilized Ti_2Ni types such as V_3Ni_2Si. In their diversity of alloying behavior manganese and zinc have much in common. Examples are the pairs: (perovskites) Mn_3MnN/Mn_3ZnN; (eta) $Ti_3Mn_3O/Nb_3Zn_3O_{0.4}$, Mn_3Ni_2Si/Zn_3Ni_2Si; (tau) $Ni_{<21}Mn_{>2}B_6/Ni_{20}Zn_3B_6$. It is to this ability of zinc to display different sizes and valences that the formation of such a unique structure as $ZrZn_{22}$ is attributed [154]. Another noteworthy peculiarity is the relative lack of selectivity of the tau borides in picking their stabilizing elements. Transition metal and non-transition metal atoms serve equally well, alone or in combination. The filled Ti_2Ni type is now recognized to have three theoretical metal:metalloid atomic ratios, all of which have been observed separately, namely 12:1, 12:2, and 12:3. The Ti_2Ni family, which includes carbides, nitrides, and oxides but no borides, is the most prolific producer of interstitial phases known to us.

ACKNOWLEDGMENT

A major portion of the work performed in the author's laboratory was supported by the U. S. Army Research Office-Durham during the past 10 years. This support is gratefully acknowledged. The author is also indebted to the students and associates whose names appear in the list of references. Without their numerous contributions this paper could not have been written.

REFERENCES

1. E. R. Morgan, J. Metals, 6, 983 (1954).

2. F. Morral, J. Iron Steel Inst., 130, 419 (1934).

3. L. J. Huetter and H. H. Stadelmaier, Acta Met., 6, 367 (1958).

4. R. G. Butters and H. P. Myers, Phil. Mag., 46, 132 (1955).

5. L. J. Hütter, H. H. Stadelmaier and A. C. Fraker, Metall, 14, 21 (1960).

6. L. J. Hütter, H. H. Stadelmaier and A. C. Fraker, Metall, 14, 113 (1960).

7. H. H. Stadelmaier, L. J. Hütter and N. C. Kothari, Z. Metallk., 51, 14 (1960).

8. H. H. Stadelmaier and W. K. Hardy, Metall, 8, 778 (1960).

9. L. J. Huetter, H. H. Stadelmaier and W. K. Hardy, Trans. Met. Soc. AIME, 218, 859 (1960).

10. H. H. Stadelmaier and J. M. Waller, Metall, 15, 125 (1961).

11. H. H. Stadelmaier and F. H. Hammad, Metall, 15, 124 (1961).

12. H. H. Stadelmaier and J. J. DuPlessis, 15, 763 (1961).

13. H. H. Stadelmaier and T. S. Yun, Z. Metallk., 52, 477 (1961).

14. H. H. Stadelmaier and A. C. Fraker, Z. Metallk., 53, 48 (1962).

15. R. Juza and H. Puff, Z. Elektrochemie, 61, 810 (1957).

16. J. D. Schöbel and H. H. Stadelmaier, Z. Metallk., 55, 378 (1964).

17. H. H. Stadelmaier, Z. Metallk., 52, 758 (1961).

18. S. Rosen and P. G. Sprang, Adv. X-ray Analysis, 8, 91 (1965).

19. S. Rosen and P. G. Sprang, Trans. Met. Soc. AIME, 233, 1265 (1965).

20. W. Jeitschko, H. Nowotny and F. Benesovsky, Monatsh. Chemie, 95, 1040 (1964).

21. W. Jeitschko, H. Nowotny and F. Benesovsky, Monatsh. Chemie, 95, 319 (1964).

22. W. Jeitschko, H. Nowotny and F. Benesovsky, Monatsh. Chemie, 95, 436 (1964).

23. H. von Philipsborn and F. Laves, Acta Cryst., 17, 213 (1964).

24. E. Ganglberger, H. Nowotny and F. Benesovsky, Monatsh. Chemie, 96, 1781 (1965).

25. C. Samson, J. P. Bouchaud and R. Fruchart, C. R. Acad. Sc. Paris, 259, 392 (1964).

26. R. G. Butters and H. P. Myers, Phil. Mag., 46, 895 (1955).

27. H. H. Stadelmaier, unpublished research.

28. W. Hume-Rothery and G. V. Raynor, Structure of Metals and Alloys, London, 4th edition (1962).

29. W. Ekman, Z. phys. Chemie, B12, 57 (1931).

30. B. C. Frazer, Phys. Review, 112, 751 (1958).

31. W. J. Takei and G. Shirane, Phys. Review, 119, 122 (1960).

32. W. J. Takei, R. R. Heikes and G. Shirane, Phys. Review, 125, 1893 (1962).

33. J. P. Bouchaud and R. Fruchart, Bull. Soc. Chim., 1579 (1964).

34. J. B. Goodenough, Magnetism and the Chemical Bond, Interscience, New York (1963).

35. G. E. Bacon, J. H. Smith and R. Street, Proc. Roy. Soc., A241, 223 (1957).

36. L. Howe and H. P. Myers, Phil. Mag., 2, 554 (1957).

37. M. Asanuma, J. Phys. Soc. Japan, 15, 1136 (1960).

38. N. S. Ibraimov and R. N. Kuz'min, Soviet Physics JETP, 21, 70 (1965).

39. H. Nowotny in P. A. Beck, editor, Electronic Structure and Alloy Chemistry of the Transition Elements, Interscience, New York, 179 (1963).

40. R. Juza, K. Deneke and H. Puff, Z. Elektrochemie, 63, 551 (1959).

41. M. Mekata, J. Phys. Soc. Japan, 17, 796 (1962).

42. H. H. Stadelmaier and L. J. Huetter, Acta Met., 7, 415 (1959).

43. A. E. Dwight and P. A. Beck, Trans. Met. Soc. AIME, 215, 976 (1959).

44. L. J. Hütter and H. H. Stadelmaier, Z. Metallk., 50, 199 (1959).

45. P. S. Rudman, Trans. Met. Soc. AIME, 233, 864 (1965).

46. W. Jeitschko, H. Nowotny and F. Benesovsky, Monatsh. Chemie, 94, 332 (1963).

47. W. Jeitschko, H. Nowotny and F. Benesovsky, Monatsh. Chemie, 94, 672 (1963).

48. W. Jeitschko, H. Nowotny and F. Benesovsky, Monatsh. Chemie, 94, 844 (1963).

49. W. Jeitschko, H. Nowotny and F. Benesovsky, Monatsh. Chemie, 94, 1198 (1963).

50. W. Jeitschko, H. Nowotny and F. Benesovsky, Monatsh. Chemie, 94, 1201 (1963).

51. W. Jeitschko, H. Nowotny and F. Benesovsky, Monatsh. Chemie, 95, 178 (1964).

52. W. Jeitschko, H. Nowotny and F. Benesovsky, J. Less Common Metals, 7, 133 (1964).

53. C. H. Johannson and J. O. Linde, Ann. Phys., 82, 449 (1927).

54. W. B. Pearson, A Handbook of Lattice Spacings and Structures of Metals and Alloys, Vol. 2, Pergamon Press, New York (1967).

55. H. H. Stadelmaier and W. K. Hardy, Z. Metallk., 52, 391 (1961).

56. A. Westgren, Jernkont. Ann., 117, 1 (1933).

57. H. Krainer, Arch. Eisenhüttenw., 21, 39 (1950).

58. Ya. S. Umanskii and N. T. Chebotarev, Izv. Akad. Nauk SSSR, Ser· Fiz., 15, 24 (1951).

59. K. Kuo, J. Iron Steel Inst., 173, 363 (1953).

60. E. N. Kislyakova, Zhur. Fiz. Khimii, 17, 108 (1943).

61. G. A. Yurko, J. W. Barton and J. G. Parr, Acta Cryst., 12, 909 (1959).

62. J. Leciejewicz, J. Less Common Metals, 7, 318 (1964).

63. W. Jeitschko, H. Nowotny and F. Benesovsky, Monatsh. Chemie, 95, 156 (1964).

64. Z. Bojarski and J. Leciejewicz, Institute of Nuclear Research Warsaw, Poland, Rept. No. 665 (1965).

65. M. H. Mueller and H. W. Knott, Trans. Met. Soc. AIME, 227, 674 (1963).

66. M. V. Nevitt in P. A. Beck, editor, Electronic Structure and Alloy Chemistry of the Transition Elements, Interscience, New York, 101 (1963).

67. H. H. Stadelmaier and R. A. Meussner, Monatsh. Chemie, 96, 228 (1965).

68. E. Parthé, W. Jeitschko and V. Sadagopan, Acta Cryst., 19, 1031 (1965).

69. W. B. Pearson, A Handbook of Lattice Spacings and Structures of Metals and Alloys, Pergamon Press, New York (1958).

70. M. V. Nevitt in J. H. Westbrook, editor, Intermetallic Compounds, John Wiley and Sons, New York (1967).

71. L. K. Borusevich, E. I. Gladyshevskii, T. F. Fedorov and N. M. Popova, Zh. Strukt. Khim., 6, 313 (1965).

72. H. Holleck and F. Thümmler, Monatsh. Chemie, 98, 133 (1967).

73. E. Reiffenstein, H. Nowotny and F. Benesovsky, Monatsh. Chemie, 96, 1543 (1965).

74. A. C. Fraker, Dissertation, North Carolina State University, Raleigh (1967); A. C. Fraker and H. H. Stadelmaier, Trans. Met. Soc. AIME, 237 (1969) in print.

75. R. A. Meussner and C. D. Carpenter, Corrosion Science, 7, 115 (1967).

76. H. H. Stadelmaier, J. M. Brett, and G. Hofer, Z. Metallk., 59, 881 (1968).

77. Yu. B. Kuz'ma, E. I. Gladyshevskii and E. E. Cherkashin, Russian J. Inorganic Chemistry (Transl.), 9, 1028 (1964).

78. Yu. B. Kuz'ma, E. I. Gladyshevskii and D. S. Byk, Zhur. Strukt. Khim, 5, 562 (1964) (Transl.: J. Struct. Chem., 5, 518).

79. Yu. B. Kuz'ma, V. I. Lakh, Yu. V. Voroshilov, B. I. Stadnik and V. Y. Markiv, Izv. Akad. Nauk SSSR, Metally, 6, 127 (1965).

80. W. Rieger, H. Nowotny and F. Benesovsky, Monatsh. Chemie, 96, 232 (1965).

81. M. V. Nevitt, Trans. Met. Soc. AIME, 218, 327 (1960).

82. M. V. Nevitt, J. W. Downey and R. A. Morris, Trans. Met. Soc. AIME, 218, 1019 (1960).

83. C. C. Wang and N. J. Grant, Trans. AIME, 200, 200 (1954).

84. R. F. Campbell, S. H. Reynolds, L. W. Ballard and K. G. Carroll, Trans. Met. Soc. AIME, 218, 723 (1960).

85. P. Rautala and J. T. Norton, Trans. AIME, 194, 1045 (1952).

86. K. Kuo, Acta Met., 1, 301 (1953).

87. H. H. Stadelmaier and R. A. Jones, Z. Metallk., 59, 878 (1968).

88. E. I. Gladyshevskii, Yu. B. Kuz'ma and P. I. Kripyakevich, Zhur. Strukt. Khim., 4, 372 (1963) (Transl.: J. Struct. Chem., 4, 343).

89. A. Westgren, Jernkont. Ann., 117, 501 (1933).

90. H. W. Lavendel, Planseeber. f. Pulvermetallurgie, 9, 80 (1961).

91. H. H. Stadelmaier and T. S. Yun, Z. Metallk., 53, 754 (1962).

92. H. H. Stadelmaier, R. A. Draughn and G. Hofer, Z. Metallk., 54, 640 (1963).

93. H. H. Stadelmaier in J. T. Waber, P. Chiotti and W. N. Miner, editors, Nuclear Metallurgy Vol. 10, IMD Special Rept. no. 13, Met. Soc. AIME, 159 (1964).

94. G. Hofer, Dissertation, N. C. State University, Raleigh (1966).

95.   S. Samson, Final Rept. ONR and Calif. Inst. of Technology, Contract No. Nonr-220-33 (1964).

96.   E. Ganglberger, H. Nowotny and F. Benesovsky, Monatsh. Chemie, 97, 101 (1966).

97.   E. Ganglberger, H. Nowotny and F. Benesovsky, Monatsh. Chemie, 97, 718 (1966).

98.   Yu. B. Kuz'ma, Yu. V. Voroshilov and E. E. Cherkashin, Izv. Akad. Nauk SSSR, Neorg. Materialy, 1, 1017 (1965).

99.   E. Ganglberger, H. Nowotny and F. Benesovsky, Monatsh. Chemie, 96, 1144 (1965).

100. H. H. Stadelmaier, J. D. Schöbel and R. A. Jones, Metall, 21, 17 (1967).

101. J. D. Schöbel and H. H. Stadelmaier, Metall, 23 (1969) in print.

102. H. H. Stadelmaier and J. G. Avery, Z. Metallk., 56, 508 (1965).

103. H. H. Stadelmaier and J. D. Schöbel, Metall, 20, 31 (1966).

104. H. H. Stadelmaier and G. Hofer, Metall, 18, 460 (1964).

105. Yu. V. Voroshilov and Yu. B. Kuz'ma, Izv. Akad. Nauk. SSSR, Neorg. Materialy, 2, 764 (1966).

106. Yu. B. Kuz'ma, M. V. Chepiga and A. M. Plakhina, Izv. Akad. Nauk SSSR, Neorg. Materialy, 2, 1218 (1966).

107. H. Haschke, H. Nowotny and F. Benesovsky, Monatsh. Chemie, 97, 1459 (1966).

108. H. H. Stadelmaier and H. H. Davis, Monatsh. Chemie, 97, 1489 (1966).

109. H. H. Stadelmaier and J. T. Lowder, Metall, 21, 1023 (1967).

110. H. H. Stadelmaier and R. A. Gregg, Metall, 16, 407 (1962).

111. E. Fruchart and R. Fruchart, C. R. Acad. Sc. Paris, 258, 3032 (1964).

112. H. H. Stadelmaier and R. B. Fitts, Metall, 16, 773 and 1229 (1962).

113. G. Hofer and H. H. Stadelmaier, Metall, 19, 1257 (1965)

114. J. D. Schöbel and H. H. Stadelmaier, Metall, 19, 715 (1965).

115. H. H. Stadelmaier and B. B. Helms, Metall, 19, 121 (1965).

116. H. H. Stadelmaier and C. A. Shumaker, Metall, 20, 1056 (1966).

117. H. H. Stadelmaier and J. B. Ballance, Metall, 21, 691 (1967).

118. J. D. Schöbel and H. H. Stadelmaier, Metall, 18, 1285 (1964).

119. H. H. Stadelmaier, M. Kotyk and G. Hofer, Metall, <u>18</u>, 1065 (1964).

120. H. H. Stadelmaier and B. E. Miller, Metall, <u>23</u>, in print (1969).

121. H. H. Stadelmaier, J. D. Schöbel and J. R. Sagmuller, Metall, <u>18</u>, 23 (1964).

122. H. H. Stadelmaier, J. D. Schöbel and L. T. Jordan, Metall, <u>16</u>, 752 (1962).

123. H. H. Stadelmaier and A. C. Fraker, Metall, <u>16</u>, 212 (1962).

124. H. H. Stadelmaier and F. M. Lee, Metall, <u>18</u>, 111 (1964).

125. H. H. Stadelmaier and L. T. Jordan, Z. Metallk., <u>53</u>, 719 (1962).

125a. H. H. Stadelmaier and C. B. Pollock, Z. Metallk., to be publ.

125b. H. H. Stadelmaier and M. L. Fiedler, Z. Metallk., to be publ.

126. G. Hofer and H. H. Stadelmaier, Metall, <u>18</u>, 963 (1964).

127. H. H. Stadelmaier and R. A. Gregg, Metall, <u>17</u>, 412 (1963).

128. H. H. Stadelmaier and J. B. Ballance, Z. Metallk., <u>58</u>, 449 (1967).

129. E. Ganglberger, H. Nowotny and F. Benesovsky, Monatsh. Chemie, <u>97</u>, 494 (1966).

130. H. H. Davis, unpublished research, North Carolina State University, Raleigh.

131. Yu. B. Kuz'ma and T. F. Fedorov, Poroshkovaya Metallurgiya, No. 11, 62 (1965).

132. H. Rassaerts, F. Benesovsky and H. Nowotny, Planseeber. f. Pulvermetallurgie, <u>14</u>, 178 (1966).

133. H. Rassaerts, F. Benesovsky and H. Nowotny, Planseeber. f. Pulvermetallurgie, <u>13</u>, 199 (1965).

134. H. Rassaerts, F. Benesovsky and H. Nowotny, Planseeber. f. Pulvermetallurgie, <u>14</u>, 23 (1966).

135. C. B. Pollock, M. S. Thesis, North Carolina State University, Raleigh (1968).

136. H. Hirota and A. Yanase, J. Phys. Soc. Japan, <u>20</u>, 1960 (1965).

137. S. Samson, personal communication.

138. R. Kiessling, J. Electrochem. Soc., <u>98</u>, 166 (1951).

139. F. Laves in ASM, Theory of Alloy Phases, Cleveland, 124 (1956).

140. J. D. Schöbel and H. H. Stadelmaier, Z. Metallk., <u>57</u>, 323 (1966).

141. H. H. Stadelmaier, R. E. Burgess and H. H. Davis, Metall, $\underline{20}$, 225 (1966).

142. G. Hägg and R. Kiessling, J. Inst. Metals, $\underline{81}$, 57 (1952-53).

143. H. H. Stadelmaier, J. D. Schöbel and R. E. Burgess, Metall, $\underline{17}$, 781 (1963).

144. J. M. Leitnaker, M. G. Bowman and P. W. Gilles, J. Electrochem. Soc., $\underline{108}$, 568 (1961).

145. E. I. Gladyshevskii, T. F. Fedorov, Yu. B. Kuz'ma and R. V. Skolozdra, Poroshkovaya Metallurgiya, No. 4, 55 (1966).

146. Yu. B. Kuz'ma, V. I. Lakh, B. I. Stadnik and Yu. B. Voroshilov, Poroshkovaya Metallurgiya, No. 6, 73 (1966).

147. A. Wittmann, H. Nowotny and H. Boller, Monatsh. Chemie, $\underline{91}$, 608 (1960).

148. M. C. Cadeville and A. J. P. Meyer, C. R. Acad. Sc. Paris, $\underline{255}$, 3391 (1962).

149. M. C. Cadeville and E. Daniel, J. de Physique, $\underline{27}$, 449 (1966).

150. M. C. Cadeville, Dissertation, Strasbourg (1965).

151. T. Shinjo, F. Itoh and H. Takaki, J. Phys. Soc. Japan, $\underline{19}$, 1252 (1964).

152. G. Fischer and A. J. P. Meyer, C. R. Acad. Sc. Paris, $\underline{265}$, 521 (1967).

153. M. C. Cadeville, R. Jesser and A. Schwab, presented at Second Int. Conf. on Solid Compounds of Transition Metals, Enschede, The Netherlands, to be published (1967).

154. S. Samson, Acta Cryst., $\underline{14}$, 1229 (1961).

ALLOY CHEMISTRY OF THORIUM, URANIUM, AND

PLUTONIUM COMPOUNDS

A. E. Dwight

Metallurgy Division, Argonne National Laboratory

Argonne, Illinois

ABSTRACT

This paper is concerned with intermetallic compounds between the A elements thorium, uranium, and plutonium and X elements from the chromium group through the Si group. Thorium, U, and Pu differ from the transition elements in the crystal structure of their compounds, in that Th, U, and Pu form compounds with definite stoichiometric ratios and never form sigma phases or other compounds which are characterized by variable composition. Furthermore, the thoride elements tend to form compounds with an average atomic volume greater than that predicted from the atomic volumes of the elements, while transition elements generally form compounds which show a contraction in the average atomic volume.

Thoride-rich compounds crystallize in uncommon crystal structures, but thoride-poor compounds have structures which are commonly found in transition element compounds.

Crystal structure data are presented for: U_2RhNi, NaTl-type, $a_0 = 6.419$ Å and five isostructural compounds; $ThRhAl$, Fe_2P-type, $a_0 = 7.183$ Å, $c_0 = 4.120$ Å and 28 isostructural compounds; and UCo_3, $PuNi_3$-type, $a_0 = 4.849$ Å, $c_0 = 24.317$ Å.

Thoride-rich compounds with nickel exhibit larger average atomic volumes than those with cobalt. The relation of this anomalous behavior to similar observations on titanium compounds is discussed.

INTRODUCTION

A quarter of a century has elapsed since the metallurgy of Th, U, and Pu became the subject of intensive scientific research, and now the quantity of data available makes information retrieval a problem. At the same time, the large amount of data presents an opportunity to discover trends in alloying behavior, particularly in the occurrence of crystal structures in intermetallic compounds.

The elements Th, U, and Pu are among those which, by long established custom, are called actinides. For reasons to be presented later in this paper, the writer prefers the designation thoride for the elements Th, U, and Pu.

A number of investigators have discussed alloy chemistry of the thorides, and probably the most complete work is that of E. S. Makarov[1], which dealt with material published prior to June 1957. The present paper will constitute an attempt to collect and to discuss data which have appeared to July 1967, but will not contain a review of equilibrium diagrams. Discussions of the equilibrium diagrams may be found in books by W. D. Wilkinson[2], M. Hansen[3], and R. P. Elliott[4]. For a complete listing of intermetallic compounds, the two volumes by W. B. Pearson[5] include most of the data through 1964.

The position of the thoride elements in the periodic table, which has varied in several published versions, is basic to the discussion of their alloy chemistry. An older version of the periodic table placed Th, Pu, and U in the Ti-, V-, and Cr-groups, respectively. The evidence which supported this arrangement was mainly the crystal structural resemblance among the elements, for example, the group IV elements β-Ti, β-Zr, and β-Hf, and β-Th all are body centered cubic; and the groups V and VI elements Cr, Mo, W, and γ-U are also all body centered cubic. Furthermore, γ-U and Mo exhibit extensive solid solubility, which usually indicates a similarity in electron configuration.

The periodic table now in general use is the "collapsed" form, in which the actinide series is appended below the lanthanide series. This sometimes has created the impression that the actinides are analogs of the lanthanides. There is evidence to support this belief, but only for the trans-plutonium elements. The alloy chemistry of uranium shows little resemblance to that of the lanthanides.

The radial form of the periodic table in Figure 1 overcomes some of the problems which beset the earlier versions. The lanthanides are inserted in their proper place between Ba and Hf. Actinium is shown as a member of the Sc group but not as the parent

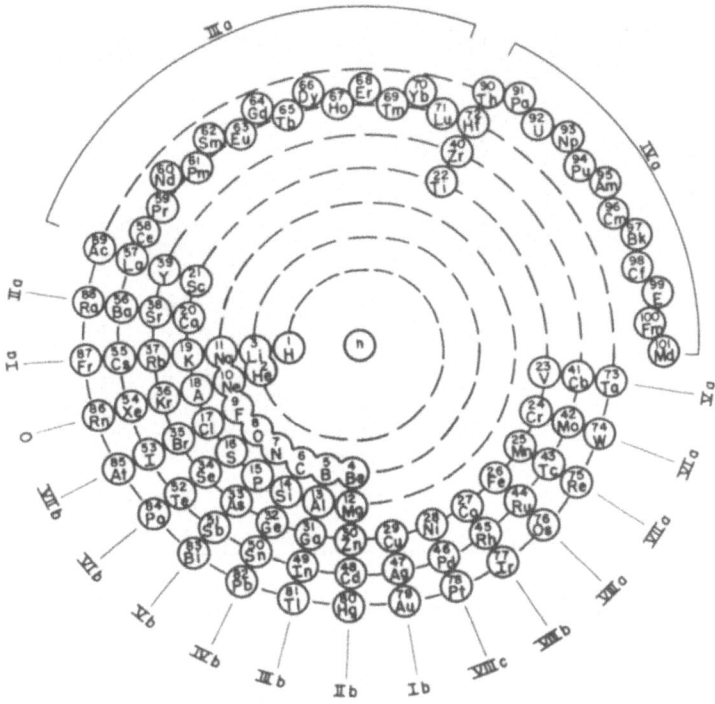

Figure 1. Periodic table in radial form.

of the actinide series. In the author's opinion, therefore, the incorrect terms "actinide series" and "actinides" should be replaced by "thoride series" and "thorides" respectively. (It is recognized that the trans-plutonium elements show less resemblance to Th and more to the lanthanides.) In the present paper, it is considered that Th is a member of the Ti group with four valence electrons, and that U and Pu have a somewhat higher effective valence electron concentration.

COMPARISON OF THORIDE COMPOUNDS WITH TRANSITION ELEMENT COMPOUNDS

It is logical to ask, first of all, in what respect do the thorides differ from the Ti-group, the V-group, and the Cr-group elements in their compound-forming tendencies? One very obvious characteristic is that compounds formed with the thorides occur at rigidly fixed stoichiometric ratios, while compounds formed by Ti, V, Mo, etc., may exhibit extensive solid solubility. The three equilibrium diagrams in Figure 2 illustrate this point. The V-Au[4] system (Figure 2a) contains the compounds VAu_2 and VAu_4, which are formed at a relatively low temperature from a terminal solid solution (congruent maximum), and have a substantial range of

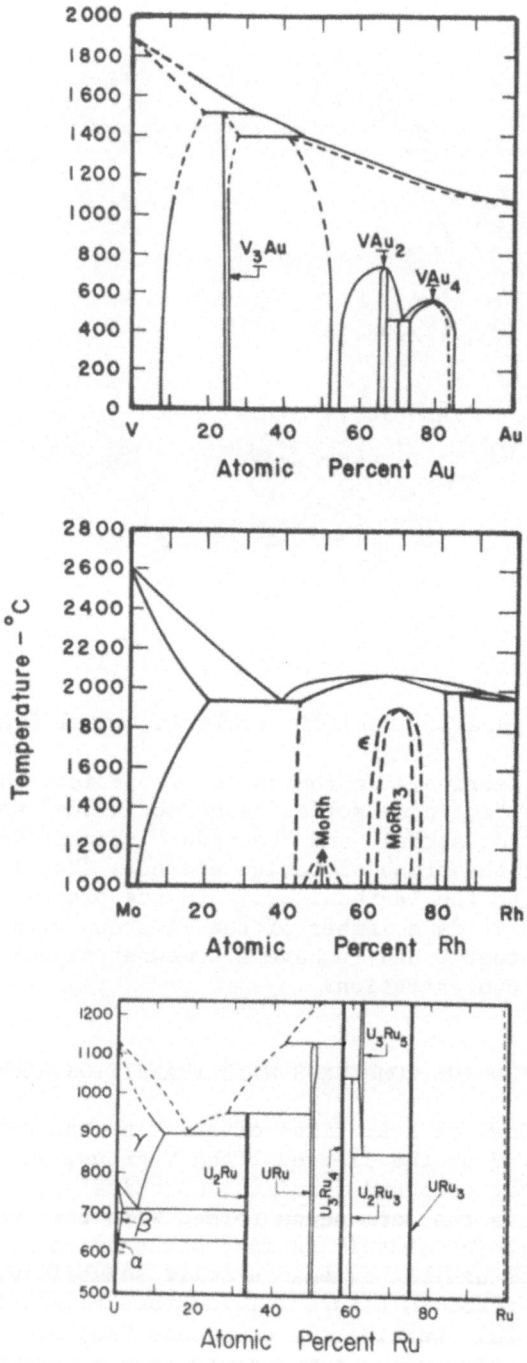

Figure 2. The V–Au, Mo–Rh, and U–Ru equilibrium diagrams.

homogeneity. This type of phase relationship is absent in Th, U, and Pu diagrams, with the single exception of U–Pd. The Mo–Rh system[4] (Figure 2b) contains an epsilon phase which is formed from the liquid and extends from 43 to 82 atomic percent Rh. A similar example is the sigma phase in the W–Os system. Transition elements of the V and Cr group form many phases of variable composition, but they are absent in Th, U, and Pu systems, with the exception of the U–Pu system.

The type of diagram typical of Th, U, and Pu systems is represented by the U–Ru system[6] (Figure 2c). All of these compounds are characterized by narrow ranges of homogeneity at specific stoichiometric ratios. These compounds are usually formed from the liquid, but occasionally by a peritectoid transformation (e.g., U_2Ru_3 in Figure 2c).

A second characteristic in which thorides differ from Ti-, V-, and Cr-group elements is the ability to undergo the essential atomic size adjustments. Some crystal structures have a definite relative atomic size requirement for the constituent elements. If a pair of elements do meet the size requirement, or if one or both can adjust atomic size, by giving up or acquiring electrons, then that particular crystal structure will be stable, provided the elements are present in the proper stoichiometric ratio and that electron concentration requirements, if any, are met. A prominent example of a crystal structure with a definite size requirement is the Laves phase, $MgCu_2$-type. In this structure, a radius ratio r_A/r_X equal to $\sqrt{3}/\sqrt{2}$ (1.23) is necessary for close contact between both A-A and X-X atoms. However, this is the ratio of the atomic size in the compound, not in the pure elements. When the radius ratio of the elements is used as an approximation, it is found[7] that Laves phases exist with radius ratios from 1.05 to 1.68. To form a Laves phase from a pair of elements whose radius ratio is 1.05, it appears that the A element must expand, and/or the X element contract so that in the compound, a ratio closer to the ideal 1.23 is attained. If an element is able to undergo size alterations, it will readily participate in compound formation, such as $MgCu_2$, which require definite size ratios. If an element is not able to undergo size changes, or only to a limited extent, then of necessity the element would avoid size-sensitive structure types and form others which will accommodate its effective size. To express the idea in a simpler way, rubber balls of different sizes will fit into a lattice array, easier than billiard balls because the rubber balls can mutually adjust their sizes. The writer believes, based on Figures 3 through 12, that the thorides are much less able to adjust their sizes than Ti-, V-, or Cr-group elements, and this prohibits the formation of the more common structure types, especially in uranium compounds.

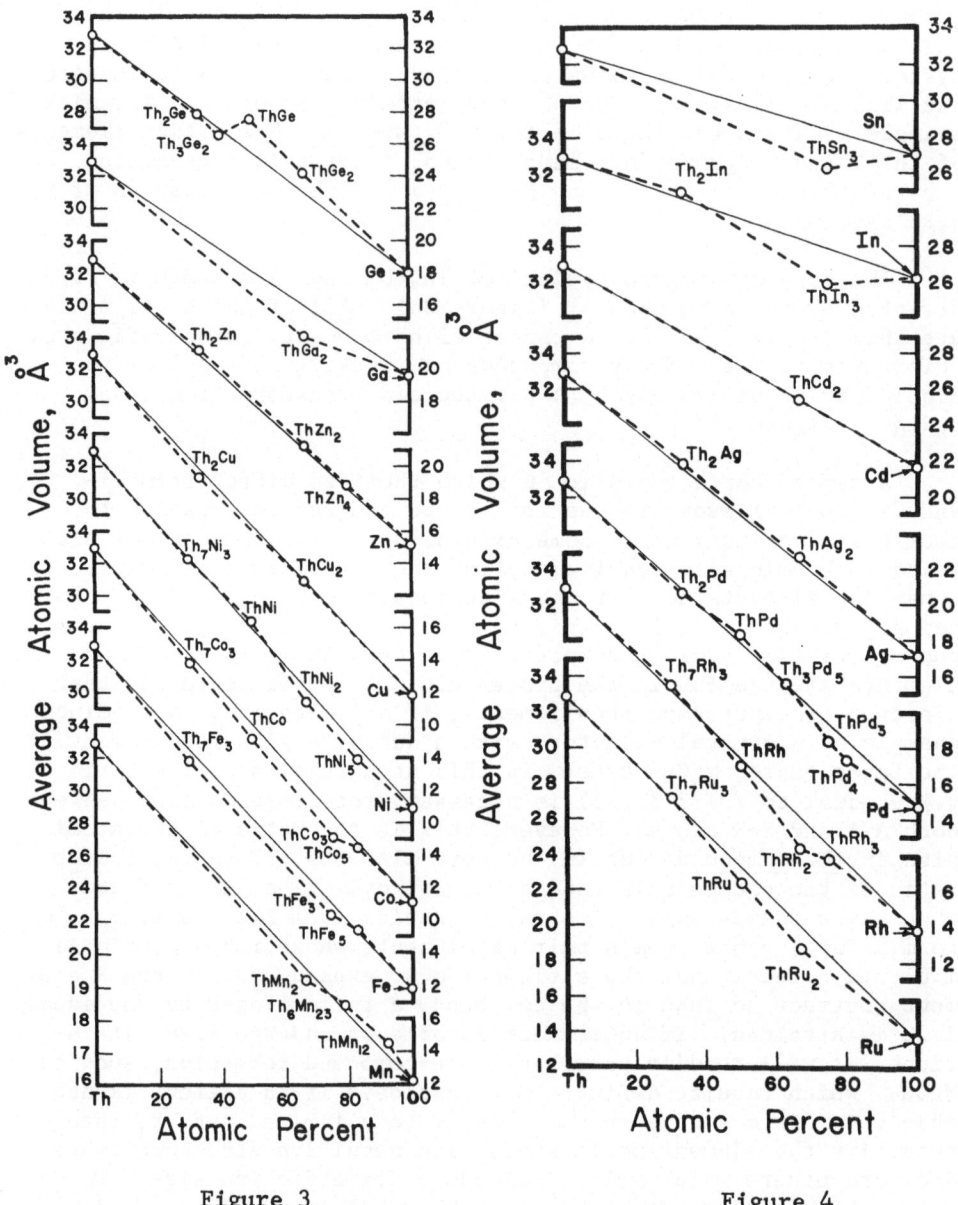

Figure 3

Figure 4

Figure 3.   Average atomic volume in compounds of Th with first
            l.p. elements.

Figure 4.   Average atomic volume in compounds of Th with second
            l.p. elements.

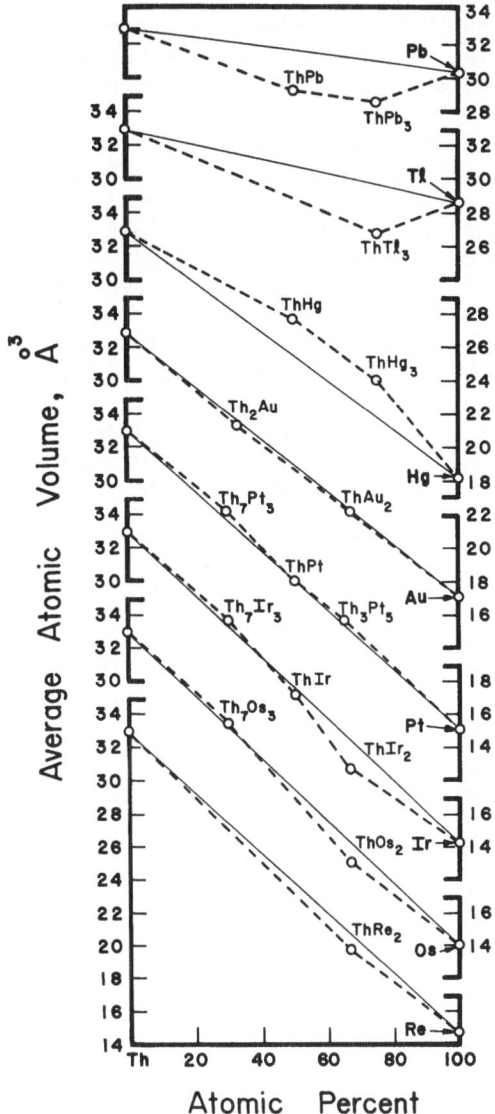

Figure 5.   Average atomic volume in compounds of Th with third
            l.p. elements.

        The average atomic volume (a.a.v.), defined as the volume of
the unit cell in $\overset{\circ}{A}{}^{3}$ divided by the number of atoms in the cell, is
a quantity which is useful for comparison of relative atomic sizes.
The atomic volumes of metallic elements, published by Rudman[8], are
used in this paper, except that new values have been computed for

U and Pu. The volumes stated by Rudman for U (20.8 $\overset{\circ}{A}^3$) and Pu
(20.3 $\overset{\circ}{A}^3$) imply that U has a larger atomic volume than Pu, which is
clearly not the case in intermetallic compounds. For example, U_6Fe
has an a.a.v. of 19.89 $\overset{\circ}{A}^3$, and Pu_6Fe has an a.a.v. of 20.93 $\overset{\circ}{A}^3$.

The atomic volume of U was computed from published data on
U-Mo alloys[9]. The lattice parameters of γ-U-Mo solid solutions,
measured at room temperature after quenching from the γ range, were
extrapolated to pure U and an a_o of 3.4808 $\overset{\circ}{A}$ and atomic volume of
21.1 $\overset{\circ}{A}^3$ (compared with Rudman's value 20.8) were obtained. Since
our value for U was obtained from a bcc structure, an attempt was
made to obtain a comparable value for the bcc allotrope of Pu, the
ε phase. The data of Ball et al., quoted by Pearson[5], give an a_o
of 3.642 $\overset{\circ}{A}$ and an atomic volume of 24.154 $\overset{\circ}{A}^3$ for ε-Pu at 530°C. A
comparable value for γ-U at 530°C was obtained by extrapolating the
data of Chiotti et al. (quoted by Pearson) down to 530°C which gave
an a_o = 3.5135 $\overset{\circ}{A}$ and an atomic volume of 21.687 $\overset{\circ}{A}^3$. The ratio of
the atomic volumes of ε-Pu to γ-U at 530°C is 1.114. Assuming that

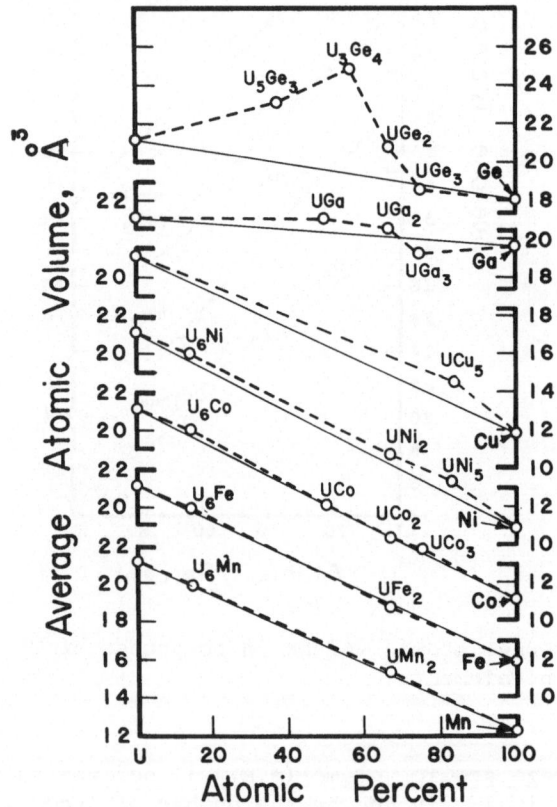

Figure 6. Average atomic volume in compounds of U with first
           l.p. elements.

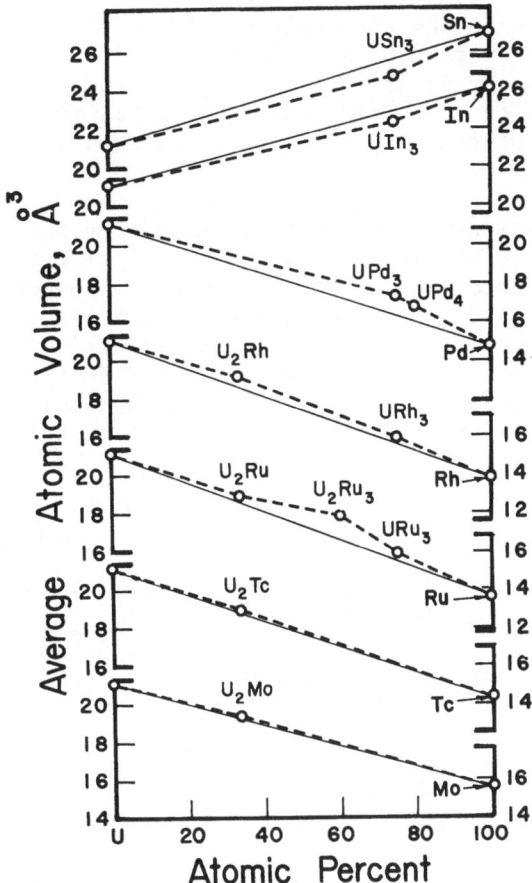

Figure 7. Average atomic volume in compounds of U with second
           l.p. elements.

the same ratio exists at room temperature, the atomic volume for
ε-Pu is the product of the Pu/U ratio and the atomic volume of
γ-U. This results in an atomic volume of 23.5 Å^3 for ε-Pu (compared with Rudman's value of 20.3). An alternative method was
used to calculate a Pu/U ratio, by plotting the Pu/U ratios of isostructural intermetallic compounds, and extrapolating back to the
elements Pu/U. This method gave a Pu/U ratio of 1.06. Due to un-
certainties in the extrapolation, the calculated ratio of 1.114 is
preferred. However, both methods indicate that Pu has a larger
atomic volume than U.

The larger apparent size of the Pu atom is of interest in relation to Zachariasen's plot[10] of estimated atomic radii. This

plot shows that U would have an atomic radius slightly larger than
Pu, if both were in the same valence state, but that pentavalent
Pu would be larger than hexavalent U. Since Pu in compounds is in
fact larger than U, it is reasonable to assume that the Pu electron configuration is $f^{n-2}d^3s^2$ and that U has $f^{n-3}d^4s^2$ (n is the
quantum number of the last full noble gas shell, and is 6 for U and
Pu).

It was mentioned that the thorides differ from other Ti-group
elements in their ability to undergo mutual size adjustments. The
expansion or contraction of Th, U, and Pu compounds is shown in
Figures 3 through 12, in which the average atomic volume is plotted
against composition. In these figures, a solid line connects the
atomic volumes of two elements (e.g. Th and Mn in Figure 3), and a
dashed line connects the a.a.v. of the intermetallic compounds.
The feature of major interest is the deviation from the solid line.

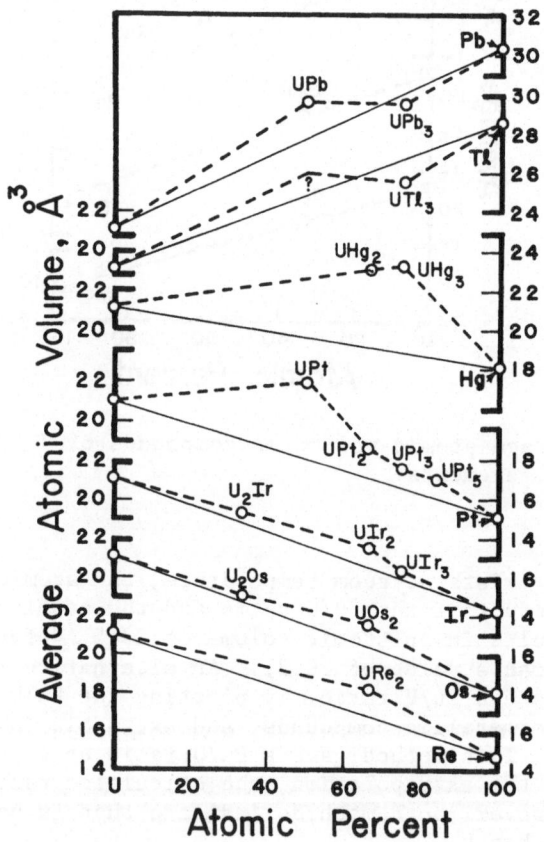

Figure 8.   Average atomic volume in compounds of U with third
            l.p. elements.

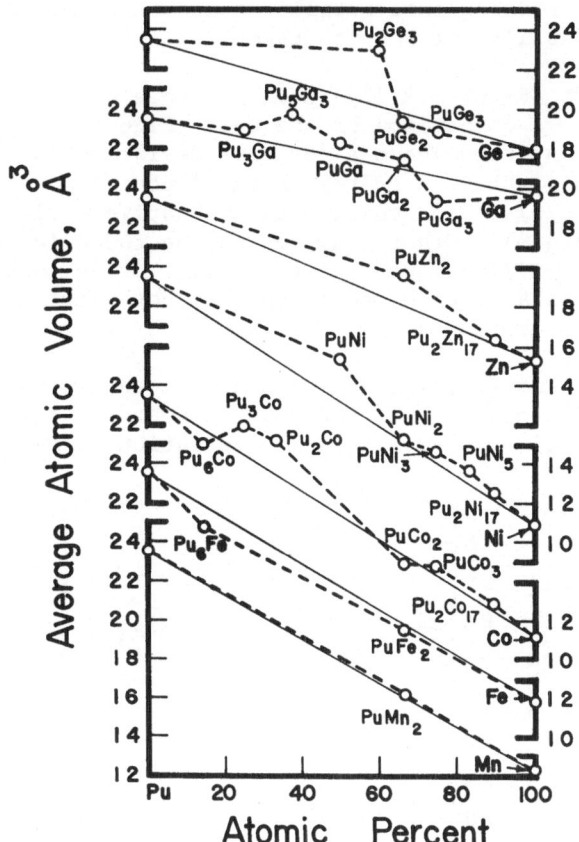

Figure 9.   Average atomic volume in compounds of Pu with first
            l.p. elements.

The following observations can be made:

1. Thorium compounds (Figures 3, 4, 5, and 12) generally display
    a negative deviation from the solid line. In this respect they
    are similar to Ti, Zr, and Hf compounds, but the negative de-
    viation for Th compounds is not as large.

2. Uranium compounds (Figures 6, 7, 8, and 12) generally exhibit
    a positive deviation from the solid line (i.e., U compounds
    have a higher a.a.v. than would a mechanical mixture of their
    component atoms). This is in contrast to similar curves for
    Ti, Zr, and Hf compounds published by Raman and Schubert[11].
    In nine Ti, Zr, and Hf binary systems, a negative deviation
    was found. This indicates that Ti, Zr, and Hf contract more

than U when forming intermetallic compounds, which reflects the relative inability of U to contract.

3. Pu compounds (Figures 9-12) resemble those of U, but to a lesser degree. A comparison of the U-Ir (Figure 8) and Pu-Ir (Figure 11) systems shows a positive deviation in both, slightly larger in UIr_2 than in $PuIr_2$. A small positive deviation in U_6Fe (Figure 6) is in contrast to a negative deviation in Pu_6Fe (Figure 9). Other examples can be found in Figures 3 through 12 to illustrate that the a.a.v. of Pu compounds resembles the a.a.v. of U compounds, but both differ from the a.a.v. of Th compounds.

4. Thorium, U, and Pu compounds with Al all show a positive deviation (Figure 12), but compounds of U and Pu with Si exhibit first a positive, then a negative deviation. The deviation

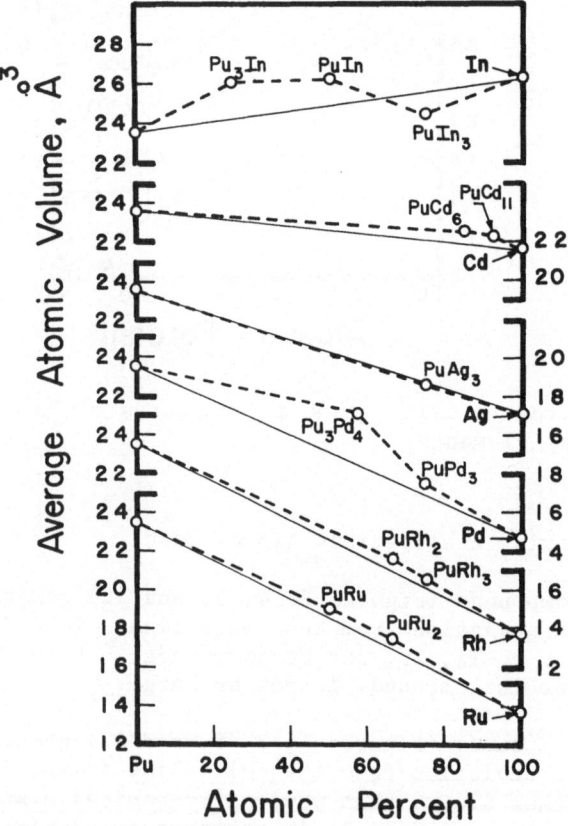

Figure 10.   Average atomic volume in compounds of Pu with second l.p. elements.

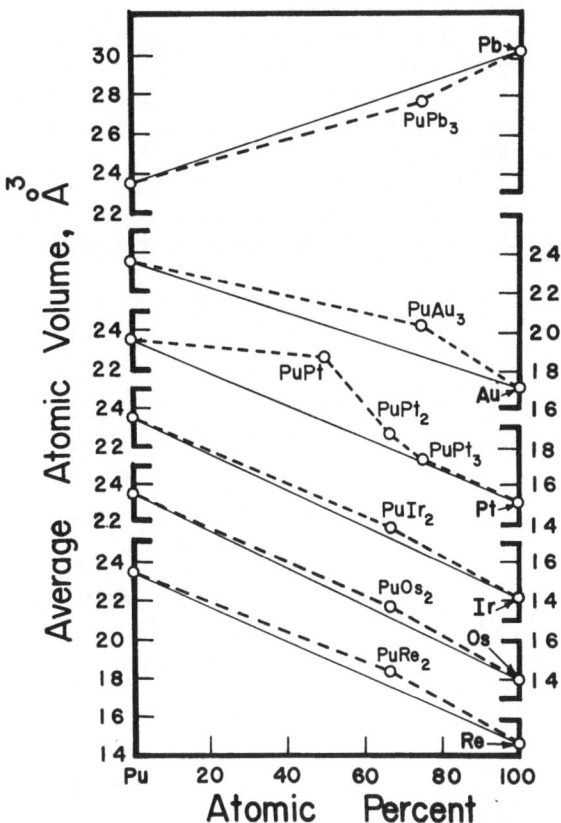

Figure 11.   Average atomic volume in compounds of Pu with third
             l.p. elements.

would be positive if an atomic volume smaller than 15.5 $\mathring{A}^3$[8]
were used for metallic Si. Yoshioka and Beck[12] have noted that
the effective atomic volume of Si in transition element sol-
vents is much smaller than the metallic atomic volume. An effective atomic volume of 11.0 $\mathring{A}^3$ is given for Si in V alloy
systems[12]. It seems probable, therefore, that Si has an atomic
volume less than 15.5 $\mathring{A}^3$ in compounds with Th, U, and Pu, and
that the a.a.v. curves in Figure 12 actually should exhibit
a positive deviation. Similar arguments indicate that Ga and
Ge have smaller effective atomic volumes than the values in the
elemental state[8].

     A third manner in which the thorides differ from elements of
the Ti-, V-, and Cr-group is in the crystal structures of the intermetallic compounds. Here a distinction must be made between

thoride-rich and thoride-poor compounds. Thoride-rich compounds
generally have crystal structures that are uncommon, i.e. not
found among compounds of other elements. As examples, U_2Ru_3, UCo,
and U_6Co have no analogs among transition element compounds. Con-
versely, the structure types that are common among Ti-, V-, and
Cr-rich compounds, such as the Cr_3Si and Ti_2Ni structure types are
rarely found in thoride-rich compounds. Common structure types
that have only one thoride representative are $MoSi_2$ (U_2Mo) and CsCl
(PuRu), which are discussed in the sections on A_2X and AX compounds,
respectively.

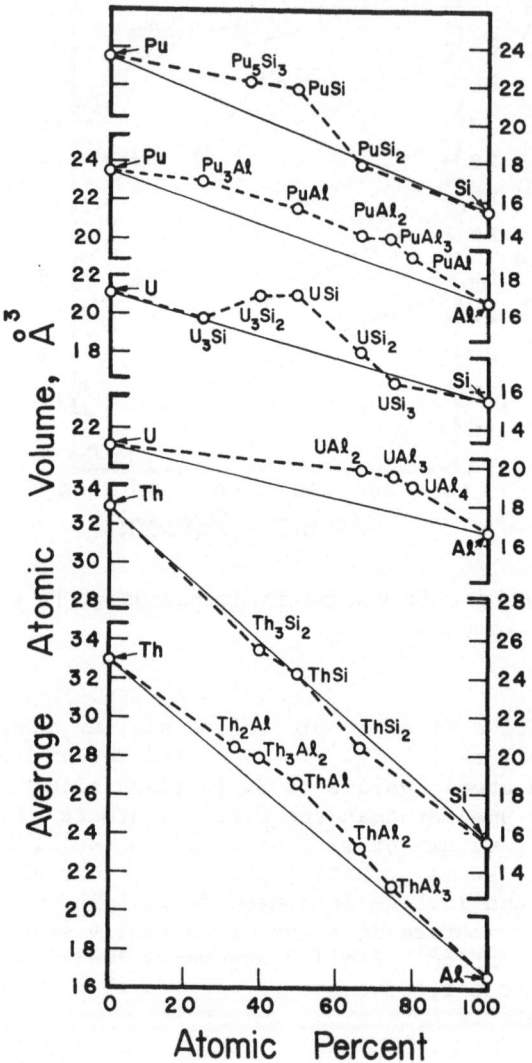

Figure 12. Average atomic volume in compounds of Th, U, and Pu
            with Al and Si.

Although there are marked differences between the thorides and the transition elements, two similar characteristics in alloying behavior are noted. First, the thoride-poor compounds often select well-known structure types, in common with the transition elements. The $MgCu_2$ and Cu_3Au structure types are well represented in thoride compounds of AX_2 and AX_3 stoichiometry. It was noted in an earlier publication[13] that Ti, Zr, Hf, Th, and U form isostructural compounds with Rh, Ir, and Pd. A second resemblance is the extensive solid solution which usually indicates a similar electronic structure. The Ti-U, Zr-U, Hf-U, Nb-U, and Zr-Pu systems exhibit at high temperatures, a complete solid solubility range. Thorium is excluded from forming an extensive range of solid solutions by its large atomic size.

Thorides differ from Ti-, V-, and Cr-group elements in the following characteristics:

1.   Thoride compounds occur at rigidly fixed stoichiometric ratios, but Ti-, V-, and Cr-group compounds may exhibit an extensive range of solid solubility.

2.   Thorides are less able to undergo size adjustments than are the Ti-, V-, and Cr-group elements.

3.   Thoride-rich compounds usually crystallize in uncommon crystal structures, which are not found in compounds of Ti, V, or Cr.

Thorides resemble the Ti-, V-, and Cr-group elements in the following characteristics:

1.   Thoride-poor compounds often have the same crystal structures as Ti- and V-group compounds.

2.   Thorides exhibit extensive solid solutions with Ti- and V-group elements, when relative size permits.

STRUCTURE DATA FOR THE THORIDE COMPOUNDS

The organizational scheme of this section is to arrange intermetallic compounds into eight families that have a definite stoichiometric ratio: A_6X, A_3X, A_7X_3, A_2X, AX, AX_2, AX_3, and AX_5. The A elements are Th, U, and Pu; the X elements are the transition elements from the Cr group through the Ni group, and elements of the Cu, Zn, Al, and Si groups.

A_6X and A_3X Compounds

The thoride compounds, richest in the thoride elements, are

those of the A_6X composition. All known examples are listed in Table I. There are no Th_6X compounds and no counterparts among transition element compounds. No new examples of A_6X compounds have been reported since Makarov's book[1] was published, and the reader is referred to Makarov for crystallographic details and coordination numbers. Two points to be noted from Table I are the following: The Pu compounds have a larger a.a.v. than their U analogs, and the a.a.v. of the U_6X series increases from U_6Mn to U_6Ni, although the atomic volume of the X elements decrease from Mn to Ni. The significance of the latter observation is discussed in a later section.

There are a small number of A_3X compounds, which are listed in Table II. The thoride A_3X compounds fall into two natural subdivisions, in one the X element is from the Al or Si group, and in the other the X element is from the Fe or Co group.

A_2X and A_7X_3 Compounds

The thoride compounds of A_2X composition are listed in Table III. The absence of $MoSi_2$-type compounds (except U_2Mo) is probably due to a size effect. The geometry of the $MoSi_2$-type structure requires that there be Mo–Mo and Mo–Si contacts, which means that the ideal atomic volume ratio is unity. Nevitt[15] has noted that in forming $MoSi_2$-type compounds "the A and B atoms make cooperative changes in their apparent radii as large as 10% in the nearest neighbor directions in order to fulfill the geometric requirements." It appears that the U atom can make the necessary size changes only with Mo, and that Pu and Th atoms, being larger than U, cannot adjust sufficiently with any element.

The occurrence of the $CuAl_2$-type structure in Th_2X compounds may be explained in terms of the relative atomic size. Nevitt[15] has noted that the $CuAl_2$ structure has less restrictive atomic-size requirements than the Ti_2Ni and $MoSi_2$ structures and that when the critical limits for the other two are exceeded, the $CuAl_2$ type is formed.

The occurrence of the monoclinic U_2Ru-type structure[16] is not explained by the previous arguments. This structure has been found only among U compounds, although the compound Pu_2Rh with an unknown crystal structure[17] may prove to be isostructural with U_2Ru. The U_2Ru-type structure is an example of a trend noted in an earlier section, that U and Pu select uncommon crystal-structure types in U and Pu-rich compounds. Only Th_2X compounds resemble transition-element compounds.

While no ternary A_2X compounds are known, data are available

Table I.   Tetragonal U_6Mn-type compounds

A_6X	a(Å)	c(Å)	c/a	a.a.v.(Å³)
U_6Mn*	10.29	5.24	0.51	19.82
U_6Fe	10.31	5.24	0.51	19.89
U_6Co	10.36	5.21	0.50	19.97
U_6Ni	10.37	5.21	0.50	20.01
Pu_6Fe	10.405	5.349	0.51	20.68
Pu_6Co	10.475	5.340	0.51	20.93

*Unit cell constants from Pearson[5] unless otherwise noted.

Table II. A_3X thoride compounds

A_3X	a(Å)	b(Å)	c(Å)	a.a.v.(Å³)	Structure Type
Th_3Ge					unknown
U_3Si*	6.029		8.696	19.76	b.c. tet.
Pu_3Al	4.499		4.538	22.96	$SrPb_3$
Pu_3Ga[14]	4.507			22.89	Cu_3Au
Pu_3In	4.705			26.04	Cu_3Au
U_3Os					unknown
Pu_3Co	3.475	10.976	9.220	21.98	ortho.
Pu_3Ru					unknown
Pu_3Os					unknown

*Unit cell constants from Pearson[5] unless otherwise noted.

Table III. A_2X thoride compounds

A_2X	a(Å)	b(Å)	c(Å)	c/a	a.a.v.(Å³)	Structure Type
Th_2Al*	7.614		5.857	0.77	28.35	$CuAl_2$
Th_2Cu	7.28		5.74	0.79	25.35	$CuAl_2$
Th_2Zn	7.60		5.64	0.74	27.15	$CuAl_2$
Th_2Pd	7.33		5.93	0.81	26.55	$CuAl_2$
Th_2Ag	7.56		5.84	0.77	27.81	$CuAl_2$
Th_2In	7.787		6.113	0.79	30.89	$CuAl_2$
Th_2Au	7.42		5.95	0.79	27.30	$CuAl_2$
U_2Mo	3.427		9.854	2.88	19.29	$MoSi_2$
U_2Tc[16]	13.407	3.271	5.213		18.93	U_2Ru
U_2Ru	13.106	3.343	5.202		18.88	U_2Ru
U_2Rh[16]	13.122	3.421	5.159		19.18	U_2Ru
U_2Os[16]	13.366	3.335	5.167		19.07	U_2Ru
U_2Ir[16]	13.210	3.457	5.095		19.27	U_2Ru
Pu_2Mg	7.34					f.c.c.
Pu_2Co	7.762		3.649	0.47	21.127	Fe_2P
Pu_2Ru[17]						unknown

*Unit cell constants from Pearson[5] unless otherwise noted.

on the solid-solubility range of the binary compounds in several ternary systems. These data could be presented as vertical or isothermal sections of a ternary-equilibrium diagram, but the num- ber of ternary systems for which data are available would require an excessive number of figures. In order to expedite the compari- son of binary and ternary data in a large number of systems, a schematic method called the area-of-stability plot has been devel- oped. Figure 13 is an area-of-stability plot for U_2X alloys and a major portion of the periodic table is included. The point desig- nated by a chemical symbol (e.g., Fe) refers to an alloy [U_2Fe] and since this specific alloy is a mixture of two phases, this point is occupied by a partially shaded triangular symbol to represent a two phase, binary alloy. The Ru position in Figure 13 has no symbol other than the chemical symbol, an indication that the alloy U_2Ru is a single-phase intermetallic compound. A typical ternary alloy

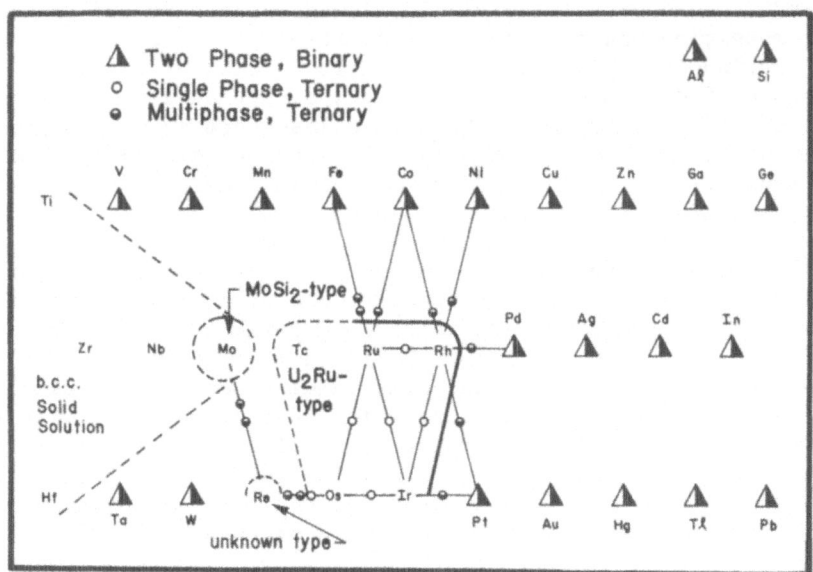

Figure 13. U_2X area-of-stability plot.

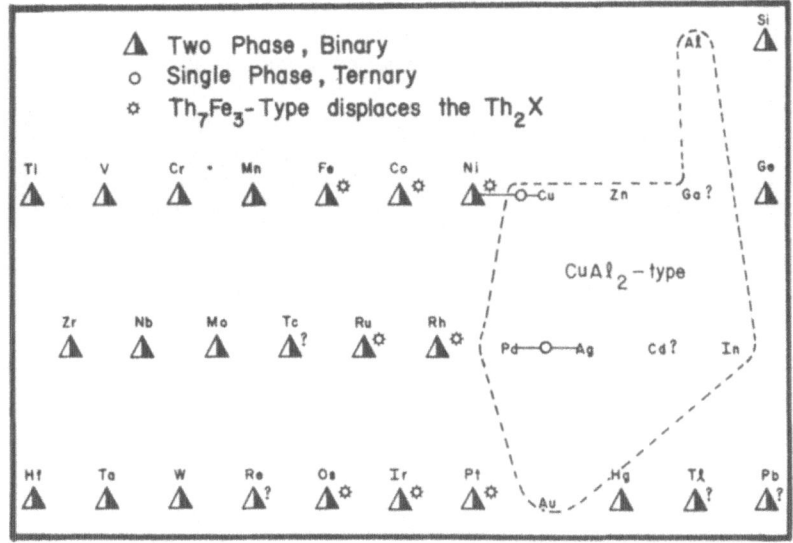

Figure 14. Th_2X area-of-stability plot.

is U_4RuRh, shown by a small circle midway between Ru and Rh, and this composition has the same crystal structure as U_2Ru and U_2Rh, i.e., continuous solid solubility exists between U_2Ru and U_2Rh at some unspecified temperature. However, less than 10 at.% Fe or Co can be substituted for Ru in U_2Ru, and less than 12 at.% Ni or Pd can be substituted for Rh in U_2Rh. The area-of-stability plot is not a substitute for an equilibrium diagram, because no attempt is made to specify temperature dependence.

In Figures 13 and 14, the elements of the second long period (l.p.) are displaced one-half space to the right relative to their counterparts of the first and third long periods. Thus, Pd is located half a space to the right of Ni and Pt. The reason for this displacement is that a compound containing a second l.p. element often selects a crystal structure which is isostructural with that of a compound containing a first or third l.p. element, one group to its right. For example, in Figure 13, the U_2Ru-type structure[16] occurs in U_2Tc but not in U_2Mn or U_2Re; thus, the alloying behavior of Tc is similar to Ru rather than to Mn or to Re. In Figure 14, Th_2Pd and Th_2Ag crystallize in the $CuAl_2$-type structure, but the compounds Th_2Ni and Th_2Pt do not exist; Pd displays a closer similarity in alloying behavior to Ag rather than to Ni or to Pt. In many of the area-of-stability plots, this behavior has been noted repeatedly; therefore, all plots are drawn with the second l.p. elements displaced half a space to the right. The apparent similarity of Pd with Ag rather than with Pt is strongest in compounds which are poor in Pd, such as Th_2Pd. In Pd-rich compounds, e.g. $TiPd_3$, the crystal structure is the same as in $TiNi_3$ and $TiPt_3$.

The relative size of the X elements in a series of isostructural compounds may be studied by plotting the a.a.v. against the group number. Figure 15 presents the a.a.v. for the Th_7X_3-type compounds[5] which displace both Th_2X and Th_3X compounds in eight binary systems. The a.a.v. rises or falls with the atomic volume of the X element, except for Fe, Co, and Ni, which exhibit the opposite trend. This anomaly has been noted in many structure types where isostructural Fe, Co, and Ni examples appear, and is discussed in a later section.

AX Thoride Compounds

The AX compounds (Table IV) exhibit the tendency for U compounds to crystallize in uncommon structures, while Th and Pu compounds occur in well-known structures. Most numerous of the Th compounds are those with the CrB-type structure, one of which (ThCo) is shown in Figure 16. All atoms are situated in alternate layers that lie parallel to the a and b axes, at z = 1/4 and 3/4. The shortest Th-Co interatomic distance is 2.86 Å which is drawn in

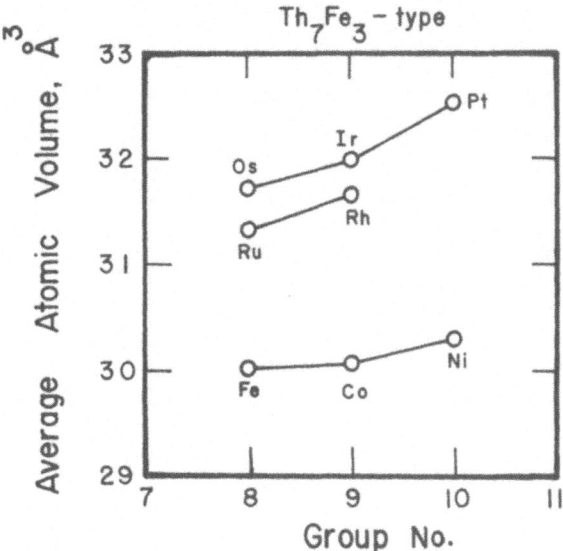

Figure 15.   Average atomic volume in Th_7Fe_3-type compounds.

Figure 16 from a Co atom to its four nearest neighbor Th atoms.
When another species is substituted for Co, the unit cell constants
change (Figure 17). Substitution of a larger atom simply increases
all parameters, but substitution of an atom from the Fe- or Ni-
groups has a peculiar effect. As the X element is changed from Ru
to Rh to Pt, the a.a.v. and c_o increase, b_o decreases, and a_o remains nearly constant. The c_o parameter, which is the spacing between every second layer, is the most sensitive to the group of the
X element. The spacing between layers is controlled by the strength
of the bond between Th and X element. The four shortest interatomic
distances should affect the a_o and c_o parameters, yet a_o remains
nearly constant; increase or decrease in Th-X bond-length appears
mainly in the c_o parameter.

When the X element is Pd, Si, and probably Ni, the FeB-type
structure is found in ThX compounds. The change from CrB-type to
FeB-type cannot be explained on the basis of atomic size, and,
therefore, is probably due to the increased number of electrons
available from Pd, Si, and Ni. The FeB- and CrB-type structures
are closely related. Hohnke and Parthé[18] have recently summarized
the structural similarities between the two types. In some of the
earlier work on the FeB-type compounds, various authors have employed different crystal settings, with the result that the designation of the axes have been interchanged in some publications. In
this paper, the setting Pnma is accepted and the unit cell constants
are in the order $a_o > c_o > b_o$.

Table IV.   AX thoride compounds

AX	a(Å)	b(Å)	c(Å)	a.a.v.(Å³)	Structure Type
ThAl*	4.42	11.45	4.19	26.51	CrB
ThCo	3.74	10.88	4.16	21.16	CrB
ThRu	3.878	11.29	4.071	22.28	CrB
ThRh	3.866	11.24	4.22	22.92	CrB
ThIr	3.894	11.13	4.226	23.11	CrB
ThPt	3.900	11.09	4.454	24.08	CrB
ThNi	14.51	4.31	5.73	22.40	ortho.
ThPd	7.249	4.571	5.856	24.25	FeB
ThSi	7.88	4.148	5.896	24.09	FeB
ThGe	6.046			27.63	NaCl
ThHg	4.80			27.64	f.c.c.
ThPb	4.545		5.644	29.15	tet.
UCo	6.355			16.04	b.c.c.
USi	7.65	3.90	5.65	21.07	FeB
UPt[17]	3.721	10.772	4.410	22.10	CrB
UGa	7.40	7.60	9.42	21.03	ortho.
UPb	11.04		10.60		tet.
$U_5Ni_2Rh_3$[21]	6.438			16.53	NaTl
U_5Ni_4Ir[21]	6.316			15.75	NaTl
U_5Ni_4Pd[21]	6.316			15.75	NaTl
U_5Ni_4Pt[21]	6.361			16.09	NaTl
$U_2Pd_3Rh_2$[21]	6.585			17.85	NaTl
U_2PdPt[21]	6.648			18.36	NaTl
PuNi	3.59	10.21	4.22	19.34	CrB
PuRu	3.363			19.02	CsCl
PuPt[17]	3.816	10.694	4.428	22.59	CrB
PuSi	7.933	3.847	5.727	21.85	FeB
PuAl	10.769			21.53	b.c.c.
PuGa[14]	6.64		8.066	22.23	tet.
PuIn	4.811		4.538	26.26	CuAu

*Unit cell constants from Pearson[5] unless otherwise noted.

    The unit cell constants for ThSi, USi, and PuSi, and for comparison, also those of TiSi, ZrSi, HfSi, ErSi, TbSi, and PrSi are shown in Figure 18. Data points for TiSi, ZrSi, HfSi, and ThSi can be fitted to a smooth curve, which supports the earlier suggestion that Th is a member of the Ti group. A curve through the a.a.v. points for ErSi, TbSi, and PrSi lies below the curve for the Ti-group silicides, and a curve through the USi and PuSi points lies above the curve. This sequence of curves is

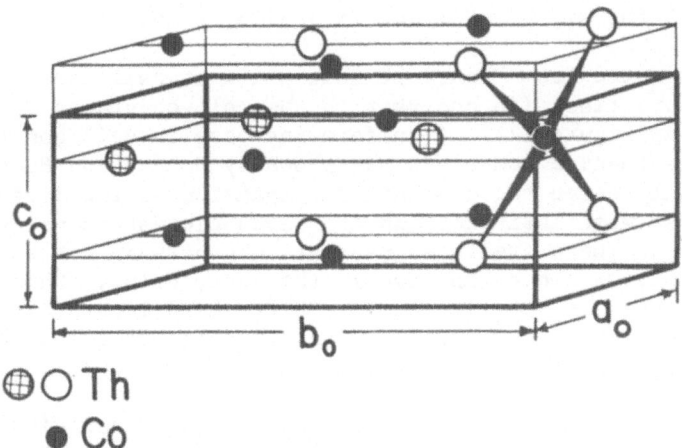

Figure 16.   The crystal structure of ThCo, a CrB–type structure.

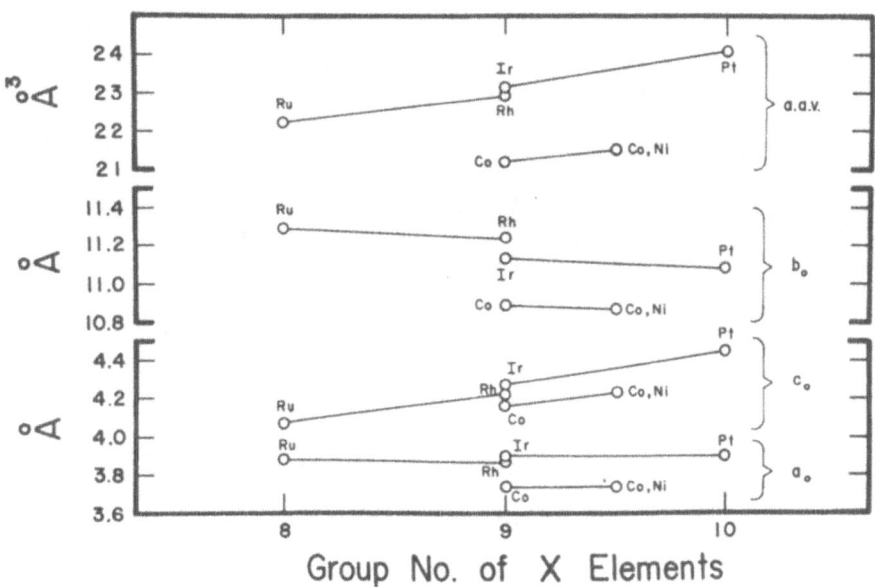

Figure 17.   Unit cell constants of ThX compounds, CrB–type structure.

compatible with average electron concentration of 3 for the lanthanides, 4 for Ti, Zr, Hf, and Th, and 5 or more for U and Pu.

It is noted from Figure 18 that all unit cell constants rise
as the a.a.v. rises (in contrast to the cell constants for CrB-type
compounds in Figure 17). A uniform increase in cell constants for
FeB-type lanthanide compounds was shown by Hohnke and Parthé[18].
The apparent reason for the uniform expansion is the direction of
the bonds between a Si atom and its seven closest Th neighbors
(Figure 19). There are three nearest neighbors in the same layer,
two in the layer above, and two in the layer below. When a larger
or smaller atom is substituted for Th the bond lengths increase or
decrease in all three directions, rather than predominantly in one
direction as in the CrB-type compounds.

The ThX area-of-stability plot in Figure 20 contains nine
CrB- and FeB-type compounds and only three compounds of other structure types. The prevalence of the CrB- and FeB-type is due to the
large volume ratio of A to X elements, which in turn is due to the
large atomic volume of Th. As previously mentioned, the choice between the CrB- and FeB-types is due to electron concentration. The
ternary systems of Th-Al-X are unexplored for the most part, and
the possibility exists that ternary $Th_{50} (Al,X)_{50}$ compounds will be
found.

The UX compounds (Table IV) also crystallize in uncommon
structure-types, just as in U_6X, U_3X, and U_2X compounds. A single
exception is USi with the FeB-type structure. An example of the
uncommon structure types is UCo, which is shown in perspective in
Figure 21. The structure is bcc[5], with a distortion which shifts
all atoms away from the corners and centers of the subcell. It is
a layer structure, but each layer is rippled. The most significant
feature of the structure is that a Co atom is linked to eight
nearest-neighbor U atoms, but one bond length is appreciably shorter
than the other seven. The structure, therefore, may be visualized
as a collection of diatomic "molecules" in which the U–Co distance
is 2.65 Å. It is this short U–Co distance which controls the unit
cell dimensions.

The effect of partial substitutions of Ni or Fe for Co in UCo
is shown in Figure 22. Only a very small substitution of Fe is
possible, but nearly 15 at.% Ni can be substituted. A point of
interest in Figure 22 is that the Fe substitution decreases the cell
constant, but the Ni substitution increases it. As noted in the
Th_7Mn_3-type compounds, this is in contrast to the decreasing atomic
size of the pure elements Fe, Co, and Ni.

Table IV includes six compounds with the NaTl-type structure.
This structure is stable for U compounds, but is not found in

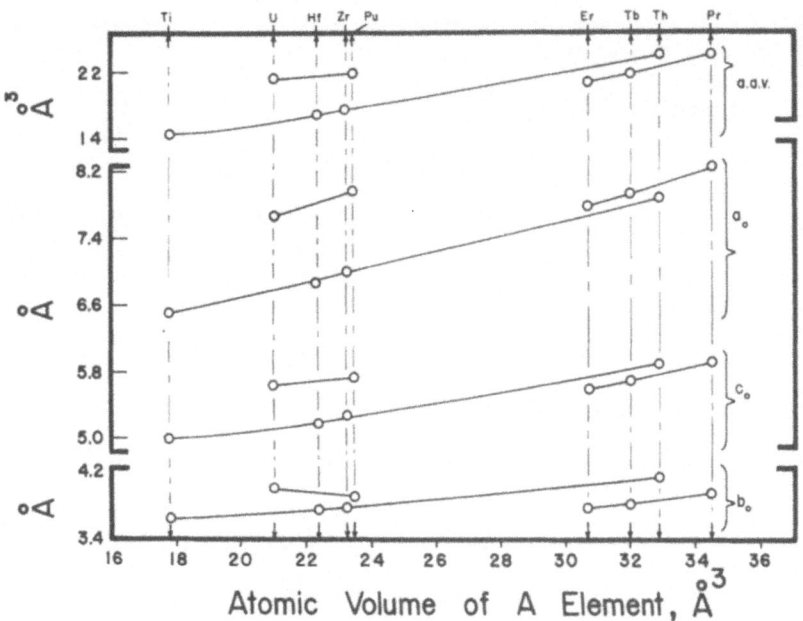

Figure 18. Unit cell constants of ASi compounds, FeB-type struc-
            ture.

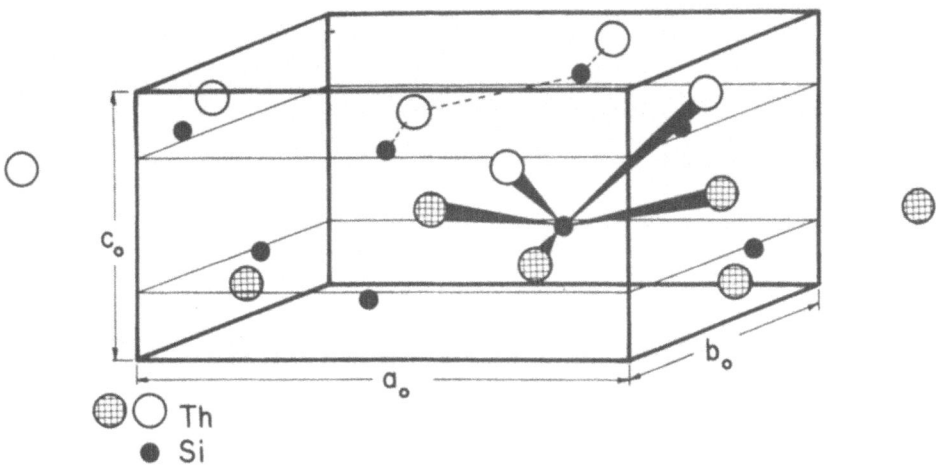

Figure 19. The crystal structure of ThSi, and FeB-type structure.

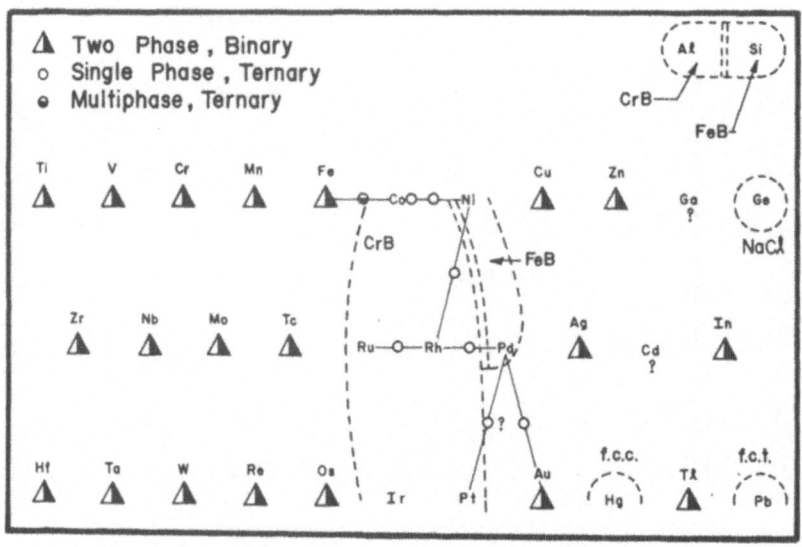

Figure 20.   ThX area-of-stability plot.

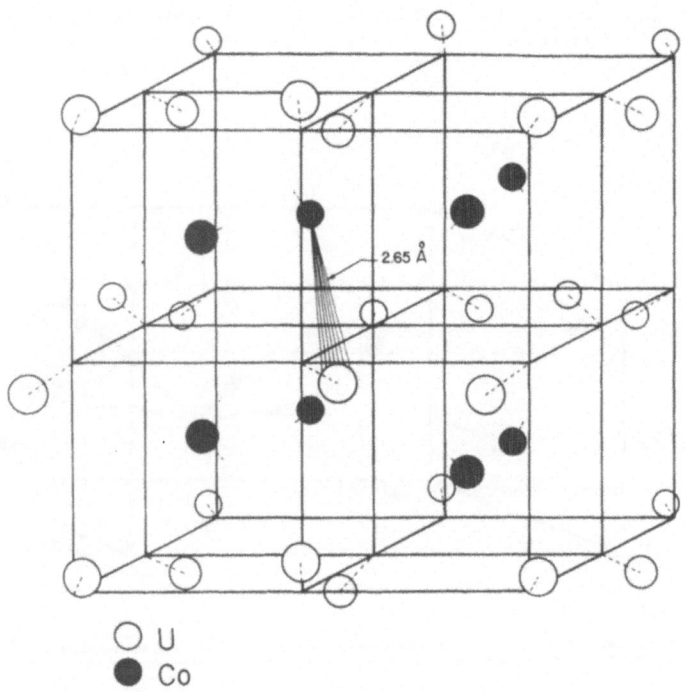

Figure 21.   The crystal structure of UCo.

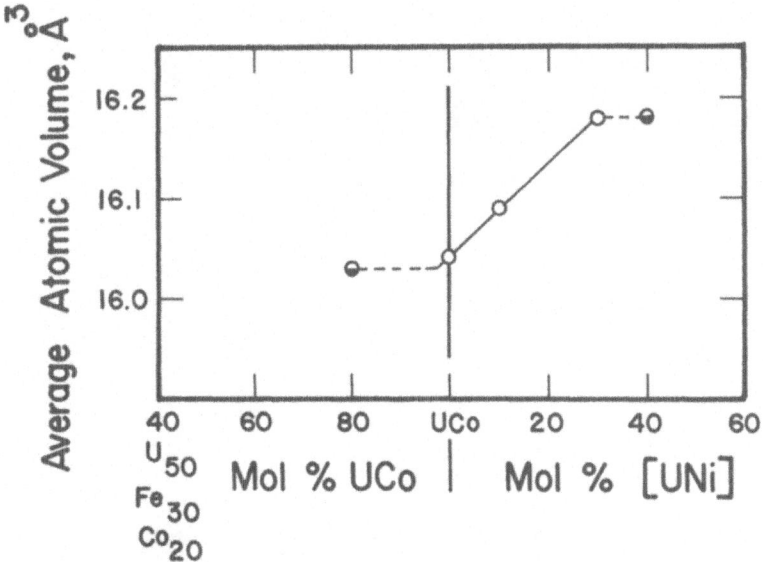

Figure 22. Average atomic volume in UCo in U(Fe, Co) and U(Co, Ni)
            vertical sections.

compounds of transition elements or thorium. The known binary
examples[5] of the NaTl-type are similar to NaTl and to LiAl in that
an extreme difference exists in the electronegativity of the components, which some authors have considered a necessary requirement for the stability of the NaTl-type structure. In the compounds of which U_2RhNi is representative, no unusual difference in
electronegativity exists.

The crystal structure of U_2RhNi (Figure 23) has a mixture of
bcc and fcc characteristics, as described in an earlier paper[19].
The essential feature of the structure is that each atom has eight
equidistant nearest neighbors; four of its own kind and four of the
other species. This means that the shortest U-U, U-Rh(Ni), and
Rh(Ni)-Rh(Ni) distances are all equal. Since U has a larger atomic
size than Rh(Ni), the U-U bond strength should be stronger than the
U-Rh(Ni) and the Rh(Ni)-Rh(Ni) the weakest of the three.

In Figures 25 and 26, the approximate compositional ranges are
shown in which the NaTl-type structure is stable in the U-Rh-Pd and
U-Rh-Ni systems, respectively. Extensive substitution of Pd and Rh
for Ni is possible, but very little substitution for U is permitted.
This indicates that the dominant feature which controls the stability is the diamond complex formed by the U atoms. Laves[20] has
pointed out that in the NaTl structure, the space-filling principle
would be satisfied best if the radius of the A element equalled
that of the X element, but that several examples exist, e.g., LiZn,
where the proposed equality does not hold. In U_2RhNi, for example,

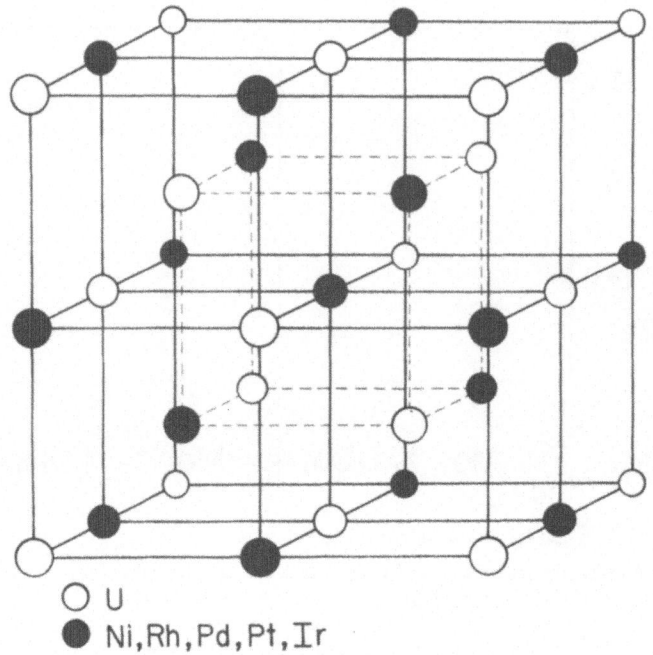

Figure 23.   The crystal structure of U_2RhNi, the NaTl-type structure.

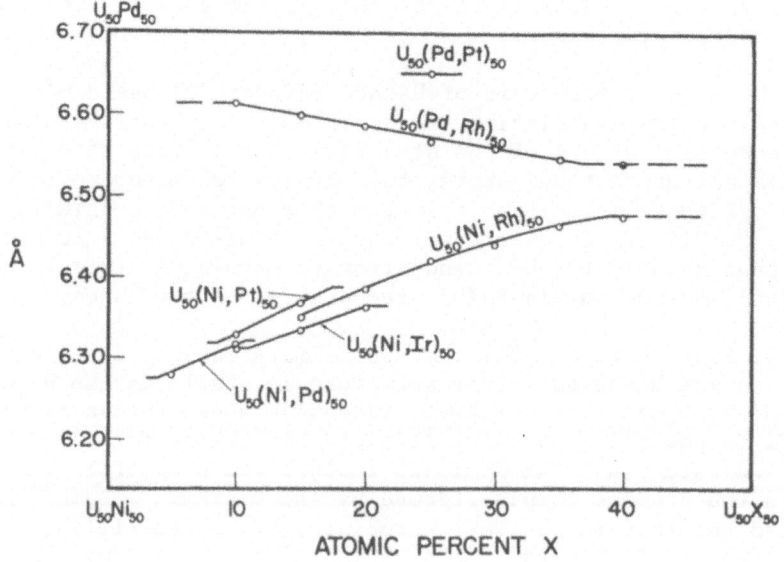

Figure 24.   Lattice parameter of NaTl-type UX compounds.

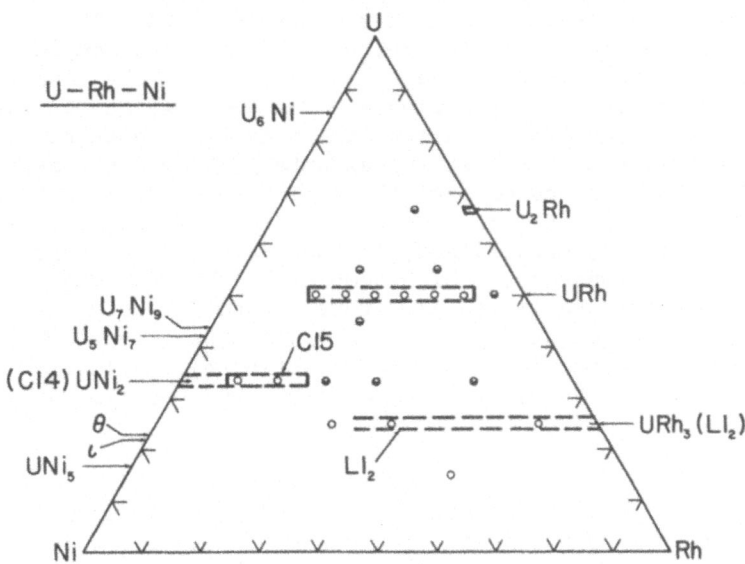

Figure 25.   Partial equilibrium diagram of the U-Rh-Ni system.

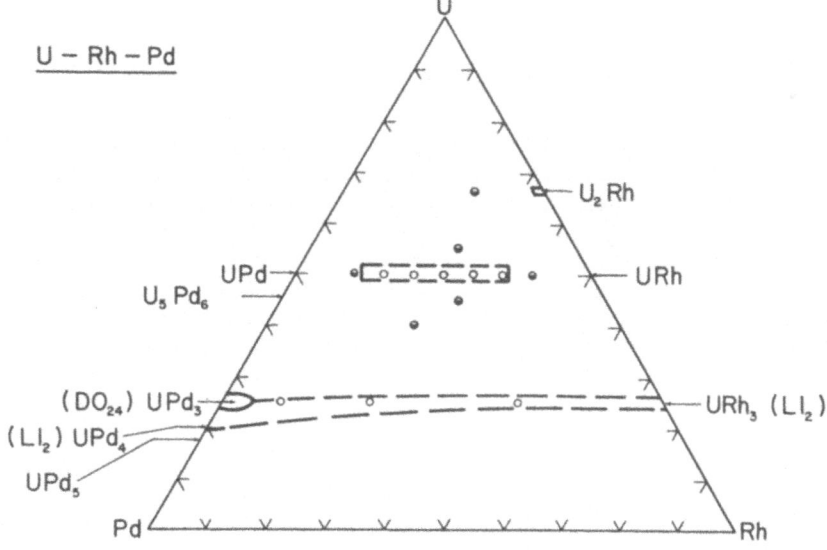

Figure 26.   Partial equilibrium diagram of the U-Rh-Pd system.

it is obvious that three atomic sizes are involved. The structure
of U_2RhNi is formed by two interpenetrating diamond complexes, one
of which (U) cannot be replaced by substitution. The other complex
is widely variable in composition[21]. The U-U distance in the U
diamond complex does not determine the unit cell constant. The cell
constant varies with the relative amounts of Ni and Rh, as shown in
Figure 24. This indicates that the cell constant is dependent upon
the length of the U-(Rh, Ni) bond. A feature of the UX NaTl-type
compounds, which is unexplained, is that none of these extend into
a binary system (unless the high temperature UPd compound should be
found to have the NaTl-type structure). Another point is that in
an alloy of $U_{50}Rh_{15}Ni_{35}$, $a_0 = 6.35$ Å (NaTl-type), and in UCo, $a_0 =$
6.355 (bcc); but, our data show very little solid solubility between the two compounds, although as mentioned UCo will dissolve
15% Ni.

The compounds which occur in UX alloys are indicated in an
area-of-stability plot (Figure 27). In addition to the UCo-, NaTl-
and FeB-type compounds, there are four compounds for which the crystal structure is unknown. These are URu, URh, UIr, and UPt.

There are five PuX compounds in Table IV. Only PuAl has an
uncommon structure; the other four crystallize in well-known structure types. In the AX compounds, Th and Pu select well-known crystal structures, but U has only one representative (i.e., USi with
the FeB-type).

AX_2 Thoride Compounds

Thoride compounds of the AX_2 stoichiometric ratio are more
numerous than for any other composition. The known representatives
(Table V) include nearly as many ternary as binary compounds. The
large number of AX_2 compounds is probably related to the fact that
Th, U, and Pu select well-known crystal structures rather than uncommon ones in the thoride-poor compounds. The only uncommon structure in Table V is that of UPt_2, which is a distorted Ni_2In-type,
and may be isostructural with other distorted Ni_2In-type structures.
The $ThSi_2$-type, which appears in a few examples in Table V, is not
uncommon as there are numerous lanthanide representatives.

The area-of-stability plot for ThX_2 compounds is given in
Figure 28. Two common compound families, Laves phases and the
AlB_2-type, dominate this plot. There is a scarcity of data on the
Th ternary systems and some uncertainty exists as to the exact
boundary of the areas of Laves phase and AlB_2-type stability. The
structure of $ThPt_2$ is unknown. $ThZn_2$ has been reported both as an
AlB_2-type and as $CeCu_2$-type, and since the latter is a distorted
variant of the AlB_2-type, both structures may be stable in $ThZn_2$.

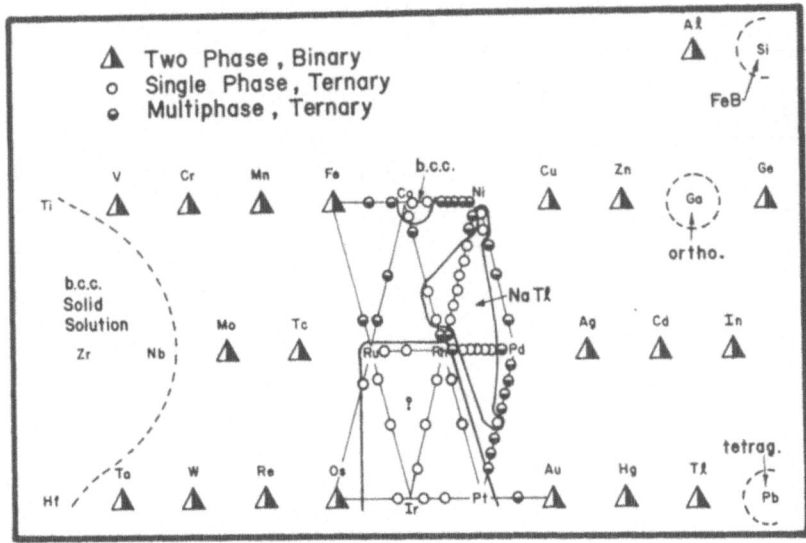

Figure 27. UX area-of-stability plot.

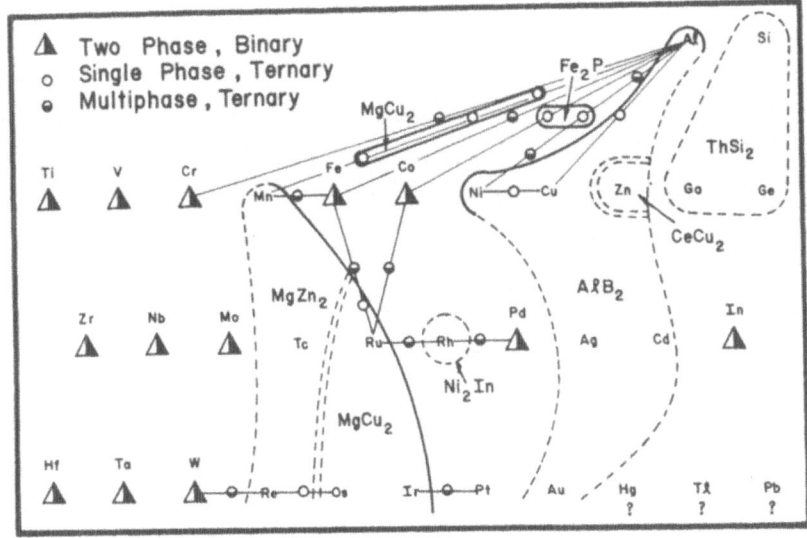

Figure 28. ThX$_2$ area-of-stability plot.

Table V. AX_2 thoride compounds

AX_2	a(Å)	b(Å)	c(Å)	c/a	a.a.v. (Å³)	Structure Type
$ThMn_2$*	5.479		8.931	1.630	19.35	$MgZn_2$
$ThTc_2$[22]	5.394		9.222	1.710	19.36	$MgZn_2$
$ThRe_2$	5.492		9.097	1.656	19.80	$MgZn_2$
$ThRu_2$	7.649				18.65	$MgCu_2$
$ThOs_2$	7.705				19.06	$MgCu_2$
$ThIr_2$	7.662				18.74	$MgCu_2$
$ThMnAl$						$MgCu_2$
$ThRh_2$	4.629		5.849	1.26	18.09	Ni_2In
$ThNi_2$	3.960		3.844	0.97	17.40	AlB_2
$ThCu_2$	4.387		3.472	0.79	19.29	AlB_2
$ThZn_2$	4.20		4.17	0.99	21.23	AlB_2
$ThAg_2$	4.837		3.353	0.693	22.65	AlB_2
$ThCd_2$	5.005		3.514	0.70	25.41	AlB_2
$ThAu_2$	4.740		3.402	0.72	22.06	AlB_2
$ThAl_2$	4.388		4.162	0.95	23.13	AlB_2
$ThSi_2$ l.t.	4.136		4.126	1.00	20.38	AlB_2
$ThSi_2$ h.t.	4.135		14.375	3.48	20.48	$ThSi_2$
$ThGa_2$	4.243		14.690	3.46	22.04	$ThSi_2$
$ThGe_2$	4.223	16.911	4.052		24.11	$ZrSi_2$
UMn_2	7.163				15.31	$MgCu_2$
UFe_2[6]	7.061				14.67	$MgCu_2$
UCo_2[6]	7.002				14.30	$MgCu_2$
UOs_2	7.512				17.66	$MgCu_2$
UIr_2	7.509				17.64	$MgCu_2$
UAl_2[23]	7.766				19.52	$MgCu_2$
$UNiRu$[24]	7.227				15.73	$MgCu_2$
U_2Ni_3Rh[24]	7.163				15.31	$MgCu_2$
URe_2 h.t.	5.433		8.561	1.576	18.24	$MgZn_2$
UNi_2[6]	4.963		8.249	1.662	14.66	$MgZn_2$

Table V. AX_2 thoride compounds (contd)

AX_2	a(Å)	b(Å)	c(Å)	c/a	a.a.v. (Å³)	Structure Type
$U_4Mo_5Si_3$	5.37		8.582	1.598	17.86	$MgZn_2$
U_2Cr_3Si[36]	5.125		8.265	1.613	47.00	$MgZn_2$
$U_4Mn_5Si_3$[36]	5.168		7.908	1.530	45.7	$MgZn_2$
U_2Fe_3Si	5.145		7.717	1.500	44.23	$MgZn_2$
U_2Co_3Si[36]	5.139		7.585	1.476	43.37	$MgZn_2$
UMnAl	5.25		8.408	1.601	16.73	$MgZn_2$
$U_4Fe_5Al_3$	5.16		8.00	1.55	15.75	$MgZn_2$
U_2Co_3Al	5.120		7.693	1.503	14.55	$MgZn_2$
U_2OsAl_3	5.380		8.482	1.577	17.68	$MgZn_2$
U_2MnNi_3	4.986		16.453	3.300	14.76	$MgNi_2$
$U_{33}Fe_{12}Ni_{55}$	4.971		16.423	3.303	14.64	$MgNi_2$
$U_{33}Co_{12}Ni_{55}$	4.963		16.397	3.303	14.57	$MgNi_2$
UPt_2	5.60	9.68	4.12		18.61	Ni_2In(dist)
USi_2	4.028		3.852	0.96	18.04	AlB_2
UGa_2	4.21		4.01	0.95	20.52	AlB_2
UHg_2	4.98		3.22	0.65	23.05	AlB_2
UGe_2	4.12	15.13	3.98		20.67	$ZrSi_2$
$PuMn_2$	7.292				16.16	$MgCu_2$
$PuFe_2$	7.189				15.48	$MgCu_2$
$PuCo_2$	7.095				14.88	$MgCu_2$
$PuNi_2$	7.141				15.17	$MgCu_2$
$PuZn_2$	7.760				19.47	$MgCu_2$
$PuRu_2$	7.476				17.41	$MgCu_2$
$PuRh_2$[17]	7.488				17.49	$MgCu_2$
$PuIr_2$[17]	7.512				17.66	$MgCu_2$
$PuPt_2$[17]	7.631				18.52	$MgCu_2$
$PuAl_2$	7.84				20.08	$MgCu_2$
$PuRe_2$	5.396		8.729	1.618	18.34	$MgZn_2$
$PuOs_2$	5.326		8.665	1.627	17.75	$MgZn_2$

Table V. AX_2 thoride compounds (contd)

AX_2	a(Å)	b(Å)	c(Å)	c/a	a.a.v.(Å^3)	Structure Type
PuGa$_2$[14]	4.248		4.120	0.970	21.46	AlB$_2$
PuSi$_2$	3.97		13.55		17.80	ThSi$_2$
PuGe$_2$	4.102		13.81		19.36	ThSi$_2$
ThCoAl[25]	7.047		4.036	0.573	19.28	Fe$_2$P
ThNiAl[25]	7.080		4.055	0.573	19.56	Fe$_2$P
ThRhAl[25]	7.183		4.120	0.574	20.45	Fe$_2$P
ThPdAl[25]	7.265		4.181	0.574	21.20	Fe$_2$P
ThIrAl[25]	7.176		4.142	0.577	20.52	Fe$_2$P
ThPtAl[25]	7.271		4.154	0.571	21.12	Fe$_2$P
ThCoGa[25]	7.048		3.987	0.566	19.06	Fe$_2$P
ThNiGa[25]	7.057		4.019	0.570	19.26	Fe$_2$P
ThRhGa[25]	7.214		4.044	0.561	20.25	Fe$_2$P
ThIrGa[25]	7.218		4.074	0.564	20.43	Fe$_2$P
ThPtGa[25]	7.297		4.095	0.561	20.93	Fe$_2$P
ThNiIn[25]	7.367		4.117	0.559	21.50	Fe$_2$P
ThPdIn[25]	7.541		4.190	0.556	22.93	Fe$_2$P
ThPtIn[25]	7.535		4.166	0.553	22.76	Fe$_2$P
UFeAl[25]	6.672		3.981	0.597	17.05	Fe$_2$P
UCoAl[25]	6.686		3.966	0.593	17.06	Fe$_2$P
UNiAl[25]	6.733		4.035	0.599	17.60	Fe$_2$P
URuAl[25]	6.895		4.029	0.584	18.43	Fe$_2$P
URhAl[25]	6.965		4.019	0.577	18.76	Fe$_2$P
UIrAl[25]	6.968		4.030	0.579	18.83	Fe$_2$P
UPtAl[25]	7.012		4.127	0.589	19.52	Fe$_2$P
UFeGa[25]	6.731		3.903	0.580	17.02	Fe$_2$P
UCoGa[25]	6.693		3.933	0.588	16.95	Fe$_2$P
UNiGa[25]	6.733		4.022	0.597	17.54	Fe$_2$P
URuGa[25]	7.076		3.818	0.540	18.39	Fe$_2$P
URhGa[25]	7.006		3.945	0.563	18.63	Fe$_2$P

Table V. AX_2 thoride compounds (contd)

AX_2	a(Å)	b(Å)	c(Å)	c/a	a.a.v.(Å³)	Structure Type
UIrGa[25]	7.033		3.944	0.561	18.77	Fe_2P
UPtGa[25]	7.063		4.065	0.575	19.51	Fe_2P
UPdIn[25]	7.414		4.096	0.553	21.67	Fe_2P
UPtIn[25]	7.413		4.058	0.547	21.46	Fe_2P

*Unit cell constants from Pearson[5] unless otherwise noted.

The area-of-stability plot for UX_2 compounds is shown in Figure 29. On this plot, the Laves phases are stable in seven areas. The AlB_2-type area is assumed to be continuous. The hexagonal Laves phases of U show a greater variation in c/a ratio than for any other element. This is especially noticeable when a comparison is made between UNi_2 (c/a = 1.664) and U_2Co_3Al (c/a = 1.503) from Steeb and Petzow[26]. (Our unpublished data show c/a = 1.488, the smallest ratio ever reported for a $MgZn_2$-type Laves phase.) A need exists for an investigation into the bond lengths and location of atoms in crystallographic sites in U_2Co_3Al.

The atomic size of Fe, Co, and Ni in UX_2 compounds shows a trend similar to that noted earlier in UX compounds. In Figure 30, the a.a.v. decreases from UFe_2 to UCo_2, then increases for UNi_2. Our data confirm the existence of a ternary $MgNi_2$-type compound[5] in the U-Co-Ni system.

Two ternary equiatomic compounds, ThCoAl and ThNiAl, are shown in Figure 28, the ThX_2 area-of-stability plot, and are enclosed in the Fe_2P type area. The structure of Fe_2P is hexagonal, space group P$\bar{6}$2m, number 189. The structure of Fe_2P reported by Hendricks and Kosting[27] is incorrect, and a corrected structure was reported by Rundqvist and Jellinek[28]. The assignment of elements to atomic sites is as follows:

Site		Fe_2P	ThNiAl
1(b)	$0,0,\frac{1}{2}$	P	Ni
2(c)	$\frac{1}{3},\frac{2}{3},0$	P	Ni
3(f)	$X,0,0$	Fe	Al
3(g)	$X,0,\frac{1}{2}$	Fe	Th

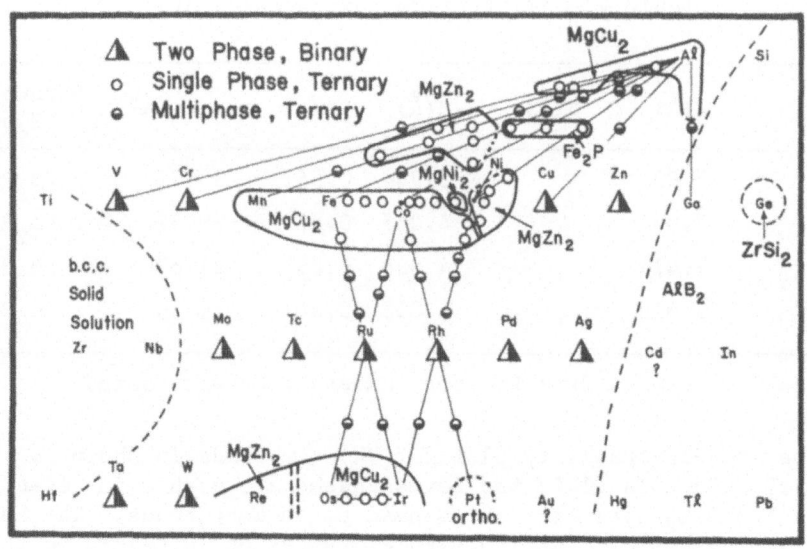

Figure 29. UX_2 area-of-stability plot.

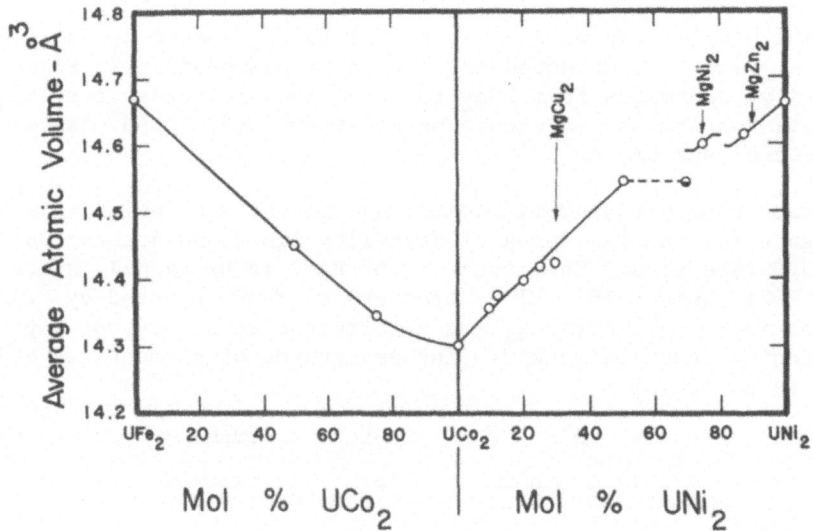

Figure 30.   Average atomic volume in UCo_2 in $U(Fe, Co)_2$ and
$U(Co, Ni)_2$ vertical sections.

A projection of the structure of ThNiAl on the base plane is shown in Figure 31. The essential feature of this configuration is that Ni atoms are in the two sites (b) and (c), which are not crystallographically equivalent. The Ni atom in the 1(b) site has six nearest Al neighbors, and each Ni atom in the 2(c) site has six nearest Th neighbors. The Ni-Al distances differ from the Ni-Th distances. This characteristic favors the formation of a ternary rather than a binary compound, and we have found more than 60 ternary compounds compared with thirteen binary compounds given in a recent compilation[29]. A recently published paper[25] gives details of the crystal structure and compositions. The ternary compounds in which a thoride element is included are listed in Table V. The only binary compound which is comparable to ThNiAl is Pu_2Co_5, in which Co probably occupies the 1(b) and 2(c) sites and Pu probably occupies both the 3(f) and 3(g) sites.

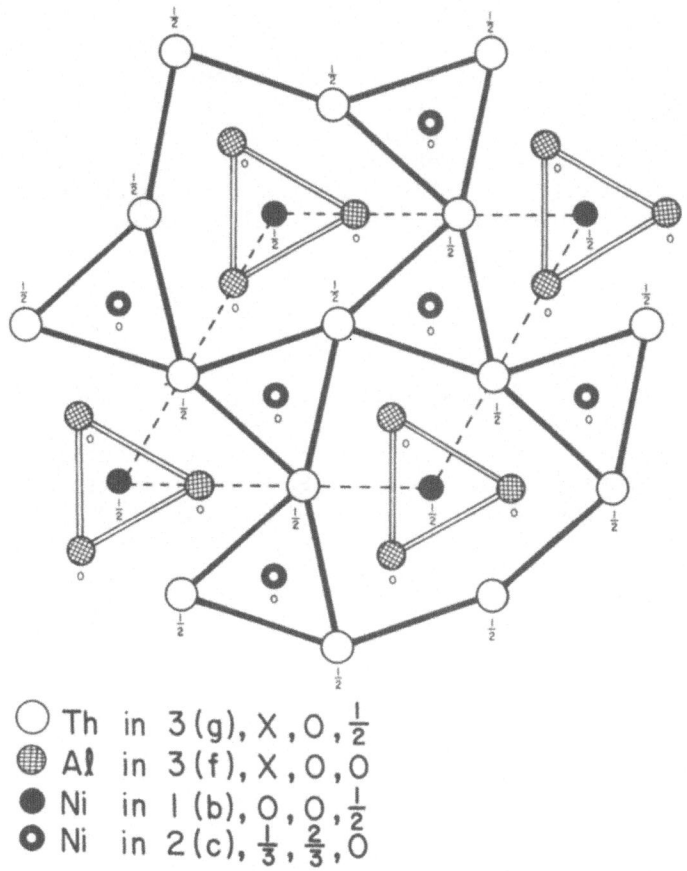

$\bigcirc$ Th in 3(g), $X, O, \frac{1}{2}$
$\oplus$ Al in 3(f), X, O, O
$\bullet$ Ni in 1(b), $O, O, \frac{1}{2}$
$\odot$ Ni in 2(c), $\frac{1}{3}, \frac{2}{3}, O$

Figure 31. Projection of ThNiAl crystal structure onto base plane. The structure is hexagonal D_{3h}^3 - $P\bar{6}2m$, No. 189.

The unit cell constants and a.a.v. for ThXAl and UXAl compounds (Fe_2P-type) plotted against atomic volume of the X element are shown in Figures 32 and 33. It is noted that the a.a.v. of the Ni compound is larger than that of the Co compound and the a.a.v. points of ThPdAl and ThRhAl lie above the curve. This reflects the reduced compressibility and correspondingly greater atomic volume of second long period elements.

The trend for thoride compounds to have uncommon crystal structures has nearly ended before the AX_2 composition was reached, and compounds richer in X than AX_2 show few uncommon crystal structures.

AX_3 Thoride Compounds

More than half of the thoride AX_3 compounds have the fcc Cu_3Au-type structure (Table VI). The X component is from one of two distinct areas of the periodic table, as shown for UX_3 compounds in Figure 34. One area-of-stability is centered around URh_3; the other includes most of the Al and Si groups. UPd_3 and UPt_3 have the hexagonal $TiNi_3$ and $MgCd_3$-type structures, respectively. These two structures are closely related to the Cu_3Au-type in that the same basic layer is stacked in an abc sequence to form Cu_3Au, in an ab sequence to form $MgCd_3$, and in an abac sequence to form $TiNi_3$. The stacking sequences in AX_3 compounds were discussed in an earlier paper[32].

The $PuNi_3$-type family, with five representatives in Table VI, has a structure that is unrelated to the Cu_3Au-, $TiNi_3$-, or $MgCd_3$-type structures. The crystal structure is rhombohedral but is ordinarily indexed as hexagonal. (Schubert[29] lists $NbBe_3$ as the prototype, although $PuNi_3$ has been widely accepted.) The known representatives are characterized by a large size ratio, i.e., a large A element and a small X element. An example of the $PuNi_3$-type structures, shown in Figure 35, is UCo_3. Our unpublished work has shown that almost no substitution of Ni for Co in UCo_3 can be tolerated.

It is noted from Figure 34 that URu_3 is isostructural with UIr_3, but not with $[UOs_3]$, which does not exist. The displacement of the second l.p. elements to the right on area-of-stability plots is supported by experimental data from U_2X through UX_3 compositions.

From the structures selected by binary ThX_3 and PuX_3 compounds (Table VI), it appears that ThX_3 and PuX_3 area-of-stability plots would show the same general features as Figure 34, with exceptions which could be explained by the difference in atomic volume among

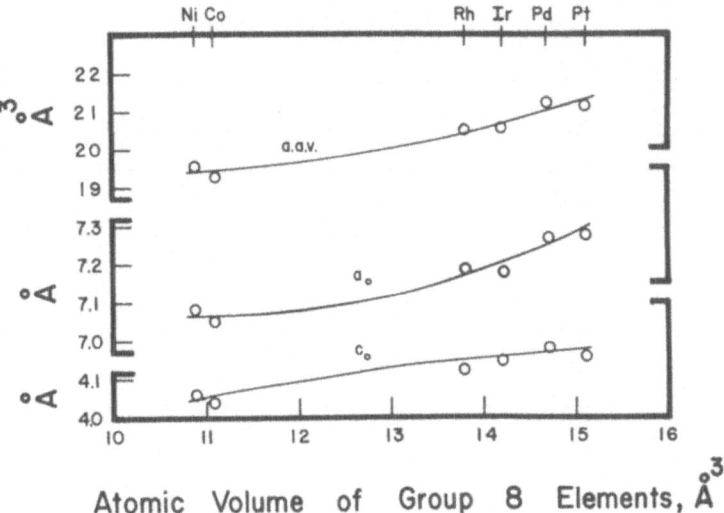

Figure 32.   Unit cell constants in ThXAl compounds, Fe_2P-type,
plotted against atomic volume of X element.

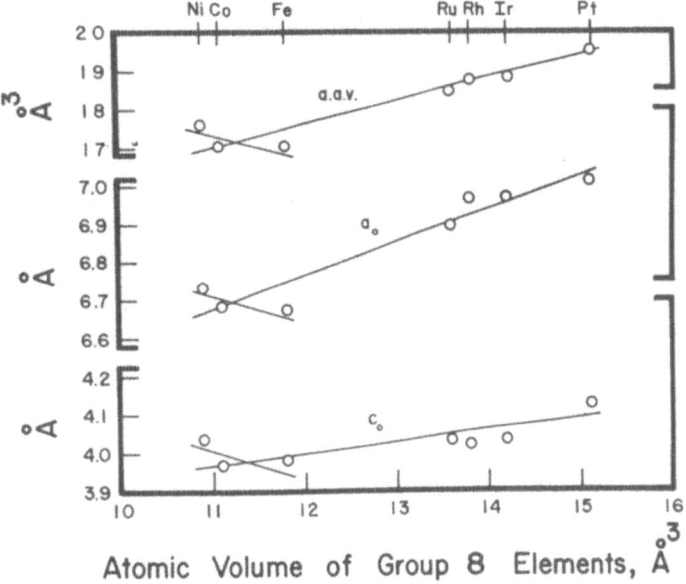

Figure 33.   Unit cell constants in UXAl compounds, Fe_2P-type,
plotted against atomic volume of X element.

Table VI. AX_3 thoride compounds

AX_3	a(Å)	c(Å)	c/a	a.a.v.(Å^3)	Structure Type
ThRh$_3$*	4.139			17.73	Cu$_3$Au
ThIn$_3$	4.695			25.87	Cu$_3$Au
ThSn$_3$	4.718			26.25	Cu$_3$Au
ThTl$_3$	4.748			26.76	Cu$_3$Au
ThPb$_3$	4.855			28.61	Cu$_3$Au
URu$_3$[6]	3.977			15.73	Cu$_3$Au
URh$_3$	3.991			15.89	Cu$_3$Au
UIr$_3$	4.023			16.28	Cu$_3$Au
UAl$_3$	4.287			19.70	Cu$_3$Au
USi$_3$	4.035			16.43	Cu$_3$Au
UGa$_3$	4.248			19.16	Cu$_3$Au
UGe$_3$	4.206			18.60	Cu$_3$Au
UIn$_3$	4.601			24.35	Cu$_3$Au
USn$_3$	4.626			24.75	Cu$_3$Au
UTl$_3$	4.675			25.54	Cu$_3$Au
UPb$_3$	4.792			27.50	Cu$_3$Au
PuRh$_3$[17]	4.009			16.11	Cu$_3$Au
PuPd$_3$[17]	4.011			16.94	Cu$_3$Au
PuPt$_3$[17]	4.107			17.32	Cu$_3$Au
PuGe$_3$	4.223			18.83	Cu$_3$Au
PuIn$_3$	4.610			24.49	Cu$_3$Au
PuSn$_3$	4.630			24.81	Cu$_3$Au
PuPb$_3$	4.808			27.79	Cu$_3$Au
ThAl$_3$	6.500	6.500	4.626	21.16	MgCd$_3$
UPt$_3$	5.764	4.898	0.85	17.62	MgCd$_3$
PuGa$_3$ l.t.					MgCd$_3$
ThPd$_3$	5.856	9.826	1.68	18.24	TiNi$_3$
UPd$_3$	5.757	9.621	1.67	17.26	TiNi$_3$
ThFe$_3$	5.22	24.96	4.78	16.36	PuNi$_3$

Table VI. AX_3 thoride compounds (contd)

AX_3	a(Å)	c(Å)	c/a	a.a.v.(Å³)	Structure Type
ThCo$_3$[31]	5.038	24.64	4.89	15.04	PuNi$_3$
UCo$_3$[35]	4.849	24.317	5.01	13.754	PuNi$_3$
PuCo$_3$	5.003	24.42	4.88	14.704	PuNi$_3$
PuNi$_3$	5.00	24.35	4.87	14.64	PuNi$_3$
ThHg$_3$	3.364	4.907	1.46	24.04	hex.
UHg$_3$	3.32	4.878	1.47	23.28	hex.
PuGa$_3$ h.t.	6.178	28.031	4.54	19.30	hex.(rhomb.)
PuAg$_3$	12.73	9.402	0.74	20.62	hex.
PuAu$_3$[17]	12.716	9.210	0.72	20.13	hex.
PuHg$_3$					UHg$_3$
PuAl$_3$	6.17	14.5	2.35	19.92	hex.

*Unit cell constants from Pearson[5] unless otherwise noted.

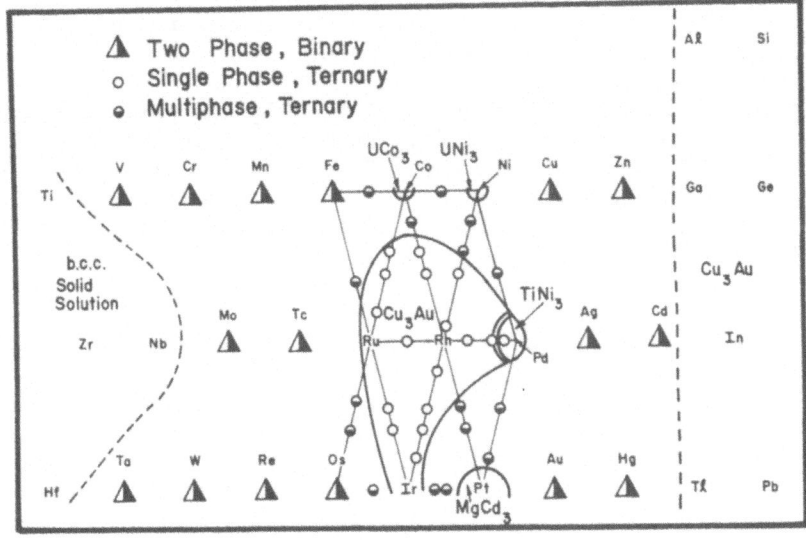

Figure 34.   UX$_3$ area-of-stability plot.

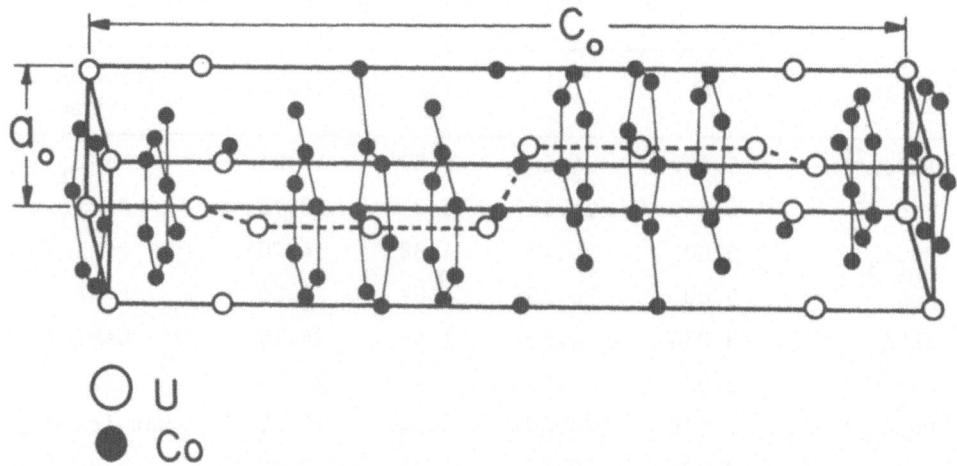

O U
● Co

Figure 35. The crystal structure of UCo_3, $PuNi_3$-type structure.

Th, U, and Pu. There are a few AX_3 compounds at the end of Table VI that are not classified as known structure types.

AX_5 Thoride Compounds

Two structure types are found among thoride AX_5 compounds, the $CaZn_5$-type has five representatives and the $AuBe_5$-type has three (Table VII). Both structure types are closely related to the $MgZn_2$- and $MgCu_2$-type Laves phases. The structural similarities have been discussed in an earlier paper[7].

From Table VII, it is noted that $ThNi_5$ has a lower a.a.v. than $ThCo_5$. Our unpublished data show a similar trend for UNi_5 in that an alloy of UNi_5 in which 5 at.% of the Ni was replaced by Co has a lattice constant of 6.786 Å compared with 6.783 for UNi_5. Co apparently exhibits an atomic volume larger than Ni in UX_5 as well as in ThX_5 compounds.

ATOMIC VOLUMES IN INTERMETALLIC COMPOUNDS OF THORIDE ELEMENTS

In the preceding sections, the comparative a.a.v. of Fe, Co, and Ni compounds has been noted. In thoride-rich compounds, e.g., U_6Mn-type, the a.a.v. increases in order Fe to Co to Ni, but elemental Fe to Co to Ni shows a decrease. This tendency persists through the AX_2 compounds, but disappears in the AX_3 and AX_5 compounds, in which the a.a.v. decreases. For convenience in discussion, the decrease in a.a.v. from Fe to Co to Ni will be called

Table VII. AX_5 thoride compounds

AX_5	a(Å)	c(Å)	c/a	a.a.v. (Å^3)	Structure Type
ThFe$_5$*	5.13	4.02	0.78	15.27	CaZn$_5$
ThCo$_5$	5.005	3.987	0.80	14.42	CaZn$_5$
ThNi$_5$	4.921	3.990	0.81	13.94	CaZn$_5$
ThIr$_5$	5.315	4.288	0.81	17.48	CaZn$_5$
PuNi$_5$	4.872	3.980	0.82	13.64	CaZn$_5$
UNi$_5$	6.783			13.00	AuBe$_5$
UCu$_5$	7.033			14.49	AuBe$_5$
UPt$_5$[17]	7.421			17.03	AuBe$_5$

*Unit cell constants from Pearson[5] unless otherwise noted.

normal behavior, and the increase from Fe to Co to Ni will be called abnormal behavior. The abnormal behavior was first noted by Philip and Beck[33], in the compounds TiFe, TiCo, and TiNi. It was noted that the decreased strength of A-B bonding from TiFe to TiNi is opposite to what would be expected from an electronegative valency effect. The abnormal behavior was also discussed by Nevitt[34] who suggested that the quasi-ionic contribution to the bonding results from an electron transfer associated with the redistribution of electrons in d states. The changeover from abnormal to normal behavior comes between AX_2 and AX_3 compositions in the thoride compounds. However, in Sc and Zr compounds which are typical of groups 3 and 4, the changeover comes between AX and AX_2 compositions. It may be that the compressibility of the Sc and Zr atoms permit the Ni atom to manifest its normally small volume at a lower Ni concentration than in the thoride-Ni compounds.

Another observation on relative atomic volume concerns compounds of Al and Ga. When the a.a.v. of a pair of isostructural Al and Ga compounds are compared, the Al compound has the larger volume. The atomic volumes[8] of the elements are 16.6 Å^3 for Al and 19.6 Å^3 for Ga, and on this basis one would expect a larger a.a.v. for Ga compounds. However, the reverse is found. Data on UFeAl and UFeGa (Table V) and on UAl$_3$ and UGa$_3$ (Table VI) are typical examples.

ACKNOWLEDGMENTS

The author wishes to acknowledge advice and assistance from many members of the Alloy Properties Group, Metallurgy Division, Argonne National Laboratory. Thanks are due to Dr. J. B. Darby, Jr. for a review of the manuscript. This work was performed under the auspices of the U. S. Atomic Energy Commission.

REFERENCES

1. E. S. Makarov, Crystal Chemistry of Simple Compounds Consultants Bureau, New York (1959).

2. W. D. Wilkinson, Uranium Metallurgy (Interscience Publishers, New York, 1962) 2.

3. M. Hansen, Constitution of Binary Alloys (McGraw-Hill Book Co., New York, 1958).

4. R. P. Elliott, Constitution of Binary Alloys, First Supplement (McGraw-Hill Book Co., New York, 1965).

5. W. B. Pearson, Handbook of Lattice Spacings and Structures of Metals (Pergamon Press, New York, 1958, 1967) 1 and 2.

6. A. E. Dwight, to be published.

7. A. E. Dwight, Trans. A.S.M., 53, 477 (1961).

8. P. S. Rudman, Trans. Met. Soc. AIME, 233, 864 (1965).

9. A. E. Dwight, J. Nucl. Mat., 2, 81 (1960).

10. W. H. Zachariasen, modified by E. A. Kmetko and J. T. Waber, "Electronic Band Structures of Plutonium and Its Monocarbide," in Plutonium 1965, A. E. Kay and M. B. Waldron, Eds. (Chapman and Hall, 1965).

11. A. Raman and K. Schubert, Z. Metallk. 55, 704 (1964).

12. T. Yoshioka and P. A. Beck, Trans. Met. Soc. AIME, 233, 1788 (1965).

13. A. E. Dwight, Nature, 187, 505 (1960).

14. F. H. Ellinger, C. C. Land, and V. O. Struebing, J. Nucl. Mat., 12, 226 (1964).

15. M. V. Nevitt,"Miscellaneous Structures of Fixed Stoichiometry," in Intermetallic Compounds, J. Westbrook, Ed. (John Wiley and Sons, New York, 1967).

16. A. F. Berndt and A. E. Dwight, Trans. Met Soc. AIME, 233, 2075 (1965).

17. V. I. Kutaitsev, et al., "Phase Diagram of Pu with the Metals of Groups IIA, IVA, VIIIA, and IB," in Plutonium 1965, A. E. Kay and M. B. Waldron, Eds. (Chapman and Hall, 1965).

18. D. Hohnke and E. Parthé, Acta Cryst., 20, 572 (1966).

19. A. E. Dwight, "Body Centered Cubic Derivative Structures," in Intermetallic Compounds, J. Westbrook, Ed. (John Wiley and Sons, New York, 1967).

20. F. Laves, "Crystal Structure and Atomic Size," in Theory of Alloy Phases, A.S.M. (1956).

21. A. E. Dwight, in Annual Progress Report for 1964, Metallurgy Division, ANL-7000.

22. J. B. Darby, Jr., A. F. Berndt, and J. W. Downey, J. Less-Common Metals, 9, 466 (1965).

23. D. J. Lam, J. B. Darby, Jr., J. W. Downey, and L. J. Norton, J. Nucl. Mat., 22, 22 (1967).

24. A. E. Dwight, in Annual Progress Report for 1963, Metallurgy Division, ANL-6868.

25. A. E. Dwight, M. H. Mueller, H. Knott, J. W. Downey, and R. A. Conner, Jr., Trans. Met. Soc. AIME, 242, 2075 (1968).

26. S. Steeb and G. Petzow, Z. Metallk. 55, 453 (1964).

27. S. B. Hendricks and P. R. Kosting, Z.Krist., 74, 511 (1930).

28. S. Rundqvist and F. Jellinek, Acta Chem. Scand., 13, 425 (1959).

29. K. Schubert, Kristallstrukturen zweikomponentiger Phasen, Springer-Verlag, Berlin (1964).

30. J.H.N. Van Vucht, J. Less-Common Metals, 11, 308 (1966).

31. J. R. Thomson, J. Less-Common Metals, 10, 432 (1966).

32. A. E. Dwight and P. A. Beck, Trans. Met. Soc. AIME, 215, 976 (1959).

33. T. V. Philip and P. A. Beck, Trans. AIME, 209, 1269 (1957).

34. M. V. Nevitt, "Alloy Chemistry of Transition Elements," in Electronic Structure and Alloy Chemistry of the Transition Elements, P. A. Beck, Ed. (Interscience Publishers, New York (1963).

35. A. E. Dwight, Acta Cryst., B24, 1395 (1968).

36. J. B. Kusma and H. Nowotny, Mh. Chem., 95, 1219 (1964).

CONSTITUTION OF NON-EQUILIBRIUM ALLOYS

AFTER RAPID QUENCHING FROM THE MELT

B. C. Giessen

Northeastern University

Boston, Massachusetts*

ABSTRACT

Following a brief description of current experimental techniques for the quenching of molten alloys at very high cooling rates, a comprehensive review of the effects of rapid quenching (especially by the splat cooling technique) on the alloy constitution is given. 133 binary and 28 ternary systems are treated; references available until December 1968 have been included.

INTRODUCTION

The development of techniques for the rapid quenching of alloys from the liquid state has resulted in a large number of non-equilibrium alloy systems being investigated in the last decade. Although the use of high cooling rates to retain metastable phases from the melt had been explored before, e.g. in the classic work by Falkenhagen and Hofmann[1], a systematic examination of alloy systems in a nonequilibrium state had to await the development of a very fast droplet quenching technique by Duwez, Willens, and co-workers.[2] This method and some of its variants have since come to be called "splat cooling" (see next chapter).

A number of review papers have been written in which the metastable alloying effects obtained by the splat cooling technique and related methods have been described extensively[3-12]

*And Massachusetts Institute of Technology, Cambridge, Massachusetts

(see especially Ref. 6, 9–12). In these articles, the material
is classified according to the physical phenomenon treated, such
as supersaturation of terminal solid solutions, retention of
amorphous phases, formation of metastable intermediate phases of
certain structure types, etc. Thus, the crystal chemistry of
stable and metastable B metal alloy phases is reviewed in
Reference 11. In the present paper, a survey is given of those
alloy systems in which nonequilibrium effects have been observed;
the systems have been arranged alphabetically. This approach was
considered necessary, as alloys from a large number of binary and
ternary systems have now been investigated, making it difficult to
obtain complete information about a given system from the present
literature. A total of 133 binary and 28 ternary systems are
treated here; references available until December 1968 have been
included.

In order to draw a border line between common nonequilibrium
phases and materials such as martensites, cementite, transitional
precipitates (e.g. G. P. zones), or glasses on the one hand, and
the phases to be discussed below on the other, the critical cooling
rate necessary to retain a given metastable structure by quenching
from the melt has been used as a criterion. Thus, nonequilibrium
alloy phases or solid solution ranges have been included if they
were produced by quenching from the liquid state at high cooling
rates (10^7–10^8°C/sec for splat cooling[13,14]), or intermediate
cooling rates (3.10^4°C/sec [1]; 5.10^4°C/sec;[15] $\geqslant 10^6$°C/sec [16]).
Nonequilibrium alloy phases that can be readily prepared by other
means, e.g. by conventional quenching from the solid state (e.g.
Fe–C martensite) or by slow cooling (e.g. metastable β–Ga [17]) have
not been included.

In addition, a small number of metastable binary phases are
listed that were produced by quenching from the vapor (vapor
quenching). These results were quoted mainly for purposes of
comparison, and a complete survey of this work was not attempted.

Finally, in some cases equilibrium phases prepared at high or
intermediate cooling rates from the melt were included, if they
could not be retained easily by quenching from the solid state, as
in the case of some high temperature phases, or if a particular
microstructure was desired.

Where illustrations are included, the equilibrium diagrams
are compared with a "quenched-phase plot" which shows the homogeneity
ranges of stable and metastable alloy phases retained after
quenching to the temperature given for the plot. The cooling rate
in each case is that associated with the quenching technique used
(for a discussion of cooling rates, see the next chapter).
Formation and extent of metastable phases depend on the

supercooling temperature attained during quenching[12] as first demonstrated for Au-Sb.[18] The supercooling temperature is closely related to the cooling rate; the effect of the latter on the observed structure has been described schematically for Au-Ge.[19] Existence and width of the metastable phase fields quoted in the text are therefore a function of the quenching technique used and may change considerably as the state of the art progresses. Experimentally, the composition ranges of metastable phase fields have mostly been derived from the appearance of a single phase upon rapid quenching; however, in some cases, they have been deduced from metastable multiphase alloys (see Au-Sn).

It has been shown that metastable crystalline phases, such as those produced by quenching and supercooling, can be considered as parts of the complete T-x phase diagram which includes equilibrium and nonequilibrium phases; it should ultimately be possible to incorporate all metastable crystalline phases discussed in the following in complete phase diagrams.[12] Unfortunately, at present there is only very limited information on the temperatures of metastable phase formation.

TECHNIQUES OF RAPID QUENCHING FROM THE MELT

The principal experimental methods used and referred to here have been extensively described; in the following, they will be listed only to permit easy identification.

1. Wedge solidification technique: molten metal is forced into a wedge-shaped metal mold by a vacuum; cooling rates of $5 \cdot 10^4$ °C/sec have been reported.[1]

2. Splat cooling technique: molten material is atomized by a shock wave; the resulting droplets impinge on a heat conducting metal substrate, where they are cooled. The shock wave is either produced with a shock tube using a driver gas and diaphragm,[3] (originally designated "gun" technique, see Ref. 6) or by an explosive discharge.[13] Induction heating[3] and resistance heating[13] of the melt has been used; in a recent modification, the crucible and substrate are kept in an argon atmosphere.[20] Physical measurements of the splat cooling process have been reported;[6,13,21] cooling rates were measured[13] and calculated[14] to be up to $10^8 - 10^9$°C/sec (for discussions, see Ref. 9 and 22). A modified splat cooling technique in which a molten metal stream strikes a rapidly rotating hearth has been employed to prepare homogeneous standards for electron microprobe work.[23] For an etymological discussion of the term "splat cooling" see Ref. 24.

3. Piston-and-anvil technique: a molten alloy droplet is flattened into a thin sheet between a rapidly moving metal piston and anvil,[25-27] with an initial cooling rate of 10^5-10^6°C/sec.[27]

Modifications of this method have been used[16,28] which are probably less effective than the design by Pietrokowski[25] and were estimated to yield cooling rates of $> 10^5$°C/sec.

4. High temperature splat cooling technique: this recently described method uses induction heating by an rf field concentrator combined with a shock wave to achieve rapid melting and splat cooling of refractory materials.[29] In another device, an alloy is arc-melted before being broken up and propelled against a cooling substrate by an inert gas stream.[30,31]

5. Catapult quenching technique: molten metal is catapulted against a stationary substrate; cooling rates of 5.10^4°C/sec [15,28] and 10^5-10^6°C/sec [32] were reported.

6. Plasma-jet spraying: alloy powders are molten in a plasma-jet and are spray-quenched against a heat conducting substrate. The quenching rate has been shown in several cases to equal that of the splat cooling technique.[33]

7. Vapor-quenching: the method used in the few experiments quoted here was that of co-depositing the components in vacuum on a copper substrate.[34,35]

ELEMENTAL NONEQUILIBRIUM ALLOYS

With the exception of Co, no pure element has yet been reported to have been retained in a nonequilibrium structure by splat cooling. Splat cooled Co consisted of the fcc high temperature modification α-Co; only traces of hcp ε-Co were found.[36] However, even in this case the transformation was not completely suppressed, and it can be stated that generally phase transformations in elements do not seem to be inhibited by splat cooling, at least with the presently available cooling rates. By contrast, a number of pure metals have been obtained in the amorphous state by vapor quenching.[35,37] The minimum cooling rate necessary to retain lead as a non-crystalline phase has been calculated.[22] An electron microscopic study of the vacancy concentration in splat cooled Al has been made.[38]

BINARY NONEQUILIBRIUM ALLOY SYSTEMS

The systems have been arranged in an alphabetic sequence of the chemical symbols of the components. All percentages are in atomic percent, unless noted otherwise. Metastable alloying effects are reported as well as the definitive absence of such effects under given experimental conditions; however, information on weak metastable effects in about 10 additional systems has been regarded as insignificant or at present unreliable, and has been omitted from discussion. To reduce the number of necessary references, data on the equilibrium diagrams are taken from Hansen[39] and Elliott,[40] except where more recent information was available.

Ag-Al Silver-Aluminum

In equilibrium, Ag dissolves max. 20.3% in solid solution, and Al dissolves max. 23.8% Ag; both phases coexist with hcp ζ phase.[39]

By use of a hammer-and-anvil technique, single phase alloys with 19-25% Al have been prepared. These alloys show strong faulting which results in the absence of coherent fcc or hcp regions; however, there is evidence for partial order in the layer sequence. On the Al-rich side, the solid solubility of Ag in Al has been increased in nonequilibrium to 40% Ag.[26]

Ag-Bi   Silver-Bismuth

An alloy with 90% Bi whose equilibrium constituents were Ag and Bi[39] was unchanged after splat cooling to -190°C.[40]

Ag-Cu   Silver-Copper

The eutectic equilibrium diagram Ag-Cu[39] shows limited solid solubility of max. 14.1% Cu in Ag and max. 4.9% in Cu.

It was shown in one of the first papers on metastable alloys obtained by splat cooling[2] that complete metastable solid solubility could be obtained by quenching to room temperature. Additional lattice parameter measurements on seven alloys[42] are shown plotted in Ref. 5; precision determinations gave mean atomic volumes with a positive deviation from a Vegard's law straight line $\Delta V/V \leq 1.0\%$.

An independent determination[43] of splat cooled Ag-Cu alloys using X-ray methods and electron microscopy showed a similar lattice parameter curve; however, the maximum deviation from a Vegard's law straight line was found to be + 0.8% for the lattice parameters, corresponding to + 2.4% for the atomic volumes. In this investigation, single phase alloys were not obtained; terminal solid solutions were also present, and were found to be supersaturated. In a recent study,[44] two discrete, fcc Ag-rich phases and a fcc Cu-rich phase were found in addition to the random Ag-Cu solid solution; the decomposition sequence of these phases was also treated.

This alloy system is frequently used to test or optimize newly designed rapid quenching equipment; thus, plasma-jet spraying permitted an alloy with 40% Cu to be retained as a single phase alloy.[33]

Amorphous and crystalline single phase Ag-Cu alloys have also been obtained by vapor-quenching to -190°C and subsequent annealing at 80 to 130°C, respectively.[34]

Ag-Ge   Silver-Germanium

The Ag-Ge phase diagram is simple eutectic, with a maximum solid solubility of Ge in Ag of 9.6% Ge. The Ag-Ge eutectic lies at 25.9% Ge.[39]

The production of a metastable intermediate hcp phase by splat cooling of Ag-Ge alloys was initially reported in Ref. 45; this was the first published finding of an intermediate crystalline phase prepared by this technique. This phase is designated here as ζ, in analogy to the isotypical Hume-Rothery phases of this type.[46] Subsequently, the homogeneity range and lattice parameters of ζ were established accurately;[47] ζ (hcp, Mg-A3) occurs as a single phase between 20-22% Ge, with an axial ratio c/a that decreases with increasing Ge content. On the basis of some changes in the relative intensity of diffraction lines, it has been assumed in Ref. 47 that splat cooled Ag-Ge alloys with >22% Ge contain a related phase designated as hcp'; however, this phase may be identical with ζ if an unidentified texture is present. The ζ phase is also quoted in Ref. 48. In addition, ζ has been found to be superconducting, with a transition temperature $T_c = 0.85°K$.[49] The formation of ζ by a plasma-jet spraying process has been demonstrated;[33] generally, its occurrence is a good criterion to establish the performance of a new quenching device.

The terminal solid solubility of Ge in Ag is increased from 9.6% to 13.0 ± 1% Ge by splat cooling.[47]

Ag-In   Silver-Indium

At the Ag-rich end, In and $AgIn_2$ coexist; there is a eutectic between In and $AgIn_2$ at 96.8% In.[39] By splat cooling to -190°C, a metastable phase α(fcc, Cu-Al) has been retained between < 70 and > 80% In.[50] For details on the occurrence of α, see Ga-Zn. α decomposes below 20°C.

Ag-Pb   Silver-Lead

The Ag-Pb diagram is simple eutectic.[39] After splat cooling to -190°C, Pb-rich alloys with 75% Pb were found to have a diffraction pattern with broad peaks, which were ascribed to an amorphous phase;[51] see also Au-Pb.

Ag-Pt   Silver-Platinum

Ag and Pt form a peritectic system with limited terminal solid solubilities; Ag dissolves up to 40.5% Pt and Pt dissolves up to 22.5% Ag.[39] By splat cooling to room temperature, a complete series of solid solutions was obtained;[52] a negative deviation from a Vegard's law straight line $\Delta a/a < 0.6\%$ was observed.

Ag-Sb   Silver-Antimony

Sb has a solid solubility in Ag of max. 7.2% Sb, while at high Sb concentrations ε phase (with ~ 25% Sb) is in equilibrium with Sb.[39] Splat cooling to room temperature raised the observed maximum solid solubility of Sb in Ag slightly to about 8.0% Sb.[53] Splat cooling of an alloy with 75% Sb to -190°C did not result in the formation of metastable phases.[41]

Ag-Si   Silver-Silicon

The simple eutectic system Ag-Si has a eutectic at 15.4% Si;[39] the terminal solid solubility of Ag for Si is practically nil. Splat cooling to room temperature yielded only faint diffraction lines due to a ζ phase (hcp, Mg-A3) in alloys with 10-25% Si[53] (see Ag-Ge for nomenclature); however, in another paper[54] the formation of ζ phase (designated β) + Ag in alloys with 5-25% Si is reported. ζ has an abnormally low axial ratio $c/a = 1.578$. There is strong faulting. No conclusive information concerning the metastable solid solubility of Si in Ag has been obtained.[53]

Ag-Sn   Silver-Tin

The Ag-rich and Sn-rich regions of the phase diagram for metastable phase formation are of interest; in equilibrium, Ag dissolves up to 11.5% Sn in solid solution, and Sn forms a eutectic with ε (Ag_3Sn).[39] Splat cooling to room temperature increased the terminal solid solubility of Sn in Ag to 13.0% Sn, with a linear change of the lattice parameter.[53] On the Sn-rich side, splat cooling to -190°C yielded a metastable γ phase (simple hexagonal, $HgSn_{6-10}$ - A_f) at ~ 5-6% Ag, which decomposed at 20°C.[55] For details, see Cd-Sn.

Ag-Te   Silver-Tellurium

In the Te-rich region, there is a eutectic between an intermediate phase tentatively identified as Ag_3Te_2 [39] and Te. By splat cooling to room temperature, single phase alloys consisting of π phase (simple cubic, α-Po - A_h) were retained between 20.5 and 30.5% Ag.[56] For details, see Au-Te. A discussion of relative atomic sizes in Ag-Te π phases is given in Ref. 57.

The π phase with the composition $AgTe_3$ was found to be superconducting.[49]

Al-Au   Aluminum-Gold

A considerable increase in the terminal solid solubility of Au in Al from 0.04% Au in equilibrium[150] to ~0.35% Au achieved by use of a catapult splat cooling method has been reported.[58] Mechanical properties of quenched and aged foils were also determined; the yield strength shows a four-fold increase over that of equilibrium bulk specimens.

Al-Co   Aluminum-Cobalt

The terminal solid solubility of 15.5% Al in α-Co[39] was raised to ~17.2% Al by splat cooling to room temperature.[36] The solid solubility of < 0.01% Co in Al is raised to > 0.5% Co by splat cooling.[59]

Al-Cr   Aluminum-Chromium

The terminal solid solubility of 0.37% Cr in Al[39] has been raised to 2.85% Cr in nonequilibrium by a wedge solidification technique.[1] Similar solubility increases were obtained by catapult[15] and drop-quenching techniques.[16]

Al-Cu   Aluminum-Copper

The terminal solid solubility of Cu in Al was raised from the equilibrium value of 2.5% Cu[39] to ~18% Cu by splat cooling to 20°C.[59]

Al-Fe   Aluminum-Iron

The maximum solid solubility of Fe in Al was raised from the equilibrium value of 0.026% Fe[39] to 0.082% Fe by a wedge solidification technique (see Al-V).[1] A high solubility increase to ~ 4% Fe has been found in splat cooled alloys.[59]

Al-Ga   Aluminum-Gallium

The simple eutectic phase diagram Al-Ga[40] is given in Figure 1a; the quenched-phase plot, Figure 1b, shows single phase fields covering almost the total concentration range after splat cooling to -190°C.[60] The terminal Ga solid solution range is increased from an undetermined value of < 1% Al to 8% Al. A complex metastable intermediate phase ψ is found between 12 and 14% Al, see also Ga-Zn. A second metastable intermediate phase α' (bct, In-A6 type) between 17.5 and 35% Al is followed immediately by an extension of the terminal solid solution field of Al (α). Increasing additions of Ga to α produce α' with a gradually changing axial ratio of α': c/a = 1.0 at 35% Al and c/a = 1.086 at 17.5% Al (using a fct setting for α', analogous to α). α' is

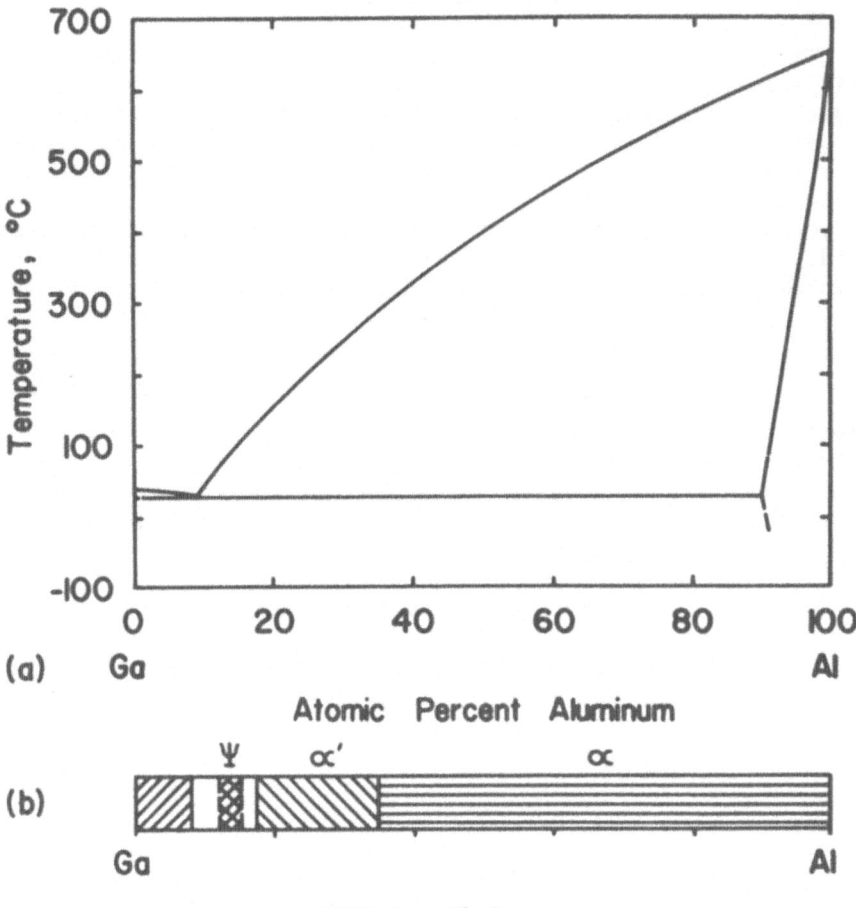

FIGURE 1.   Al–Ga

isostructural with the high pressure modification GaII.[61]   All
metastable phases decompose below 20°C.

Al–Ge   Aluminum–Germanium

Al and Ge form a simple eutectic diagram, with a eutectic at
30.3% Ge and a maximum solid solution of   2.8% Ge in Al.[39]
Alloys with about 30% Ge show complex and line-rich patterns due
to one or more metastable intermediate phases;[51] the solid
solubility of Ge in Al is substantially increased.[62]

Al-Mg   Aluminum-Magnesium

The equilibrium diagram indicates considerable mutual terminal solid solubility of Al and Mg; Al dissolves up to 18.9% Mg, and Mg dissolves up to 11.6% Al.[39] It was shown that by splat cooling to room temperature these solid solubilities could be increased to ⩾ 36.8% Mg in Al and to ⩾ 22.6% Al in Mg.[63] The lattice parameters match very well with those obtained on equilibrium alloys. At intermediate compositions, no changes in crystal structure were found. An almost identical increase of the solid solubility of Mg in Al to 36% Mg upon quenching has been found independently.[64]

Al-Mn   Aluminum-Manganese

The maximum terminal solid solubility in Al is 0.9% Mn.[40] A metastable supersaturation of 4.7% Mn was achieved by a wedge solidification technique (see Al-V).[1] A catapulting technique yielded a maximum solubility of 5.3% Mn;[15,65] a similar value was attained by a hammer-and-anvil technique[16] (see C-Co). By splat cooling, a maximum solubility of > 9% Mn was reached.[59]

Al-Mo   Aluminum-Molybdenum

The maximum solid solubility of Mo in Al has been given as 0.07% Mo.[40] By a catapulting technique, a "maximum supersaturation" of 0.18% Mo and a "homogeneous solid solution" of 0.11% Mo were obtained;[66] see Al-V for details on the measurements.

Al-Nb   Aluminum-Niobium

According to the Al-Nb phase diagram,[67] an alloy with 25% Al consists of Nb_3Al at high temperatures and contains off-stoichiometric Nb_3Al solid solution and σ at lower temperatures. After splat cooling with high temperature equipment (see C-W) a metastable structure (bcc, W-A2) was found,[29] which is probably an extension of the terminal α-Nb solid solution.

Al-Ni   Aluminum-Nickel

The maximum terminal solid solubility of Ni in Al, given as 0.023% Ni,[39] is raised to ~ 1.0% Ni by splat cooling to 20°C.[59]

Al-Pd   Aluminum-Palladium

The solid solubility of Pd in Al which is too small to be determined by X-ray diffraction in equilibrium,[39] could be raised to > 7% Pd by splat cooling to 20°C.[59]

Al-Si   Aluminum-Silicon

Al and Si form a simple eutectic system; the maximum solid
solubility of Si in Al is 1.59% Si.[39] By splat cooling to 20°C,
a metastable solubility increase to 11 ± 0.5% Si was obtained, as
found by X-ray diffraction and electron microscopy.[68] A lattice
parameter curve has been given, and the changes of the lattice
parameters of supersaturated Al-Si solid solutions on aging have
been measured. By a wedge solidification technique,[1] no
solubility increase over the equilibrium value was found, demon-
strating the higher cooling rate of the splat cooling method
(see Al-V).

The precipitation of Si from splat cooled Al-Si alloys has
been the subject of an electron microscopic study;[69] activation
energies for the start of precipitation have been calculated. The
dendrite spacings of splat cooled Al-Si alloys have been measured
and correlated with the cooling rate of the quenching process.[21]

Al-Sn   Aluminum-Tin

The Al-Sn equilibrium diagram is simple eutectic.[39] A
metastable γ phase (simple hexagonal, $HgSn_{6-10}$ - Af) with an
undetermined composition between 15 and 50% Al has been observed
after splat cooling to -190°C.[55] For details, see Cd-Sn.

Al-Ta   Aluminum-Tantalum

An alloy with 25% Al, reported to consist of σ phase in
equilibrium, was found to be bcc (W-A2) after splat cooling;[29] see
Al-Nb.

Al-Ti   Aluminum-Titanium

The maximum terminal solid solubility of Ti in Al is not
known reliably; values ranging from 0.1 to 0.7% Ti have been
quoted.[39] It has been reported that the solubility has been raised
from 0.09 to 0.19% Ti with the aid of a wedge solidification
technique.[1]

Al-V   Aluminum-Vanadium

The maximum terminal solid solubility of V in Al has been
given as 0.2% V.[39] Rapid quenching of thin wedges of molten alloy
in a chill mold produced a maximum solubility of 0.55% V.[1] This
technique is substantially slower than the splat cooling method
(see Al-Si); with the latter technique, a considerably larger
solubility increase could be expected. Catapulting of the melt
onto a copper sheet led to a supersaturation of 0.65% V;[66]

however, the authors differentiate between a "maximum supersaturation", as detected by x-rays and a "homogeneous solid solution" of 0.32% V, as found by paramagnetic susceptibility measurements. By plasma-jet spray-quenching, a considerable solubility increase has been attained.[70]

Al-W   Aluminum-Tungsten

The solid solubility of W in Al, given as 0.024% W,[40] has been raised to a "maximum supersaturation" of 0.15% W and a "homogeneous solid solution" of 0.1% W by a catapulting technique;[66] for details, see Al-V.

Au-Bi   Gold-Bismuth

Figure 2a shows the Au-Bi constitution diagram;[39] there is one compound Au_2Bi. The quenched-phase plot, Figure 2b, indicates 3 new metastable phases, Au_7Bi_8, π', and π.[41,71] Of these, π (simple cubic, α-Po - A_h) with 75-80% Bi occurs upon splat cooling to -190°C. π' (rhombohedral, β-Po - A_i) with about 60% Bi forms if splat foils prepared at -190°C are annealed briefly at -30°C; Au_7Bi_8 (rhombohedral, related to NiAs-B8)[71] with an estimated homogeneity range of < 0.3% around 53.5% Bi, forms upon annealing at -30 to 0°C. After splat cooling to -190°C, alloys in the composition ranges of π' and Au_7Bi_8 are microcrystalline. For further details, see Reference 41.

Au-Co   Gold-Cobalt

The Au-Co phase diagram is simple eutectic, with terminal solubilities of 23.5% Co in Au and 2.5% Au in α-Co.[29] By splat cooling to room temperature, up to 42% Co was dissolved in Au in single phase alloys, and up to 49% in multiphase alloys.[72] On the other side, up to 31% Au was dissolved in Co in metastable multiphase alloys that also contained Co-rich and Au-rich solid solutions. The lattice parameters of the metastable Au-Co solid solutions could be joined by a smooth curve. Solid state precipitation has been considered in order to account for the multiphase Co-rich alloys.[72]

By vapor quenching, amorphous Au-Co alloys were obtained which could be annealed to form metastable fcc single phase alloys before decomposing into Au and Co[34] (see Ag-Cu).

Au-Ge   Gold-Germanium

The simple eutectic phase diagram Au-Ge is given in Figure 3a. The quenched-phase plot, Figure 3b, shows that two metastable phases could be retained by splat cooling to room temperature:

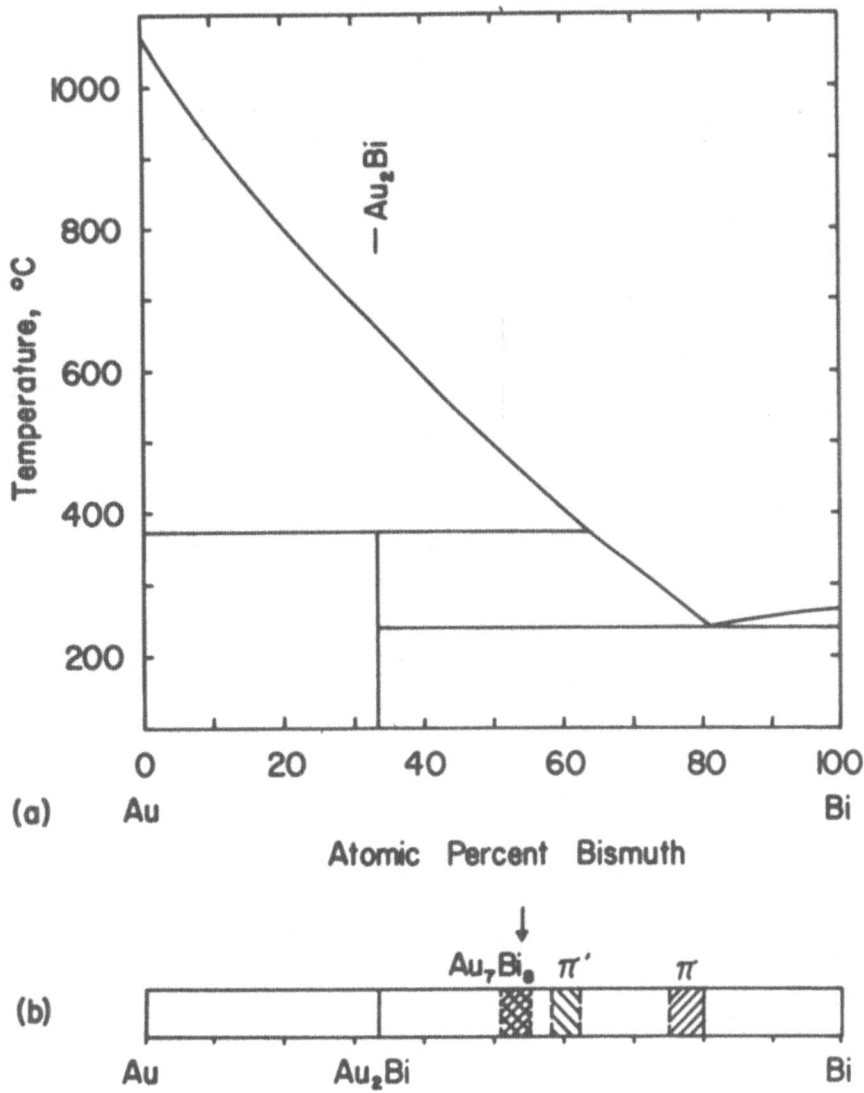

FIGURE 2. Au-Bi

the more Au-rich of these is ζ, (hcp, Mg-A3) from 20-22%
Ge.[19,48,49,51] (For the nomenclature, see Ag-Ge.) Lattice
parameters are given in References 19 and 51; it is surprising
that the axial ratios c/a given in Reference 51 increase with
increasing Ge content, contrary to observations on Ag-Ge[47]
and the general tendency of hcp Hume-Rothery phases in the valence

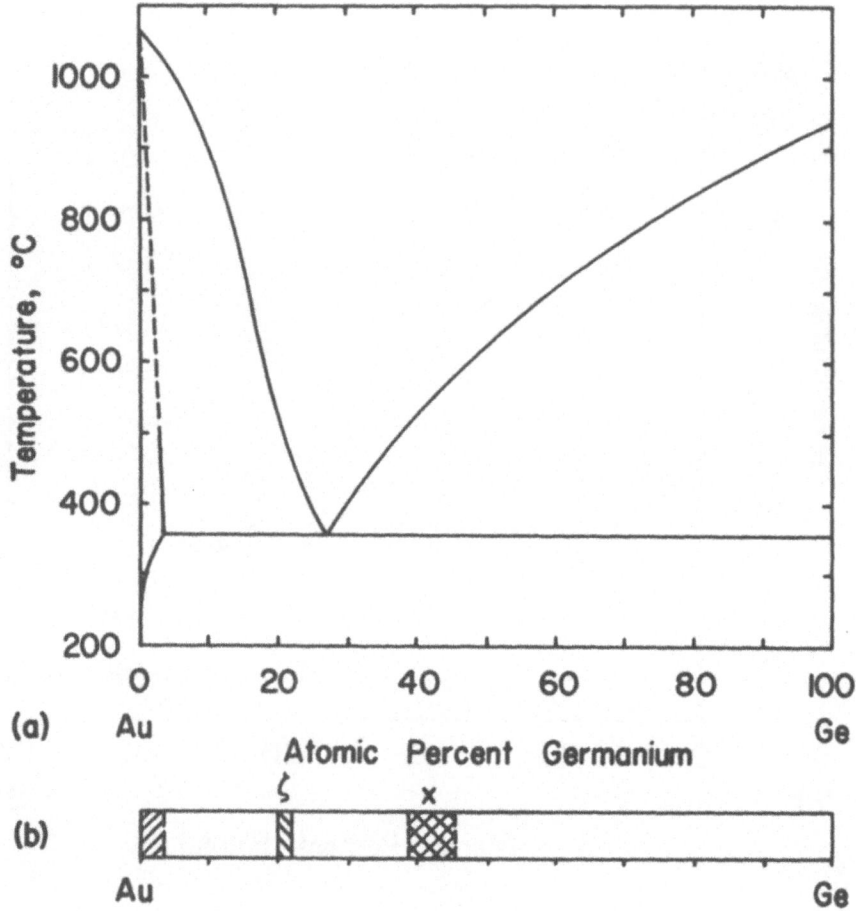

FIGURE 3.   Au–Ge

electron concentration range of ζ.

The second metastable phase X occurs as a single phase between 39 and 46% Ge[19] (see also Reference 48); however, its superconducting transition temperature was found to change linearly with composition between 27 and 50% Ge,[49] indicating a larger homogeneity range. The diffraction pattern of X has been indexed with a large tetragonal cell with 176 atoms/cell, considered as a superstructure of Au;[19] space group and atomic positions are not given.

Au-Pb   Gold-Lead

In the Pb-rich region, there are 2 equilibrium phases, $AuPb_2$[39] and $AuPb_3$,[40] and there is a eutectic at 85% Pb. Splat cooling to -190°C leads to the formation of a phase identified as microcrystalline[50] or amorphous[51] in alloys with 75 to 85% Pb. This phase transforms into the equilibrium phases at temperatures below 20°C. Diffraction peak widths and positions are given.[50,51]

Au-Sb   Gold-Antimony

The constitution diagram, Figure 4a, indicates an equilibrium phase $AuSb_2$.[39] According to the quenched-phase plot, Figure 4b, two metastable phases can be retained by splat cooling to 20°C: ζ phase (hcp, Mg-A3) at 13 to 15% Sb[51] and π phase (simple cubic, α-Po - A_h) from 72 to 84% Sb.[41] Of these, ζ can be regarded as a "3/2" Hume-Rothery phase, analogous to stable ζ (Au-Sn); π is closely related to SnSb.[41] The structure of π can also be derived from that of Sb; π decomposes slowly into the equilibrium phases at 20°C.

The heats of formation of metastable ζ and π have been studied, proving both phases to be metastable at all temperatures.[18] Considerable supercooling of 150-200°C accompanies their formation.

Au-Si   Gold-Silicon

The Au-Si system is simple eutectic, with a eutectic at 18.6% Si[40] and 370°C.[39,40]

The existence of a noncrystalline phase after rapid quenching of an alloy with 25% Si has been reported;[73] this was the first phase with a noncrystalline structure produced by splat cooling. It was assumed to be amorphous[6,73] and is highly unstable at room temperature. A diffraction pattern has been given and it was stated that complex decomposition phases exist.[73] Indirect proof for an amorphous structure was provided by work on splat cooled Au-Ge-Si alloys[74-76] (see Au-Ge-Si). The amorphous phase has also been prepared by a hammer-and-anvil technique;[26] it decomposes into "Au_5Si" and a second metastable phase with a complex diffraction pattern. Its radial distribution function has been compared to that of liquid Au.[26]

In a study of splat cooled Au-Si alloys,[51] the existence of a metastable crystalline phase γ (cubic, related to γ-brass - $D8_{1-3}$) at 20-21% Si was described. This phase exists in splat cooled as well as in as-cast alloys. It can be retained easily in supercooled melts; however, it is not an equilibrium phase at any temperature.[77] γ was also observed in Au-rich Au-Si alloys quenched by a

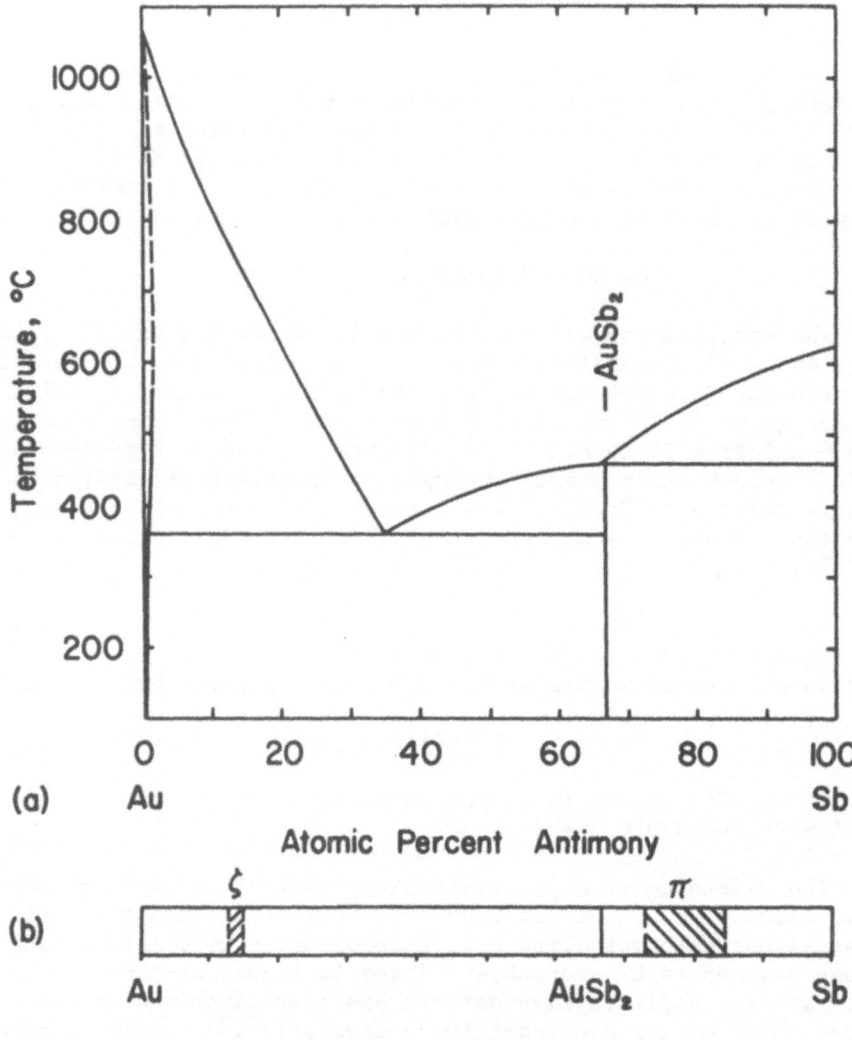

FIGURE 4. Au-Sb

hammer-and-anvil technique; the composition Au₅Si was tentatively assumed.[26] Evidence for a second metastable phase with a higher Si content than γ has been found.[77]

Two complex metastable phases β and "γ", produced by splat cooling and assumed to represent superlattices of Au, have been claimed to exist, based on a tentative x-ray analysis.[48,54] According to its composition, β is probably identical with the γ phase previously mentioned;[51] however, the diffraction patterns given in Reference 54 do not agree in detail with those observed for γ.[51]

Au-Sn   Gold-Tin

Figure 5a shows the Au-Sn equilibrium diagram, characterized by the phases ζ, AuSn, $AuSn_2$, and $AuSn_4$.[39] The quenched-phase plot, Figure 5b,[78] reveals several metastable phases and phase extensions.

The ζ phase (hcp, Mg-A3) is extended from $\sim$16% Sn to $\sim$20% Sn (indicated as ζ_m). A metastable γ phase (bcc, γ-brass – $D8_{1-3}$) occurs at 20.5% Sn. For details on the structure of this Hume-Rothery phase "missing" in equilibrium and the off-stoichiometric preparation of γ, see Reference 78. A metastable phase X with unidentified structure exists between 25 and 31% Sn. These three phases and phase extensions can be obtained by splat cooling to 20°C.

After splat cooling of alloys near 30% Sn to -190°C, a microcrystalline phase is retained[78] (see Au-Bi).

At the Sn-rich end, splat cooling to -190°C leads to the formation of γ_{hex} (simple hexagonal, $HgSn_{6-10}$ – A_f) at $\sim$ 8% Au.[55] For details on γ_{hex} see Cd-Sn.

Au-Ta   Gold-Tantalum

An alloy with 25% Au of an undetermined two phase equilibrium structure consisted of a metastable phase (bcc, W-A2) after splat cooling.[29]

Au-Te   Gold-Tellurium

The constitution diagram[39] shows a compound $AuTe_2$ melting congruently at 464°C and two eutectics of Au + $AuTe_2$ and $AuTe_2$ + Te, respectively. Splat cooling of Au-Te alloys to room temperature was very effective: a metastable π phase (simple cubic, α-Po – A_h) was found in alloys with 15.3 to 40.4% Au.[56] (The designation π was chosen in this review to indicate the relationship with isostructural simple cubic phases based on Sb or Bi, see Au-Sb). The structure of π was compared to that of $AuTe_2$.[56] At most compositions, π phase decomposes after 15 minutes at 110°C; only the alloy with 37.5% Au was less unstable. Electrical measurements were made.

Comments on the relative sizes of the constituent atoms in (Au-Te) π phases can be found in Reference 57. Mössbauer spectrum measurements[79] on the π phase confirm its cubic symmetry by the

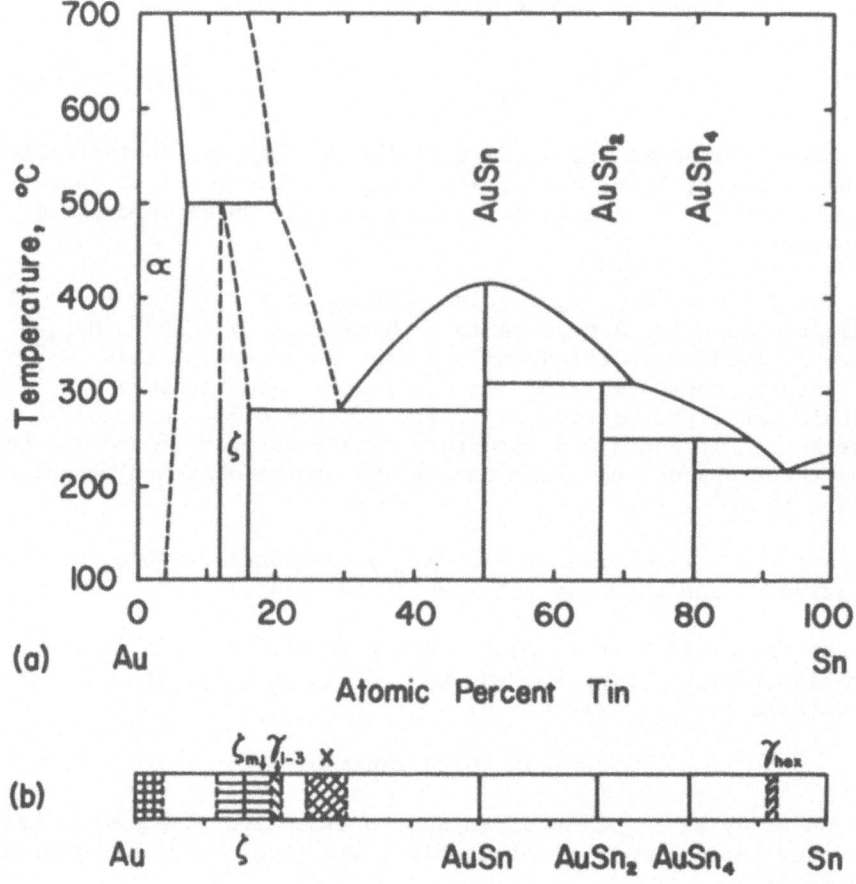

FIGURE 5. Au-Sn

absence of quadruple splitting. A sample of π phase with the composition Au_3Te_5 was found to be superconducting.[49] In a recent investigation, the variations of the lattice parameter, superconducting transition temperature, and thermoelectric power of quenched metastable simple cubic Au-Te alloys (55 to 85% Te) with composition have been measured and interpreted in terms of Fermi surface-Brillouin zone interactions.[80]

Au-Tl   Gold-Thallium

The Au-Tl system is simple eutectic.[39] By splat cooling, two metastable phases were obtained, and their superconducting transition temperatures were determined as 3.8°K – 2.5°K between 25 to 60% Au,

and 4.2°K between 28 to 33.3% Au, respectively. Au additions
stabilized β-Tl.[81] The phase with the higher transition
temperature has been identified as metastable $AuTl_2$ (bct, $CuAl_2$-
C16) with a = 7.07 Å and c = 5.55 Å.

Be-Cu   Beryllium-Copper

The Be-Cu phase diagram[39] indicates a maximum solid solubility
of 16.4% Be in Cu; there is an intermediate phase β (bcc, W-A2)
at Be contents > 23.6% Be. At higher temperatures, there is a
continuous transition β → β' (cubic, CsCl-B2).

Splat cooling to 20°C caused departure from equilibrium in
two points:[82] the Cu-rich boundary of β was moved to ~19% Be,
increasing the β range; and long range ordering corresponding to
the CsCl-B2 type of β' was detected in quenched off-stoichiometric
alloys with ⩾ 32% Be. This effect is either the result of nonequilibrium solidification or it indicates an extension of the
β' field into the region considered to be disordered β.

Bi-Cd   Bismuth-Cadmium

The constitution diagram Bi-Cd is simple eutectic.[39] By splat
cooling to -190°C, a metastable γ phase (simple hexagonal, $HgSn_{6-10}$-
A_f) could be retained between ~ 40 and ~75% Bi.[50] For details
on γ, see Cd-Sn. The γ phase decomposes below 20°C.

Bi-Cu   Bismuth-Copper

An alloy with 90% Bi which consisted of Bi and Cu in
equilibrium[39] did not change upon splat cooling to -190°C.[41]

Bi-Ga   Bismuth-Gallium

Bi and Ga form a monotectic system.[40] In two alloys with 40
and 80% Bi which were splat cooled to -190°C, traces of a new
metastable phase with an unidentified structure were found.[50]

Bi-In   Bismuth-Indium

The equilibrium diagram shown in Figure 6a follows Hansen[39],
and has been revised by the insertion of the new equilibrium phase
In_5Bi_3.[83,84] The stable and metastable phases present after
splat cooling to -190°C are shown in the quenched-phase plot in
Figure 6b.[83] Six metastable phases were found which cover a large
part of the concentration range. They are, with rising Bi contents:
$α_1$ (bct, In-A6); $α_2$ (bcc, W-A2); γ (simple hexagonal, $HgSn_{6-10}$ - A_f);
$γ_1$ (orthorhombic); β (bct, Sn-A5); "α-Bi" (not determined).
$α_2$ and γ each occur in 2 separated concentration ranges. All

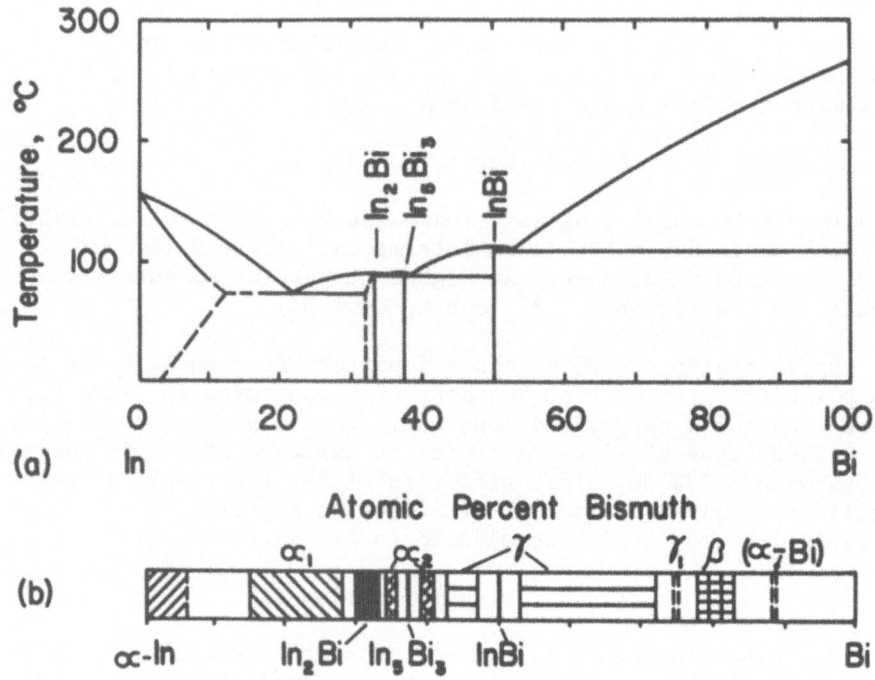

FIGURE 6.　Bi-In

metastable structures are element-like and disordered; α_2 is closely related to In_2Bi. The occurrence of β with the structure of white tin is of interest. For a discussion of the crystal chemistry of the metastable phases, see Reference 83. All metastable phases decompose below 20°C.

Bi-Ni　　Bismuth-Nickel

In the Bi-rich section, $NiBi_3$ is in equilibrium with Bi.[39] No metastable phase was found after splat cooling an alloy with 90% Bi to -190°C.[41]

Bi-Pb　　Bismuth-Lead

The equilibrium diagram Bi-Pb[39] is shown in Figure 7a. The quenched-phase plot, Figure 7a,[85] is characterized by the occurrence of two metastable phases: X-phase, between 52 and 65% Bi and with an unknown crystal structure, has a complex, line-rich powder pattern. Its decomposition into the equilibrium phases ε and Bi was found to proceed at a rapid rate at a temperature of -35 ± 5°C.

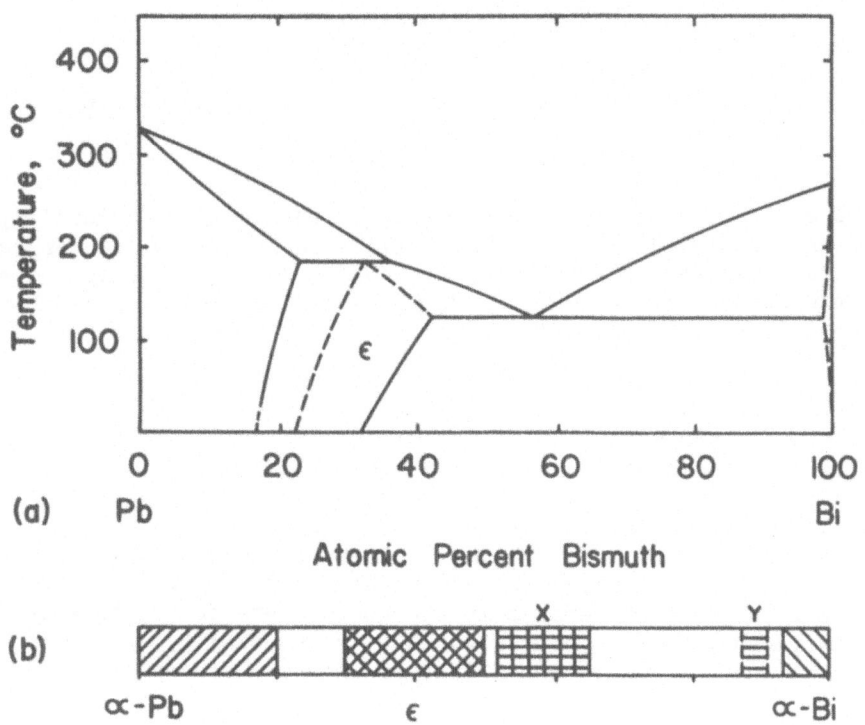

FIGURE 7. Bi-Pb

Y phase appears between 87 and 91% Bi; its structure was not
identified. In addition, there is a considerable metastable
extension of the ε phase field from 42% Bi at equilibrium to 50%
Bi, and of terminal Bi solid solution from < 1% Pb to ~ 6% Pb.
The rhombohedral angle α of α-Bi alloys decreases with increasing
Pb content. All metastable effects disappear upon heating to
20°C. The solubility range of ε had previously been investigated
by splat quenching to room temperature;[86] between 27 and 32% Bi
single phase alloys were retained.

Bi-Pd   Bismuth-Palladium

In the Bi-rich region, where $PdBi_2$ and Bi are in equilibrium,[40]
an alloy with 90% Bi did not show nonequilibrium effects after
splat cooling to -190°C.[41]

Bi–Pt   Bismuth–Platinum

An alloy with 90% Bi, consisting of $PtBi_2$ and Bi in equilibrium,[40] did not change after splat cooling to -190°C.[41]

Bi–Sn   Bismuth–Tin

The Bi-Sn system is simple eutectic; the terminal solubility of Bi in β-Sn is 13.1% Bi.[39] The solid solution range of β-Sn has been increased to > 50% Bi by splat cooling to -190°C, beyond the eutectic composition at 43% Bi.[55] The axial ratio c/a of β-Sn decreases with increasing Bi contents.

C–Co   Carbon–Cobalt

Co dissolves up to 4.5% C in solid solution.[39] By splat cooling to 20°C, a solid solubility increase to 5.3% C was achieved, as estimated from the lattice parameter of a C-rich alloy after quenching.[87] A solubility increase up to 7.8% C, achieved with the aid of a "hammer and anvil" technique (from an equilibrium value given as 3.8% C) has been reported;[16] the numerical values of Reference 16 are quoted here from Reference 88. The discrepancy between the two values is surprising, as the technique used in Reference 87 must be regarded as capable of a somewhat faster quench than that in Reference 16: the hcp ε (Fe-C) phase[89] was prepared with the same equipment used in Reference 87, whereas no metastable Fe-C phase was found according to Reference 16. See C-Fe for details.

C–Fe   Carbon–Iron

The Fe-rich portion of the constitution diagram C-Fe shows two reactions involving the liquid; one of these is peritectic, the other is eutectic, $L \rightarrow \gamma$-Fe + Fe_3C (in the metastable Fe-Fe_3C system) at 1147°C.[39] In a concentration range from 13.7 to 16.9% C (3.3 - 4.2 wt. % C) a metastable phase ε (hcp, Mg-A3) was prepared by splat cooling to 20°C.[89] The new phase is considered to be distinctly different from ε carbide. According to later work, the C atoms probably occupy ordered lattice sites.[90] The ε phase decomposes upon annealing to 120°C. In an investigation using a "hammer-and-anvil" technique, no new phase was reported[16] in alloys with 3.9 to 4.4 wt. % C.

C–Mn   Carbon–Manganese

An alloy with 50% C, reported to consist in equilibrium of Mn_7C_3 and C, was hcp, with c/a = 1.61, after splat cooling to room temperature;[29] no details on the structure were given.

C-Mo   Carbon-Molybdenum

There is an equilibrium high-temperature phase α-MoC_{1-x} with a fcc NaCl-B1 type structure[91] which can be retained by a fast conventional bulk quenching technique (quenching into a liquid tin bath). By splat cooling to room temperature, α-MoC_{1-x} (designated as "cubic MoC") has also been retained;[29,92] a superconducting transition temperature T_c = 14.3°K has been given (see also C-W). The lattice parameter a_o = 4.2777 Å given[92] corresponds to ~42.5% C for the splat cooled alloy, using the values of Reference 91.

C-Ni   Carbon-Nickel

Ni dissolves up to 2.7% C in solid solution.[39] By splat cooling to 20°C, the solid solubility could be raised to 7.4%, as determined from lattice parameter measurements.[87] A metastable solid solubility of 8.2% C in Ni has been reached by use of a "hammer-and-anvil" technique;[16] the numerical value has been quoted here from Reference 88. For comments, see C-Co.

C-W   Carbon-Tungsten

There are three equilibrium phases: W_2C, WC (hexagonal, WC-B_h), and β-WC_{1-x} (fcc, NaCl-B1); the latter is stable only at temperatures above 2525°C [93] and could be only partially retained by quenching with conventional techniques. Successful attempts to retain the cubic carbide by plasma-spraying and spark-machining have been reviewed in Reference 94.

Splat cooling with high temperature equipment[29] yielded complete retention of β-WC_{1-x} between 46 and 50% C. While hexagonal WC is not superconducting to 0.3°K, the cubic phase has a superconducting transition temperature T_c of 9 to 10°K; the lattice parameters and T_c increase with increasing C-content.[29,92]

Neither the structure nor the superconducting properties of W_2C were changed by splat cooling.[92]

Ca-Sn   Calcium-Tin

At Sn-rich compositions, Sn is in equilibrium with $CaSn_3$.[39] By splat cooling to -190°C, a metastable γ phase (simple hexagonal, $HgSn_{6-10}$ - A_f) with > 15% Ca was retained; γ decomposes below 20°C.[55] For details, see Cd-Sn.

Cd-Ga   Cadmium-Gallium

In four alloys with 15 to 60% Ga, only the equilibrium phases Cd and Ga were found after splat cooling to -190°C.[50]

Cd-In   Cadmium-Indium

The Cd-In constitution diagram is shown in Figure 8a; there are two intermediate phases "Cd_3In" and α, both of the fcc Cu-Al type.[40] (For "Cd_3In", the possibility of ordering exists.) As demonstrated by the quenched-phase plot, Figure 8b,[50] single phase α (fcc, Cu-Al) exists from ~20 to ~95% In, connecting the equilibrium phase fields of "Cd_3In" and α. The atomic volumes follow a Vegard's law straight line.[50] For details on α, see also Ga-Zn.

Cd-Pb   Cadmium-Lead

The Cd-Pb system is simple eutectic.[39] Five alloys with 30 to 75% Pb were reported to consist of Cd and Pb after splat cooling to -190°C;[50] however, there is a considerable metastable solid solubility of Cd in Pb (> 20% Cd).[62]

Cd-Sb   Cadmium-Antimony

In four alloys with 10 to 35% Sb, the phases present in the as-cast state and after splat cooling to -190°C were found to be identical.[50] It is stated that both sets of alloys solidified in the metastable system Cd-Sb,[39] forming Cd_3Sb_2 (see also Reference 12).

Cd-Sn   Cadmium-Tin

The constitution diagram Cd-Sn, Figure 9a,[39] shows a high temperature phase at high Sn contents. This phase and its isotypes in nonequilibrium and equilibrium have been the subject of several studies[55,95] and have been designated γ after their prototype, $HgSn_{6-10}$ - A_f. For Cd-Sn, there is a eutectic between γ and Cd.

The quenched-phase plot, Figure 9b, obtained at -190°C,[50] shows a large metastable extension of γ phase (simple hexagonal, $HgSn_{6-10}$ - A_f) to about 50% Cd. An extension to > 30% Sn had been reported previously;[55] in this reference, a correlation of the axial ratio c/a of γ with the valence electron concentration is given. The γ phase is thought to occur at valence electron concentrations between 3.4 and 3.9 e/a. Further, there is a metastable phase α (fcc, Cu-Al) between ~15 and ~30% Sn;[50] lattice parameters have been determined. For details on α, see also Ga-Zn. All metastable Cd-Sn phases decompose below 20°C into Cd and Sn.

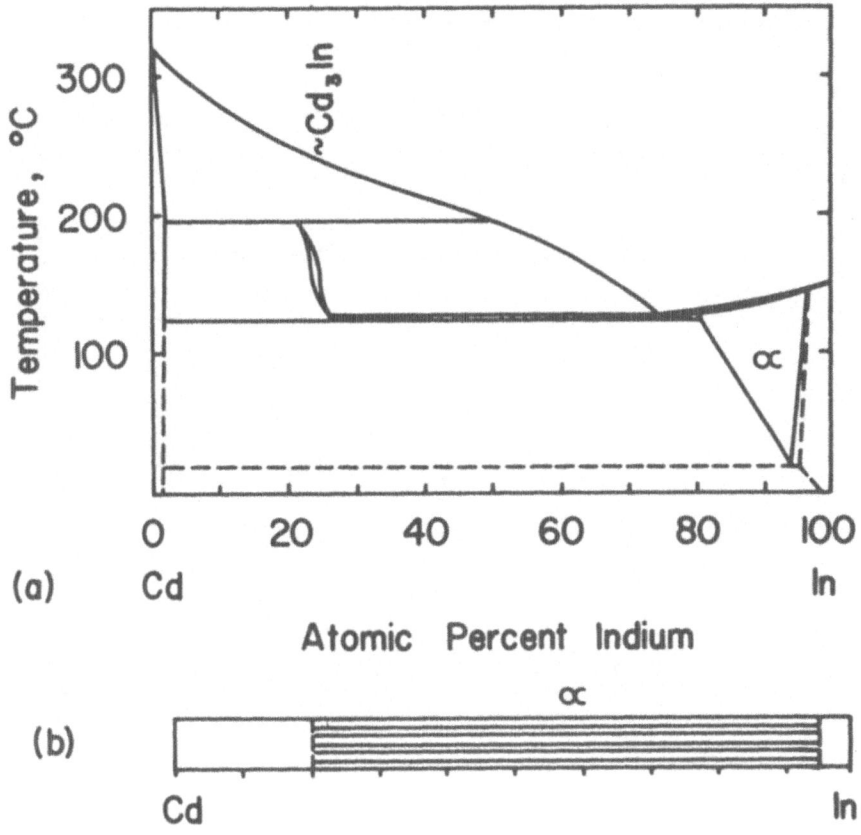

(a)

Atomic Percent Indium

(b)

FIGURE 8. Cd-In

Cd-Zn   Cadmium-Zinc

The terminal solid solubilities in the simple eutectic system
Cd-Zn are 5% Zn in Cd and 1.5% Cd in Zn (the Zn solidus line is
retrograde).[39] It had been stated on the base of lattice parameter
measurements of limited accuracy that there was no change in solid
solubility after splat cooling Cd-Zn alloys to -196°C;[6] however,
recent experiments in which the lattice parameters were evaluated
carefully indicate an increase in the Cd solubility in Zn to
~ 3% Cd.[96] This metastable solubility increase in a system with a
retrograde solidus line is of importance for the theory of
solidification. There is also a considerable solubility increase
on the Cd-rich side.[97]

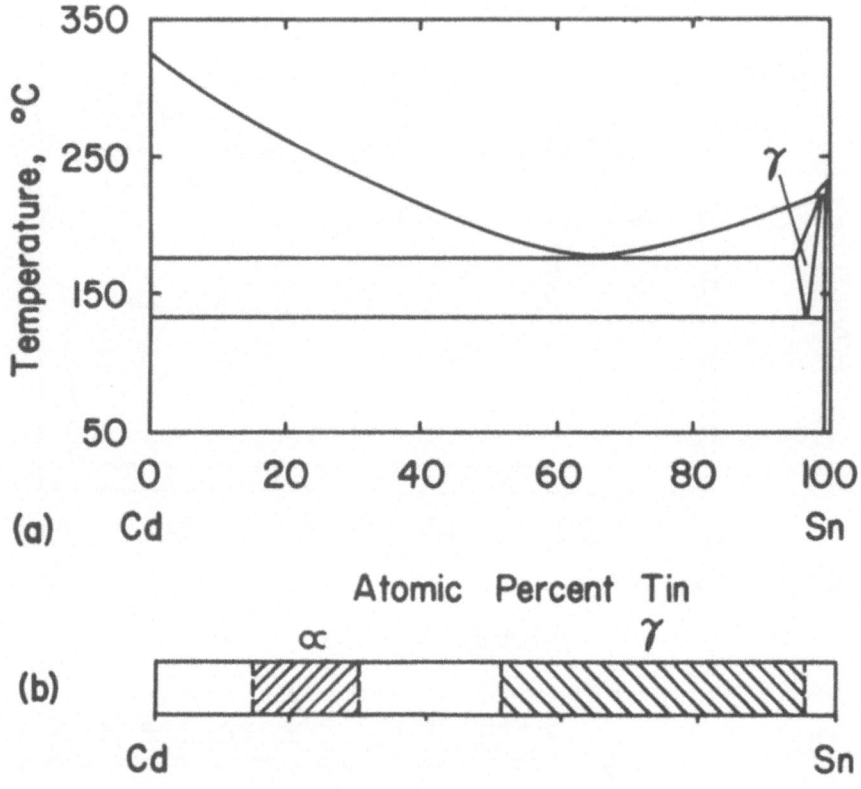

FIGURE 9.   Cd-Sn

Co-Cu   Cobalt-Copper

The peritectic Co-Cu phase diagram indicates a maximum solid solubility of 12% Cu in α-Co and 5.5% Co in Cu.[39] By splat cooling to room temperature, metastable increases in the solubilities of Cu in Co up to 25% Cu (in multiphase alloys up to ~ 35% Cu) and of Co in Cu up to 15% Co were obtained, as found by lattice parameter measurements.[72] A miscibility gap in undercooled liquids has been considered in order to explain the limited mutual solid solubilities of the isostructural elements in nonequilibrium. Metastable Co-Cu solutions have also been obtained by vapor quenching.[98,99]

Co-Ga   Cobalt-Gallium

The terminal solid solubility of Ga in α-Co, tentatively given as 11% Ga,[40] was raised to 16.2% Ga by splat cooling to room

temperature.[36] There is an unexplained change of slope in the
lattice parameter curve at 13% Ga.

Co-Ge   Cobalt-Germanium

The terminal solid solubility of Ge in α-Co, estimated to be
~15% Ge,[39] has been increased to 17.4% Ge by splat cooling to
room temperature,[36] as shown by lattice parameter measurements.

Co-Sb   Cobalt-Antimony

An alloy with 90% Sb consisting of $CoSb_3$ and Sb in
equilibrium,[39] was unchanged after splat cooling to -190°C.[41]

Co-Si   Cobalt-Silicon

The terminal solid solubility of about 14% Si in α-Co [39] was
not changed by splat cooling to room temperature.[36]

Co-Sn   Cobalt-Tin

The solid solubility of Sn in α-Co, estimated to be 2.5% Sn,[39]
has been increased to 5% Sn by splat cooling to room temperature.[36]
Lattice parameters have been given.

Co-Ti   Cobalt-Titanium

A large metastable extension of the intermediate phase TiCo
was observed in alloys which had been splat cooled to room
temperature[100] (see Fe-Ti).

Co-Zr   Cobalt-Zirconium

A splat cooled Co-Zr alloy with 28% Co had diffraction
patterns typical of a non-crystalline phase[101] (see Ni-Zr).

Cr-Cu   Chromium-Copper

The maximum solubility of Cr in Cu has been raised from the
equilibrium value of 0.8% Cr[39] to 1.8% Cr by a wedge solidification
technique.[1]

Cu-Fe   Copper-Iron

The solid solubility of Fe in Cu is 4.5% Fe at the peritectic
temperature of 1094°C.[39] Splat cooling to room temperature resulted
in the formation of single phase alloys with up to 20% Fe.[102]
X-ray measurements indicate a discontinuous change of the lattice
parameter a_o with composition: there is a linear increase of a_o

up to 7.2% Fe, followed by a non-linear, slow decrease up to 20%
Fe. A hypothetical metastable miscibility gap was invoked to
explain the partial solid solubility, see Co-Cu.

Metastable Cu-Fe solid solutions have been produced by vapor
quenching.[103]

Cu-Rh   Copper-Rhodium

Limited information suggests a simple peritectic phase diagram
with limited terminal solubilities[29] of the two fcc metals. Splat
cooling experiments resulted in the preparation of single phase
alloys throughout the diagram;[104] the lattice parameters show
positive deviations from a Vegard's law straight line of
$\Delta a/a \leqslant 0.5\%$.

Cu-Sb   Copper-Antimony

In the Sb-rich region, Cu_2Sb is in equilibrium with Sb.[39]
Splat cooling of alloys with 33 and 75% Sb to -190°C did not
produce metastable alloy phases.[41]

Cu-Si   Copper-Silicon

The highly complex Cu-Si diagram shows at least 6 equilibrium
phases in the region between 10 and 25% Si; see Reference 39 for
details. There are three high temperature phases; of these,
β has a bcc W-A2 structure, while δ and η are variants of the
γ-brass type. Splat cooling to 20°C[82] resulted in suppression of
the low temperature phases γ and ϵ. Instead, following the range
of hcp χ phase, an apparently continuous region joining δ and η
(and possibly β) exists in quenched alloys; the varying degrees
of distortion of the γ-brass-related or W-A2-related structures
were not established.

Cu-Sn   Copper-Tin

In the Sn-rich region, Sn is in equilibrium with η phase
(CuSn).[39] Splat cooling to -190°C yielded a metastable γ phase
(simple hexagonal, $HgSn_{6-10} - A_f$) with > 14% Cu,[55] which
decomposes below 20°C. For details on γ, see Cd-Sn.

Cu-Te   Copper-Tellurium

Te-rich Cu-Te alloys consist of CuTe and Te.[39] Between
26-35% Cu, an amorphous phase can be retained by splat cooling to
room temperature.[57] For information on the atomic distribution,
see Reference 79.

Cu-Ti   Copper-Titanium

In equilibrium, there is a congruently melting phase $TiCu_3$ (orthorhombic, $TiCu_3$-DO_a type).[39] By splat cooling to 20°C, another modification of $TiCu_3$ was retained;[62] an increase of the solid solubility of Ti in Cu was also observed. Splat cooled alloys with 65-70% Cu were non-crystalline[101] (see Ni-Zr).

Cu-Zr   Copper-Zirconium

In a composition range of 40-75% Cu, a non-crystalline phase has been retained by splat cooling[101] (see Ni-Zr).

Fe-Ga   Iron-Gallium

An equilibrium phase diagram for Fe-Ga[105] indicates solid solubility of Ga in α-Fe up to ~50% Ga, and formation of a phase Fe_3Ga with a cubic $AuCu_3$-Ll_2 structure. Splat cooling of alloys with 34 to 50% Ga revealed the presence of ordering of the bcc solid solution to form the cubic CsCl-B2 type. In addition, conventional powder quenching produced Fe_3Ga (fcc, BiF_3-DO_3). There is a lattice parameter maximum between 30 and 40% Ga.[106]

Fe-Ti   Iron-Titanium

In equilibrium, there are two stoichiometric phases TiFe and $TiFe_2$.[39] Splat cooling to room temperature yielded a continuous metastable solid solution range (bcc, W-A2) extending from 0 to > 50% Fe, with partial, cooling-rate dependent ordering at and near TiFe. The solid solubility of Ti in α-Fe was increased from 9.8% Ti in equilibrium to > 16% Ti in nonequilibrium.[100]

Ga-Ge   Gallium-Germanium

The phase diagram is eutectic;[39] four alloys between 15 and 30% Ge consisted of the equilibrium constituents Ga and Ge after splat cooling to -190°C.[50]

Ga-In   Gallium-Indium

Ga and In form a eutectic.[39] By splat cooling to -190°C, a Ga-rich metastable phase has been prepared[62] (see also Reference 11).

Ga-Ni   Gallium-Nickel

Ga-rich alloys consist in equilibrium of Ga and $NiGa_4$.[39] By splat cooling two alloys with 5 and 7% Ni to -190°C, traces of a new metastable phase with unidentified structure were retained; the positions of the diffraction peaks indicate a relationship to ψ (Ga-Zn).[50]

Ga-Pb   Gallium-Lead

Two alloys with 5 and 7% Ga consisted of the equilibrium phases Ga and Pb after splat cooling to -190°C.[50]

Ga-Sn   Gallium-Tin

The Ga-Sn constitution diagram is simple eutectic.[39] At Sn-rich compositions, a γ phase (simple hexagonal, $HgSn_{6-10}$ - A_f) at ~20% Ga was retained by splat cooling to -190°C;[43] for details on γ, see Cd-Sn. At Ga-rich compositions, a metastable ψ phase occurs, with a complex, unidentified structure.[62] Both phases decompose below 20°C.

Ga-Te   Gallium-Tellurium

Te-rich Ga-Te alloys are reported to consist of $GaTe_3$ and Te.[39] In alloys with 10 to 30% Ga, an amorphous phase has been retained by splat cooling to room temperature.[107] For details, see Ge-Te.

Ga-Zn   Gallium-Zinc

The Ga-Zn phase diagram is simple eutectic, see Figure 10a.[39] By splat cooling to -190°C, two metastable single phase fields can be retained, as shown in the quenched-phase plot, Figure 10b: α (fcc, Cu-Al), between > 20 and < 40 Ga; and ψ, with an unidentified and complex structure, between ~65 and 83% Ga.[50]

The fcc structure of α and of its isotypes in other B_2-B_3 systems is regarded as connected with the valence electron concentration range in which it occurs, i.e., 2.2 - 2.4 e/a. There is also a close relation to the structure of In. According to its x-ray diffraction pattern, ψ is structurally related to other Ga-based metastable phases,[11] e.g. ψ (Al-Ga)[60] and ψ (Ga-In).[62] All metastable Ga-Zn phases decompose below 20°C.

Ge-Nb   Germanium-Niobium

An off-stoichiometric phase $Nb_3(Nb_xGe_{1-x})$, where $0.12 < x < 0.4$, of the cubic $A15$-Cr_3Si structure type occurs in this system.[108,109] Alloys with 25 to 29% Ge were rapidly quenched to room temperature with an arc-furnace splat cooling technique;[30] in an alloy with 29% Ge, the solubility range of the A15-type phase was found to be increased up to the stoichiometric composition, as determined by lattice parameter extrapolation. Considerable disordering of the A15-type structure was observed. The superconducting transition temperature T_c was raised from 6.9°K for the non-stoichiometric composition to 17°K for Nb_3Ge.

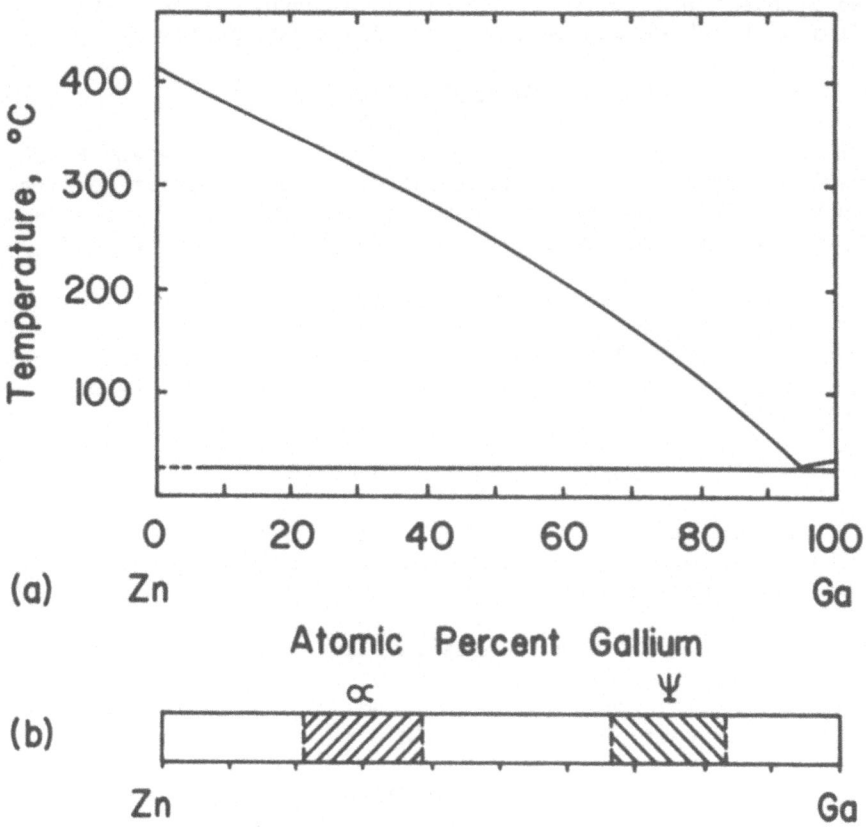

FIGURE 10.   Ga-Zn

Ge-Ni   Germanium-Nickel

The terminal solid solubility of Ge in Ni was enhanced from
the maximum equilibrium value of ~12% Ge[39] to ~20% Ge by splat
cooling to room temperature.[110] The lattice parameter plot shows
a smooth fit of equilibrium and nonequilibrium values.

Ge-Pd   Germanium-Palladium

The Pd-Ge equilibrium diagram is not known; however, from
the analogy to the Pd-Si diagram and preliminary measurements[111] a
eutectic between Pd and a compound Pd_5Ge_2,[39] has been assumed.
As shown by x-ray analysis and electron microscopy, splat cooled
alloys with 15 to 18% Ge consisted of a bcc phase not found in slow

cooled alloys; an alloy with 20% Pd was amorphous (see Pd-Si); and an alloy with 18% Ge was partly crystalline and partly amorphous.[111]

Ge-Pt   Germanium-Platinum

In equilibrium, there is a eutectic between Pt_3Ge and Pt_2Ge at ~30% Ge; Pt_3Ge melts peritectically.[112] After splat cooling, alloys with 17 and 30% Ge were found to be amorphous, whereas an intermediate alloy at 22% Ge was not amorphous[111] (see Pd-Si).

Ge-Rh   Germanium-Rhodium

An alloy with 17% Ge contained an amorphous phase after splat cooling to room temperature; a eutectic alloy with 22% Ge was crystalline.[111]

Ge-Sb   Germanium-Antimony

In the simple eutectic Ge-Sb system, a metastable intermediate phase has been retained by splat cooling to room temperature[113] (see also Reference 11).

Ge-Sn   Germanium-Tin

A 10% Ge alloy consisting of Ge and Sn in equilibrium[39] was unchanged after splat cooling to -190°C.[55]

Ge-Te   Germanium-Tellurium

Te-rich Ge-Te alloys consist of GeTe and Te.[39] Between 10 and 25% Ge, an amorphous phase was retained by splat cooling to room temperature; a diffraction pattern of a 10% Ge alloy is given.[107] A radial distribution analysis has been carried out;[114] results of this work are reported in abstract form in Reference 6 (see also Reference 115). The radial distribution functions strongly resemble those of liquid Te. Dendritic crystallization of an amorphous alloy with 15% Ge has been observed upon heating in the electron microscope.[116]

In-Ni   Indium-Nickel

The In-Ni phase diagram[39] is shown in Figure 11a; in the Ni-rich region, there are the phases Ni_3In and ε (Ni_2In).

The quenched-phase plot, Figure 11b,[117] indicates no extension of the terminal solid solubility of In in Ni, as reported e.g. for Ni-Sn;[110] instead, a metastable phase ε' (probably bcc, W-A2) occurs between 17 and 25% In. ε' is structurally distorted from the A2 type. The high temperature phase β (cubic, CsCl-B2)

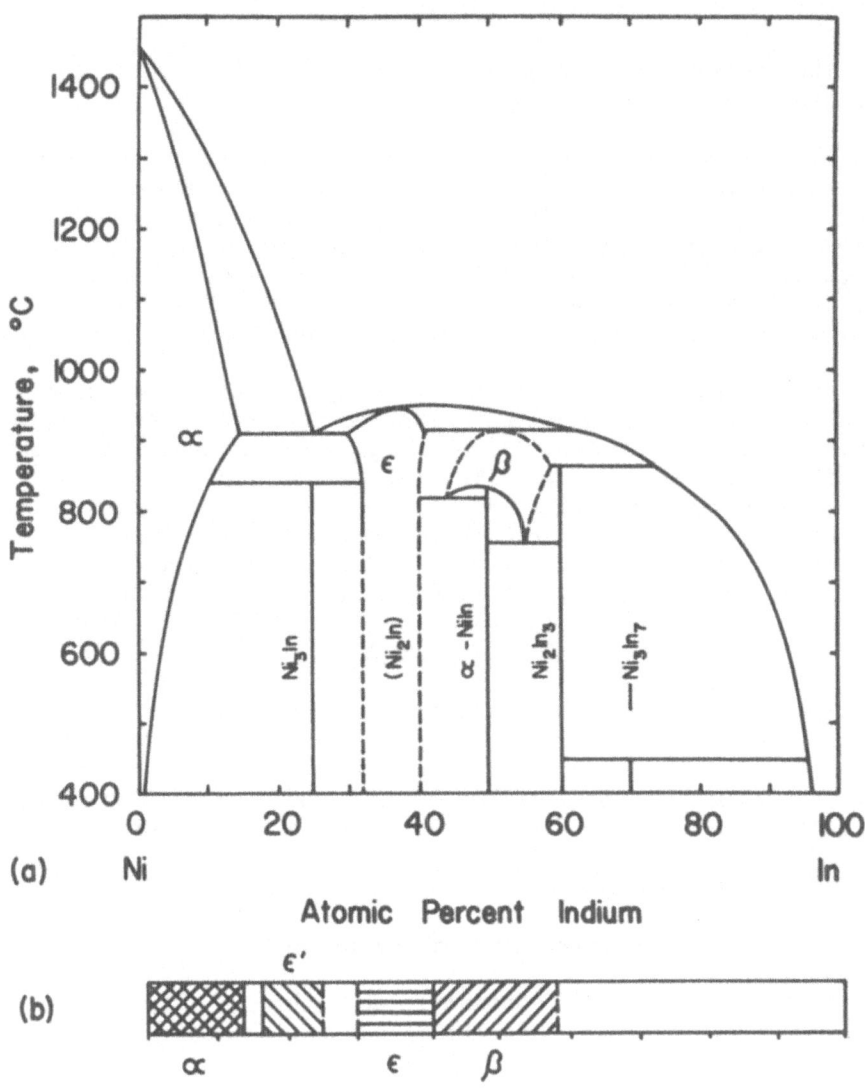

FIGURE 11.   In-Ni

was retained by splat cooling. The atomic volume curve shows
changes in slope similar to those in splat cooled metastable
Fe-Ga alloys.[106]

In-Sb   Indium-Antimony

The equilibrium diagram In-Sb given in Figure 12a[39] is characterized by the occurrence of the compound InSb.

Rapid quenching experiments have revealed the existence of two metastable phases, as shown in the quenched-phase plot, Figure 12b. In the In-InSb region, splat cooling to -190°C resulted in the formation of γ (simple hexagonal, $HgSn_{6-10}$ - A_f) at In_{1-x} with $0.35 < x < 0.45$.[118] The γ phase is isotypic with many stable and metastable γ phases containing Sn[11,55] and is isoelectronic with them; it decomposes below 20°C.

In the region InSb-Sb, splat cooling to room temperature produced a metastable phase at the eutectic composition of ~68% Sb (simple cubic, α-Po - A_h).[119] At neighboring compositions the equilibrium phases appeared. The relation of this simple cubic phase to the one obtained in Au-Te alloys[56] was recognized. In the same composition range, the finding of a metastable phase $In_{0.25}Sb_{0.75}$ has been reported,[118] which was tentatively identified as primitive rhombohedral, in analogy to SnSb.

In-Sn   Indium-Tin

The In-Sn equilibrium diagram[120] is characterized by a bct phase α_1 (designated "β" in Reference 120) and a hexagonal phase γ. Upon splat cooling of eight alloys with 20 to 52.5% Sn to -190°C, a slight increase of the maximum solid solubility of Sn in α_1 (bct, In-A6) from 44 to > 47% Sn and a large increase in the maximum solid solubility of In in γ (simple hexagonal, $HgSn_{6-10}$ - A_f) from 23 to 47.5% In were found.[50]

In-Te   Indium-Tellurium

Te-rich In-Te alloys consist in equilibrium of In_2Te_5 and Te.[39] By splat cooling to room temperature, an amorphous phase was retained between 10 and 30% In.[107] For details, see Ge-Te.

In-Tl   Indium-Thallium

Across the equilibrium phase diagram,[39,121] there are 4 phase fields with wide solubilities: In solid solution (bct); α-phase (fcc, Cu-Al); β-Tl solid solution (bcc, W-A2), extension of the high-temperature modification of Tl; and α-Tl solid solution (hcp, Mg-A3). In an investigation of superconducting transitions of In-Tl alloys,[122] alloys with 55 to 100% Tl were splat quenched to -196°C, and their structures determined by x-ray diffraction. The results were: 55 to 62% Tl: α + β-Tl solid solution; 65 to 83% Tl: β-Tl solid solution; 84.5 to 95% Tl: β-Tl solid

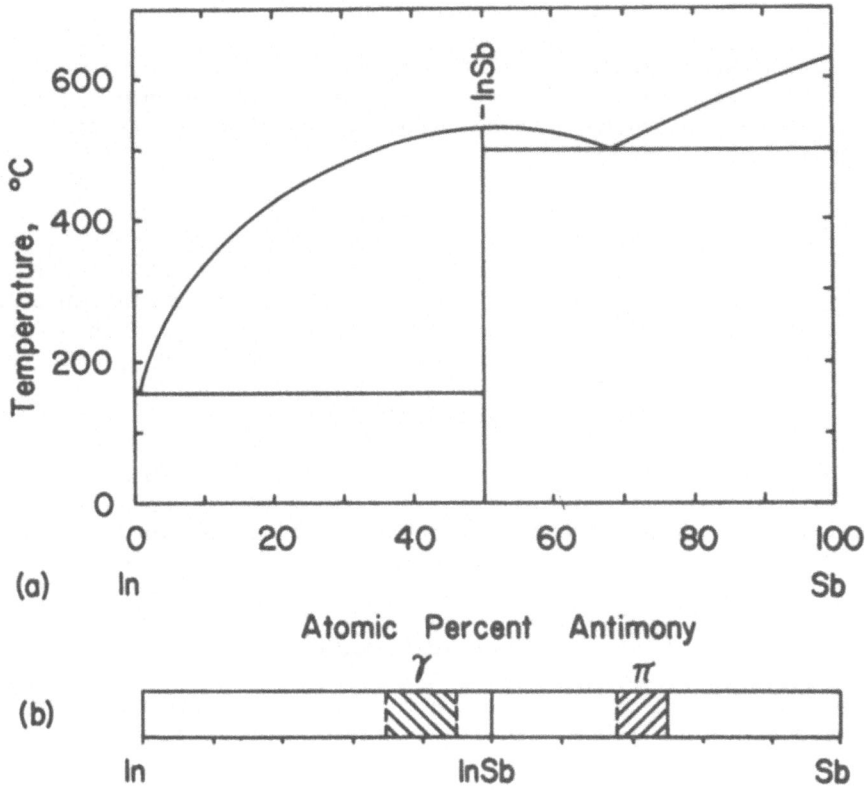

FIGURE 12.   In-Sb

solution + α-Tl solid solution; 100% Tl: β-Tl. After heating to
room temperature, β-Tl solid solution was found as a single phase
only between 75 and 83% Tl. Compared to the equilibrium diagram,
no extensions of equilibrium phases were observed at -196°C; the
maximum solubilities of α (60% Tl at 30°C) and β-Tl solid solution
(55% Tl at 170°C) were not reached in nonequilibrium. The
diffusionless β → α transformation of pure Tl was not suppressed.
The superconducting transition temperature T_c for β-Tl was derived
by extrapolation from β-Tl solid solution.

In-Zn   Indium-Zinc

In and Zn form a simple eutectic phase diagram.[39] In five
alloys with 25 to 90% In splat cooled to -190°C, no phase changes
were found.[50]

Mg–Mn   Magnesium–Manganese

The maximum solid solubility of Mn in Mg has been given as
~1.55%[39] and 0.98%.[40] By use of a catapulting technique (see
Al–V) up to 2.5% Mn have been dissolved, as shown by lattice
parameter measurements.[123]

Mg–Sn   Magnesium–Tin

Sn is in equilibrium with Mg_2Sn; they form a eutectic at
9% Mg.[39] Splat cooling to -190°C resulted in the formation of a
metastable γ phase (simple hexagonal, $HgSn_{6-10}$ - Af) with 12 to
> 15% Mg;[55] γ decomposes below 20°C. For details on γ, see Cd–Sn.

Mg–Zr   Magnesium–Zirconium

Values for the maximum solid solubility of Zr in Mg differ
widely; a solubility limit of 0.27% Zr is regarded as most
likely.[39] A solid solubility increase (from an assumed value of
0.22% Zr) to 0.33% Zr by a catapulting technique has been
reported.[123]

Mn–Sb   Manganese–Antimony

An alloy with 90% Sb which consisted in equilibrium of MnSb
and Sb[39] was found to be unchanged after splat cooling to -190°C.[41]

Mo–Ru   Molybdenum–Ruthenium

Between 1920 and 1945°C, Mo solid solution coexists with Ru
solid solution; at 1920°C, a σ phase "Mo_5Ru_3" forms.[40] By plasma-
jet spraying, the Mo content of the Ru solid solution was raised
from max. 49% to ~54% Mo and the formation of σ was suppressed.
The superconducting transition temperature of an alloy with 60%
Mo was raised by 0.7°K.[33] For details, see Ge–Te.

Na–Pb   Sodium–Lead

The maximum solid solubility of Na in Pb, given as 12% Na
in equilibrium,[39] has been increased to 23.8% by a wedge solidification technique[1] (see Al–V).

Nb–Ni   Niobium–Nickel

The Nb–Ni phase diagram, Figure 13a, is characterized by the
two equilibrium phases μ and $NbNi_3$. The quenched-phase plot at
20°C, Figure 13b,[20,124] shows several metastable phase ranges. The
terminal α–Ni solid solution range is enhanced to > 15.1% Ni.
A metastable phase ζ (hcp, Mg–A3 type) occurs at 81% Ni. $NbNi_3$ is

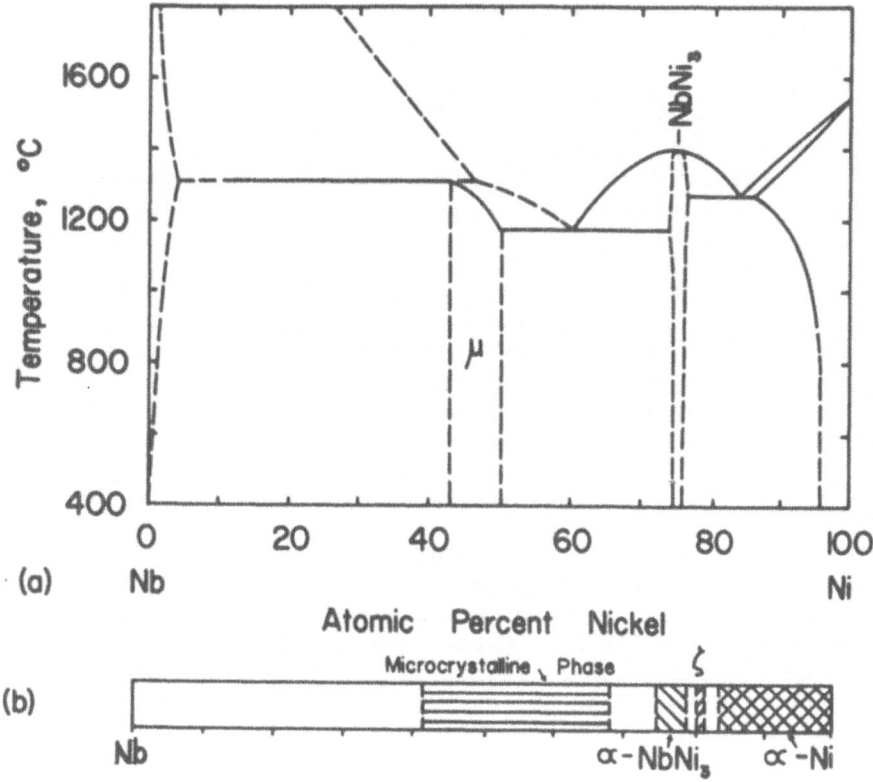

FIGURE 13.   Nb-Ni

extended in nonequilibrium to about 78% Nb; ζ is structurally
closely related to NbNi₃.[124]

     At higher Nb concentrations, splat cooled alloys display
diffraction patterns with broad peaks, which have been explained
as being due to microcrystal formation; assuming a bcc W-A2-like
atomic coordination, microcrystals contain about 150 atoms and
have a diameter of ∿ 15 Å. This microcrystalline phase was found as
a single phase between 42 and 46 and between 52 and 67% Ni; partial
retention of the microcrystalline phase was observed between 33
and 77% Ni. Electron diffraction patterns were also used.[20]

                    Ni-Rh   Nickel-Rhodium

     It had been stated that Ni and Rh mix in the solid state in
all proportions.[125] This result was confirmed by splat cooling
experiments[104] which succeeded in forming a complete series of

single phase alloys; thermal treatments of the splat foils up to 800°C produced no decomposition.

Ni-Sb   Nickel-Antimony

In the Sb-rich region, NiSb$_2$ and Sb form a eutectic at 97% Sb.[39] By splat cooling to 20°C, a metastable π phase (simple cubic, α-Po - A$_h$) was retained at 88% Sb.[41] For details, see Au-Sb.

Ni-Si   Nickel-Silicon

Splat cooling permitted the retention of Ni-rich terminal solid solutions with ⩾ 15% Si;[110] the lattice parameters show a smooth slope to 18% Si. By comparison, the maximum equilibrium solubility is 17.6% Si,[39] indicating almost no metastable solubility increase in this system.

Ni-Sn   Nickel-Tin

The terminal solid solubility of Sn in Ni was increased by splat cooling from the equilibrium value of 10.4% Sn to ⩾ 12% Sn;[110] the precision lattice parameter measurements suggest a higher supersaturation than 12% Sn. Recent experiments also showed a substantial solubility increase; in addition, more complex diffraction patterns which are not due to the equilibrium phases were obtained above 17% Sn.[62]

Ni-Ta   Nickel-Tantalum

The Ni-Ta equilibrium diagram[20,126] is characterized by the phases Ta$_2$Ni, μ, TaNi$_2$, and TaNi$_3$. Both the equilibrium diagram and the quenched-phase plot for Ni-Ta show strong similarities with the respective diagrams for Nb-Ni; two metastable Ni-Ta phases occur after splat cooling to room temperature. At Ni + 21% Ta, there is a ζ phase (hcp, Mg-A3),[124] and between 33 and 45% Ni, there is a microcrystalline phase, probably based on a bcc Mg-A2-like atomic coordination.[20] For details, see Nb-Ni.

Ni-Ti   Nickel-Titanium

Several splat cooled Ni-Ti alloys had diffraction patterns with broad peaks, indicating a microcrystalline or amorphous phase[101] (see Nb-Ni). A large metastable extension of the intermediate phase TiNi was also found[100] (see Fe-Ti).

Ni-V   Nickel-Vanadium

The phase diagram Ni-V[39] given in Figure 14a shows extended
σ phase and α-Ni solid solution ranges, as well as 3 stoichiometric
phases V_3Ni, VNi_2, and VNi_3. The quenched-phase plot at 20°C,
Figure 14b,[117] indicates 3 phase ranges: an extension of the α-V
solid solution range to ~18% Ni; an extension of the α-Ni
solid solution range from 43 to 51% V; and a metastable phase
β (bcc, W-A2) between 33 and 53% Ni. The two bcc phase fields do
not seem to be continuously connected; β phase replaces the σ
phase in nonequilibrium. There is a region from 50 to 54% Ni where
both bcc β and fcc α-Ni solid solution occur at the same concentration. Lattice parameter measurements are given. Broadened
peaks due to microcrystalline phases (see Nb-Ni) were also
observed.

Ni-Zr   Nickel-Zirconium

In this complex alloy system, splat cooling to room temperature
led to the retention of a noncrystalline phase (lacking long-range
atomic periodicity) in a composition range of 20-40% Ni, as
indicated by broad diffraction maxima.[101] A detailed radial
distribution analysis to determine whether this phase is
microcrystalline (see Nb-Ni) or amorphous was not carried out.

Pb-Sb   Lead-Antimony

Figure 15a shows the simple eutectic phase diagram Pb-Sb;[39]
the quenched-phase plot at -190°C, Figure 15b,[85] is marked by an
extension of the solid solubility of Sb in Pb from 5.8% to 17% Sb,
and of the solid solubility of Pb in Sb from between 1.5 and 2.7%
to 11% Pb. In addition, there are two metastable intermediate
phases: β (bcc, W-A2) between 34 and 38% Sb, and π (fcc, NaCl-B1
or simple cubic, α-Po - A_h) between 50 and 60% Sb. The experimental
distinction between the ordered NaCl type and the disordered α-Po
type is difficult and presently not certain, although the former
type is more likely. The metastable substitution of Pb in Sb
strongly influences the rhombohedral angle α of Sb which decreases
with increasing Pb content. All metastable effects disappear
upon heating to 20°C.

Pb-Sn   Lead-Tin

The Pb-Sn phase diagram is simple eutectic, with limited
terminal solid solubilities of 1.45% Pb in Sn and 29% Sn in Pb.[39]
By splat cooling to -190°C, a metastable γ phase (simple hexagonal
$HgSn_{6-10}$ - A_f) with ~25% Pb was retained, which is unstable at
20°C.[55] For details on γ, see Cd-Sn. Additional metastable phase
formation has been observed at higher Pb contents[62] (see also
Reference 11).

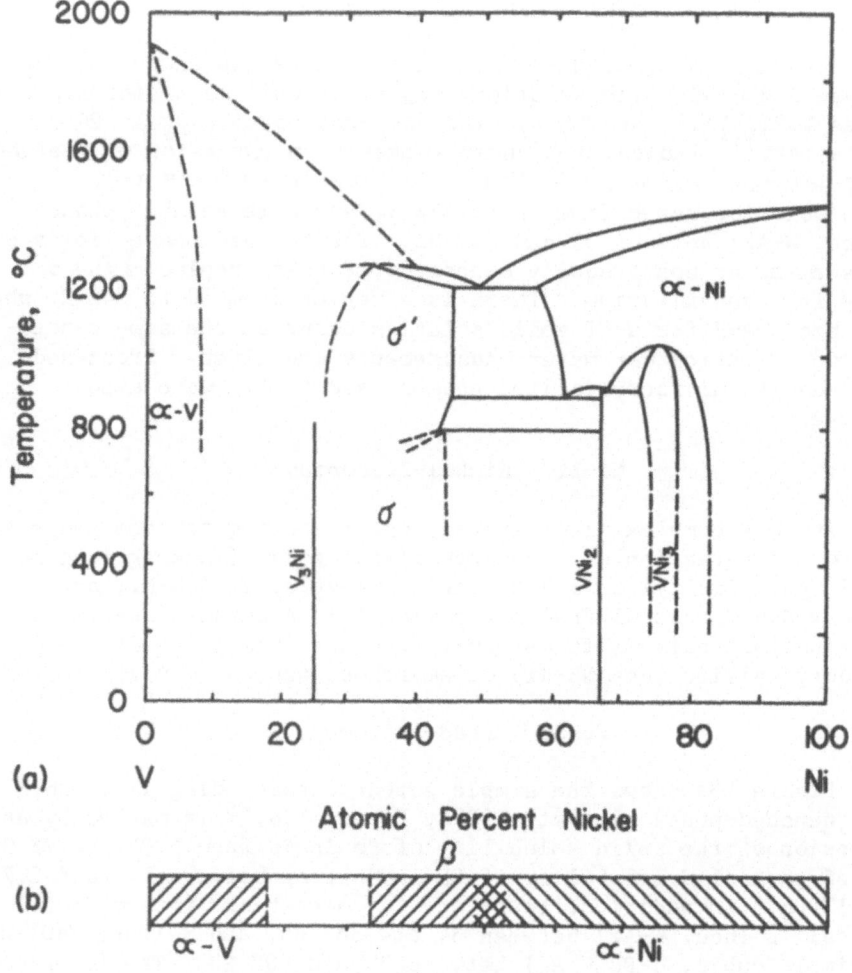

FIGURE 14.   Ni-V

Pd-Sb   Palladium-Antimony

In the Sb-rich region, PbSb$_2$ and Sb are in equilibrium and form a eutectic at 89% Sb.[39] Splat cooling to 20°C yielded a metastable π phase (simple cubic, α-Po - A$_h$) between 84.5 and 87.5% Sb.[41] For details, see Au-Sb.

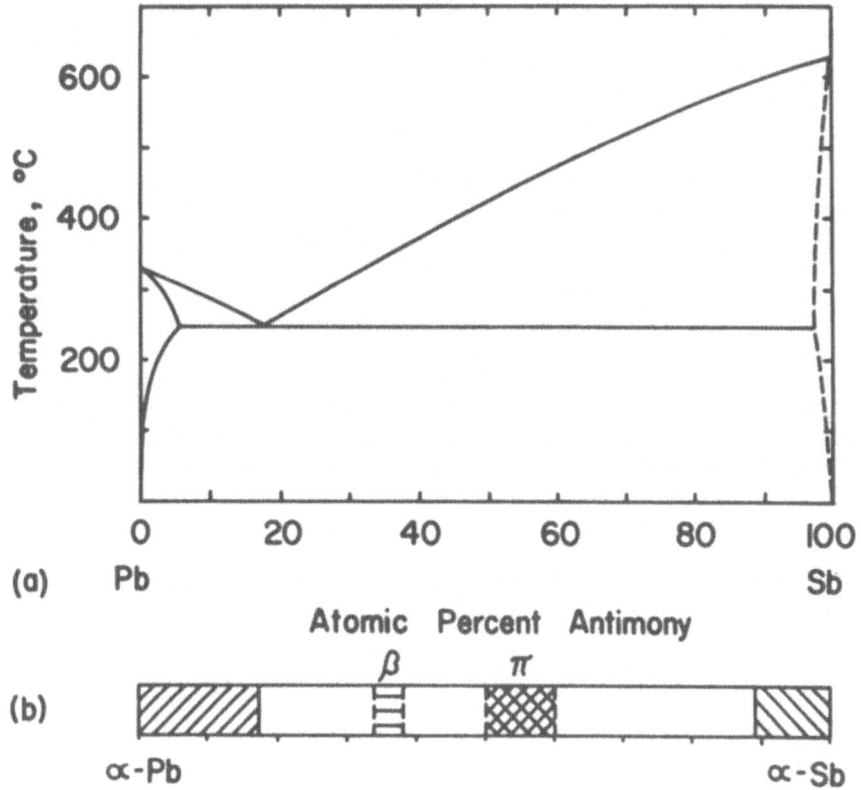

FIGURE 15.   Pb-Sb

Pd-Si   Palladium-Silicon

The equilibrium diagram Pd-Si shows the compounds Pd_3Si [40] and Pd_9Si_2; the latter probably forms a low-melting eutectic with Pd at ~17% Si.[9] After rapid quenching to room temperature, alloys with 12 to 23% Si were found to have x-ray diffraction patterns with broad peaks regarded as typical of an amorphous structure.[127] Both splat cooling and hammer-and-anvil techniques were used. The amorphous phases had a high electrical resistivity and crystallized into undetermined phases upon heating to ~400°C, for heating rates of 20-10000°C/min.[127] If an amorphous alloy with 20% Pd is heated to a temperature below 325°C, a metastable fcc phase of the Cu-Al type crystallizes.[128] A radial distribution analysis of the amorphous phase indicated 11.6 nearest neighbors at a distance of 2.79 Å, to be compared to 12 nearest neighbors at 2.77Å in crystalline Pd[129] (see also Pt-Si).

At higher Si contents, 2 further eutectics exist at ~45 and
~58% Si; however, splat cooling to -190°C yielded no metastable
phases between 45 and 61% Si.[130]

Pd-Sn   Palladium-Tin

At high Sn contents, Sn and $PdSn_4$ are in equilibrium.[39] A
metastable γ phase (simple cubic, $HgSn_{6-10}$ - A_f) with ~8% Pd
could be formed by splat cooling to -190°C; γ decomposes below
20°C.[55] For details on γ, see Cd-Sn.

Pd-W   Palladium-Tungsten

The solid solubility of W in Pd has been increased from the
equilibrium value of 22% W[131] to 44% W by an arc-furnace splat
cooling technique described in Reference 105;[31] lattice parameters
and superconducting transition temperatures were determined.[31]

Pd-Zr   Palladium-Zirconium

In alloys with 20-35% Pd, a non-crystalline phase has been
retained by splat cooling to room temperature[101] (see Ni-Zr).

Pt-Sb   Platinum-Antimony

The Pt-Sb phase diagram[39] shows a eutectic at 33.6% Sb formed
by two phases with 27 ± 2 and 40% Sb, respectively.[132] Splat
cooling to room temperature produced a phase causing broad x-ray
and electron diffraction reflections, tentatively analyzed as due
to a microcrystalline phase. In alloys with 34 and 36.5% Sb, only
this phase was found; in other alloys with 30 to 43% Sb, it occurred
together with unidentified crystalline phases.[132] In another
investigation,[111] splat cooled alloys with 33.3 and 36.9% Sb were
found to be amorphous in their thinner regions and crystalline in
the thicker regions; this agrees with the observation[132] that the
new phase cannot be easily retained with the present cooling rate.
The activation energy of the equilibration process was determined
by isothermal heat-treatments and a quantitative x-ray technique.[132]

Pt-Si   Platinum-Silicon

At 23% Si, a eutectic exists between Pt and Pt_5Si_2; there is
also a eutectic at 67.5% Si between PtSi and Si.[39] After splat
cooling to room temperature, hypereutectic alloys were amorphous
(see Pd-Si); eutectic alloys gave some evidence of being
microcrystalline, with no ~30 Å particle size (based on a fcc,
Cu-Al type structure), with primary Pt solid solution grains also
appearing at hypoeutectic concentrations.[111]

At 68% Si, an amorphous phase was retained by splat cooling to -190°C, together with an unidentified crystalline phase.[132]

Pt-Sn   Platinum-Tin

At Sn-rich compositions, Sn and $PtSn_4$ coexist.[39] No phase change was observed after splat cooling two alloys with 4 and 8% Pt to -190°C.[55]

Pt-W   Platinum-Tungsten

Pt dissolves max. 64% W in solid solution;[39] other findings[40] put the solubility limit at a lower value.

By an arc furnace splat cooling technique,[30] single phase alloys with max. 67% W were prepared and their superconducting transition temperatures were measured.[31]

Rh-Si   Rhodium-Silicon

In an alloy with 22% Si of unknown equilibrium structure, an amorphous phase was found to be present after splat cooling to room temperature;[111] a fcc crystalline phase was also observed.

Sn-Tl   Tin-Thallium

In the composition range between β-Sn and an intermediate equilibrium phase (fcc, Cu-Al type)[40] which forms a eutectic with β-Sn, three metastable phases have been retained by splat cooling to -190°C. Of these, one is isomorphous with γ (Cd-Sn) (simple hexagonal, $HgSn_{6-10}$ - A_f), another is isomorphous with $α_1$ (In-Bi) (bct, In-A6)[133] (see also Reference 11).

Sn-Zn   Tin-Zinc

The simple eutectic phase diagram Sn-Zn shows a eutectic at 15% Zn and 198°C.[39] By splat cooling to -190°C, a metastable γ phase (simple hexagonal, $HgSn_{6-10}$ - A_f) with 12 to > 15% Zn was retained; it decomposes below 20°C into Sn and Zn[55] (see Cd-Sn).

TERNARY NONEQUILIBRIUM ALLOY SYSTEMS

The component elements are generally listed alphabetically; however, where additions of varying third elements to a given binary combination have been investigated, the binary system has been listed first (e.g. C-Mo-Cr).

Ag–Au–Te   Silver–Gold–Tellurium

See Au-Cu-Te.

Ag–Cu–Te   Silver–Copper–Tellurium

See Au-Cu-Te.

Al–Cr–Si   Aluminum–Chromium–Silicon

Use of a wedge solidification technique produced a metastable ternary phase in Al-rich Al-Cr-Si alloys.[88] Mechanical properties were also determined.[134]

Al–Mg–O   Aluminum–Magnesium–Oxygen

Pseudobinary $MgO-Al_2O_3$ melts were splat cooled onto glass slides. A metastable spinel-type phase was found in the concentration range from $\gamma-Al_2O_3$ to 80% MgO, with a linear lattice parameter variation.[135]

Al–Nb–Ni   Aluminum–Niobium–Nickel

In the Al-Nb-Ni system, there is a complex ternary M phase with a structure related to that of μ (Nb-Ni).[136,137] An alloy with an equilibrium structure of M phase consisted of a microcrystalline phase after splat cooling to 20°C[20] (see Nb-Ni).

Al–Mn-Ta, Ti, W, etc.,   Aluminum–Manganese–Tantalum, etc.

The influence of third components, such as Ta, Ti, W, Zr, etc., on physical properties of supersaturated Al-Mn solid solutions produced by a catapult technique has been investigated[65] (see Al-Mn).

Au–Cu–Te   Gold–Copper–Tellurium

Metastable alloys lying on sections with 70% Te through the systems Ag-Au-Te, Ag-Cu-Te, and Au-Cu-Te were prepared by splat cooling to room temperature. A metastable π phase (simple cubic, α-Po - A_h) (see Au-Te) was found in all Ag-Au-Te alloys and in the Ag-rich and Au-rich Ag-Cu-Te and Au-Cu-Te alloys; the Cu-rich alloys were amorphous (see Cu-Te). A considerable difference in the size factors for Ag and Au was noted.[57] Mössbauer spectra of an amorphous alloy $Au_5Cu_{25}Te_{70}$ have been obtained.[79] Resistivity and thermoelectric measurements on amorphous $Au_5Cu_{25}Te_{70}$ show the amorphous alloy to be a semiconductor.[115]

Au-Ge-Si   Gold-Germanium-Silicon

The specific heat and resistivity of a metastable amorphous
Au-Ge-Si alloy produced by splat cooling has been measured;
evidence for a glass transition was found.[74,75] Viscosity
measurements have been reported.[76]

Au-Sb-Sn   Gold-Antimony-Tin

Alloys on the metastable pseudobinary section SnSb (fcc,
NcAl-Bl)[39] and π (Au-Sb 75) (simple cubic, α-Po - A_h)[41] were
prepared by splat cooling to -190°C. A continuous variation of
the lattice parameter between π and the simple cubic substructure
of SnSb was observed.[41]

B-Fe-Ni   Boron-Iron-Nickel

Martensitic phases were found in splat cooled Fe-Ni-B alloys
indicating retention of B in the metal matrix; this result is of
special interest, as solid solutions with B could not be retained
in the binary systems B-Fe and B-Ni.[138]

Bi-Pb-Sb   Bismuth-Lead-Antimony

Ten ternary Bi-Pb-Sb alloys were examined after splat
cooling to -190°C; metastable β (Pb-Sb) (bcc, W-A2) and stable
ε (Bi-Pb) (hcp, Mg-A3) extend into the ternary, while metastable
π (Pb-Sb) (simple cubic, α-Po - A_h) and X (Bi-Pb) have no
noticeable ternary solid solution range[85] (see Bi-Pb and Pb-Sb).

C-Fe-P   Carbon-Iron-Phosphorus

Ternary alloys of Fe with C and P were retained in an
amorphous state (see Pd-Si) after rapid quenching by a "piston
and anvil" technique. The resulting phase was ferromagnetic.[139,140]
The temperature dependence of the bulk magnetization and the
magnetic hyperfine field obtained from Mössbauer spectra were
measured; a magnetic moment per iron atom of 2.10 ± 0.01 μ_B and a
Curie temperature T_c = 586 ± 2°K were found.[140] Hall coefficient
measurements showed no temperature dependence up to the Curie
temperature; this was interpreted as a confirmation of the
amorphous structure of the Fe-C-P phase.[141] There is a minimum in
the electrical resistivity *vs*. temperature curve.[142]

C-Fe-Si   Carbon-Iron-Silicon

Small Si additions were shown to lead to increased retention
of the metastable phase (up to 100%) in splat cooled Fe-rich
Fe-C alloys[89,90] (see C-Fe).

C-Mo-Cr, Fe, Mn   Carbon-Molybdenum-Chromium, Iron, Manganese

The effect of paramagnetic impurities on the superconducting properties of fcc Mo carbide (see C-Mo) prepared by splat cooling has been measured.[143] It had been found previously that an addition of 2% "MnC" to superconducting fcc MoC prepared by splat cooling makes the alloy normal to ≤ 1.7°K.[29]

C-Mo-Nb   Carbon-Molybdenum-Niobium

The lattice parameters and superconducting properties of alloys from the six pseudobinary systems formed by the carbides NbC, TaC, MoC, and WC, such as NbC-MoC, TaC-WC, etc., were investigated. Splat cooled alloys were utilized and the fcc NaCl-B1 type carbide modifications were retained in every case.[144]

C-Mo-Ta, W   Carbon-Molybdenum-Tantalum, Tungsten

See C-Mo-Nb.

C-Nb-Ta, W   Carbon-Niobium-Tantalum, Tungsten

See C-Mo-Nb.

C-Ta-W   Carbon-Tantalum-Tungsten

See C-Mo-Nb.

Co-Cr-Fe   Cobalt-Chromium-Iron

See Co-Fe-V.

Co-Fe-V   Cobalt-Iron-Vanadium

The magnetic properties of splat cooled Co-Fe-V alloys with 52% Co and up to 20% V were measured; a reversible change of the magnetic moment with temperature was observed in quenched alloys with 14% V.[145] A similar behavior is found in splat cooled Co-Fe-Cr alloys, i.e. with Cr replacing V. An inductance thermometer utilizing this property of splat cooled Co-Fe-V and Co-Fe-Cr foils to measure temperatures has been described.[146]

Co-Pd-Si   Cobalt-Palladium-Silicon

The addition of Co, Fe and Ni to the metastable amorphous Pd-Si phase (see Pd-Si) has produced ferromagnetic amorphous phases.[147] A later measurement of the magnetic moments of splat cooled amorphous alloys with the compositions $Pd_{80-x}Co_xSi_{20}$ showed a change from paramagnetism to ferromagnetism at $x=10$.[148]

Fe-Pd-Si   Iron-Palladium-Silicon

See Co-Pd-Si.

Ga-Ge-Sb   Gallium-Germanium-Antimony

The pseudobinary system Ge-GaSb is simple eutectic with small mutual solid solubilities. By splat cooling to room temperature, a complete series of solid solutions has been obtained; the lattice parameters decrease linearly with increasing Ge concentration.[149]

Ni-Pd-Si   Nickel-Palladium-Silicon

See Co-Pd-Si.

ACKNOWLEDGMENTS

This work has been supported under NSF Grant GK 1374 and NASA Contract NsG-117-61. Thanks are due to Professor P. Duwez, Dr. R. Willens and Dr. H. L. Luo for making unpublished results available; to Professor N. J. Grant for his continued interest; to the author's collaborators at M.I.T., especially Dr. P. Predecki, Dr. R. Kane, Dr. R. Ruhl, Dr. P. Srivastava, Dr. R. Wang, Mrs. C. Jansen, Mr. J. Muratet, and Mr. R. Ray, on whose work much of the foregoing is based; and to Miss L. Demerjian for preparing the typescript.

REFERENCES

1. G. Falkenhagen and W. Hofmann, Z. Met. 43, 69 (1952).

2. P. Duwez, R. H. Willens, and W. Klement, Jr., J. Appl. Phys. 31, 1136 (1960).

3. P. Duwez and R. H. Willens, Trans. TMS-AIME 227, 362 (1963).

4. P. Duwez, in "Energetics in Metallurgical Phenomena", Vol. I, W. M. Mueller, Ed., Gordon and Breach, New York, (1965).

5. P. Duwez, in "Alloying Behavior and Effects in Concentrated Solid Solutions", T. B. Massalski, Ed., Gordon and Breach New York, 420 (1966).

6. P. Duwez, Progr. in Solid State Chemistry, 3 (8), 377 (1966).

7. P. Duwez, in "Intermetallic Compounds", J. H. Westbrook, Ed., J. Wiley and Sons, New York, 340 (1967).

8. P. Duwez, in "Phase Stability in Metals and Alloys",
    P. S. Rudman, J. Stringer, and R. I. Jaffee, Eds., McGraw-Hill
    Book Co., New York 523 (1967).

9. P. Duwez, Trans. ASM 60, 607 (1967).

10. B. C. Giessen, in "Strengthening Mechanisms, Metals and
    Ceramics", Proc. of the 12th Sagamore Army Materials Research
    Conference, 1965, J. J. Burke, N. L. Reed, V. Weiss, Eds.,
    Syracuse University Press, Syracuse, New York, 273 (1966).

11. B. C. Giessen, in "Advances in X-Ray Analysis", Vol. 12,
    C. S. Barrett, G. R. Mallett, and J. B. Newkirk, Eds.,
    Plenum Press, New York, 23 (1969).

12. B. C. Giessen and R. H. Willens, in "The Use of Phase Diagrams
    in Ceramic, Glass, and Metal Technology", A. M. Alper, Ed.,
    Academic Press, New York (1969).

13. P. Predecki, A. W. Mullendore, and N. J. Grant, Trans. TMS-
    AIME, 233, 1581 (1965).

14. R. C. Ruhl, Mat. Sc. and Eng. 1, 313 (1967).

15. N. I. Varich and K. Y. Kolesnichenko, Izv. Vysshikh Uchebn.
    Zavedenii, Tsvetn. Met. 4, 131 (1960).

16. I. S. Miroshnichenko, Krist. i Fazovye Perechody, Otd. Fiz.
    Tverd. Tela i Poluprov. Akad. Nauk Belorussk. SSR 133 (1962).

17. A. Defrain and I. Epelboin, J. Phys. Rad. 21, 76 (1960).

18. A. K. Jena, B. C. Giessen, M. B. Bever, and N. J. Grant,
    Acta Met. 16, 1047 (1968).

19. T. R. Anantharaman, H. L. Luo, and W. Klement, Jr.,
    Trans. TMS-AIME 233, 2014 (1965).

20. R. C. Ruhl, B. C. Giessen, M. Cohen, and N. J. Grant,
    Acta Met. 15, 1693 (1967).

21. H. Matyja, B. C. Giessen, and N. J. Grant, J. Inst. Met.
    96, 30 (1968).

22. P. T. Sarjeant and R. Roy, Mat. Res. Bull. 3, 265 (1968).

23. J. I. Goldstein, F. J. Majeske, and H. Yakowitz, in
    "Advances in X-Ray Analysis" Vol. 10, J. B. Newkirk and G. R.
    Mallett, Eds., Plenum Press, New York, 431 (1967).

24. N. J. Grant and P. Duwez, J. Met. <u>17</u>, 1167 (1965).

25. J. Pietrokowski, Rev. Sc. Instr. <u>34</u>, 445 (1963).

26. J. Dixmier and A. Guinier, Mem. Sc. Rev. Met. <u>64</u>, 53 (1967).

27. D. H. Harbur, J. W. Anderson, and W. J. Maraman, Abstr.
     Bull. IMD-AIME <u>3(2)</u>, 70A (1968).

28. I. V. Salli, <u>Structure Formation in Alloys</u>, Translation by
     Consultants Bureau Enterprises, Inc., New York, 64 (1964).

29. R. H. Willens and E. Buehler, Trans. TMS-AIME <u>236</u>, 171
     (1966).

30. B. T. Matthias, T. H. Geballe, R. H. Willens, E. Corenzwit,
     and G. W. Hull, Jr., Phys. Rev. <u>139</u>, A1501 (1965).

31. H. L. Luo, J. Less. Comm. Met. <u>15</u>, 299 (1968).

32. R. Roberge and H. Herman, Mat. Sc. and Eng. <u>3</u>, 62
     (1968).

33. M. Moss, D. L. Smith, and R. A. Lefever, Appl. Phys. Lett. <u>5</u>,
     120 (1964).

34. S. Mader, A. S. Novick, and H. Widmer, Acta Met. <u>15</u>, 203 (1967).

35. R. Hilsch, in 'Non-Crystalline Solids', V. C. Frechette, Ed.,
     J. Wiley and Sons, New York (1960).

36. H. L. Luo and P. Duwez, Can. J. Phys. <u>41</u>, 758 (1963).

37. W. Buckel, Z. Phys. <u>138</u>, 136 (1954).

38. G. Thomas and R. H. Willens, Acta Met. <u>12</u>, 191 (1964).

39. M. Hansen, <u>Constitution of Binary Alloys</u>, McGraw-Hill Book
     Company, New York (1958).

40. R. P. Elliott, <u>Constitution of Binary Alloys</u>, <u>First Supplement</u>,
     McGraw-Hill Book Co., New York (1965).

41. B. C. Giessen, U. Wolff, and N. J. Grant, Trans. TMS-AIME <u>242</u>,
     597 (1968).

42. R. K. Linde, J. Appl. Phys. <u>37</u>, 934 (1966).

43. S. Nagakura, S. Toyama, and S. Oketani, Acta Met. <u>14</u>, 73 (1966).

44. R. Stoering and H. Conrad, Abstr. Bull. IMD-AIME 3(1), 89A (1968).

45. P. Duwez, R. H. Willens, and W. Klement, Jr., J. Appl. Phys. 31, 1137 (1960).

46. T. B. Massalski and H. W. King, "Alloy Phases of the Noble Metals", Progr. Mat. Sc. 10 (1), 30 (1961).

47. W. Klement, Jr., J. Inst. Met. 90, 27 (1961).

48. T. R. Anantharaman, H. L. Luo, and W. Klement, Jr., Nature 210, 1040 (1966).

49. H. L. Luo, M. F. Merriam, and D. C. Hamilton, Science 145, 581 (1964).

50. P. K. Srivastava, B. C. Giessen, and N. J. Grant, Acta Met. 16, 1199 (1968).

51. P. Predecki, B. C. Giessen, and N. J. Grant, Trans. TMS-AIME 233, 1438 (1965).

52. W. Klement, Jr., and H. L. Luo, Trans. TMS-AIME 227, 1253 (1963).

53. W. Klement, Jr., Trans. TMS-AIME 233, 1182 (1965).

54. H. L. Luo, W. Klement, Jr., and T. R. Anantharaman, Trans. Indian Inst. Met., 214 (December, 1965).

55. R. H. Kane, B. C. Giessen, and N. J. Grant, Acta Met. 14, 605 (1966).

56. H. L. Luo and W. Klement, Jr., J. Chem. Phys. 36, 1870 (1962).

57. C. C. Tsuei and P. Duwez, California Institute of Technology Report CALT-221-33 (1967).

58. T. Toda, R. Maddin, and H. Herman, Abstr. Bull. IMD-AIME 3 (2), 48A (1968).

59. C. Jansen, B. C. Giessen, and N. J. Grant, Trans. TMS-AIME, to be published.

60. B. C. Giessen, U. Wolff, and N. J. Grant, J. Appl. Cryst. 1, 30 (1968).

61. L. F. Vereshchagin, S. S. Kobalkina, and Z. V. Troitskaya, Sov. Phys. Doklady 9, 894 (1965).

62. B. C. Giessen, unpublished results.

63. H. L. Luo, C. C. Chao, and P. Duwez, Trans. TMS-AIME 230, 1488 (1964).

64. I. S. Miroshnichenko, Dokl. Akad. Nauk SSSR 164 (1), 137 (1965).

65. N. I. Varich, C. M. Burov, and K. Y. Kolesnichenko, Fiz. Metal. Metalloved. 18, 396 (1964).

66. N. I. Varich, L. M. Burov, K. Y. Kolesnichenko, and A. P. Maksimenko, Fiz. Metal. Metalloved. 15, 292 (1962).

67. C. E. Lundin and A. S. Yamamoto, Trans. TMS-AIME 236, 863 (1966).

68. M. Itagaki, B. C. Giessen, and N. J. Grant, Trans. Quart. ASM 61, 330 (1968).

69. H. Matyja, B. C. Giessen, K. C. Russell, and N. J. Grant, Trans. TMS-AIME, to be published.

70. M. Moss, Acta Met. 16, 321 (1968).

71. B. C. Giessen and R. Wang, Acta Cryst., to be published.

72. W. Klement, Jr., Trans. TMS-AIME 227, 965 (1963).

73. W. Klement, R. H. Willens, and P. Duwez, Nature 187, 869 (1960).

74. H. S. Chen and D. Turnbull, Appl. Phys. Lett. 10, 284 (1967).

75. H. S. Chen and D. Turnbull, J. Appl. Phys. 38, 3646 (1967).

76. H. S. Chen and D. Turnbull, J. Chem. Phys. 48, 2560 (1968).

77. G. Kiessler, B. C. Giessen, and N. J. Grant, unpublished results.

78. B. C. Giessen, Z. Metallk. 59, 805 (1968).

79. C. C. Tsuei and E. Kankeleit, Phys. Rev. 162, 312 (1967).

80. C. C. Tsuei and L. R. Newkirk, California Institute of Technology Report CALT-221-60 (1968).

81. H. L. Luo, Abstr. Bull. IMD-AIME 2 (1), 44 (1967), and private communication.

82. J. C. Muratet, B. C. Giessen, and N. J. Grant, unpublished results.

83. B. C. Giessen, M. Morris, and N. J. Grant, Trans. TMS-AIME 239, 883 (1967).

84. R. Wang, B. C. Giessen, and N. J. Grant, Z. Krist., to be published.

85. C. Borromee-Gautier, B. C. Giessen, and N. J. Grant, J. Chem. Phys. 48, 1905 (1968).

86. W. Klement, Jr., J. Chem. Phys. 38, 298 (1963).

87. R. C. Ruhl and M. Cohen, Scripta Met. 1, 73 (1967).

88. P. Esslinger, Z. Met. 57, 12 (1966).

89. R. C. Ruhl, Acta Met. 15, 159 (1967).

90. R. C. Ruhl and M. Cohen, Trans. TMS-AIME 245, 241 (1968).

91. E. Rudy, St. Windisch, A. J. Stosick, and J. R. Hoffman, Trans. TMS-AIME 239, 1247 (1967).

92. R. H. Willens and E. Buehler, Appl. Phys. Lett. 7, 25 (1965).

93. R. V. Sara, J. Am. Cer. Soc. 48, 251 (1965).

94. E. Krainer and J. Robitsch, Planseeber. Pulvermet. 15, 46 (1967).

95. G. V. Raynor and J. A. Lee, Acta Met. 2, 616 (1954).

96. J. Baker and J. W. Cahn, Acta Met., 17, 575 (1969).

97. J. Baker and B. C. Giessen, unpublished results.

98. S. Mader, H. Widmer, F. M. d'Heurle, and A. S. Novick, Appl. Phys. Lett. 3, 201 (1963).

99. E. F. Kneller, J. Appl. Phys. 33, 1355 (1962).

100. R. Ray, B. C. Giessen, and N. J. Grant, unpublished results.

101. R. Ray, B. C. Giessen, and N. J. Grant, Scripta Met. 2, 357 (1968).

102. W. Klement, Jr., Trans. TMS-AIME 233, 1180 (1965).

103. E. F. Kneller, J. Appl. Phys. 35, 2210 (1964).

104. H. L. Luo and P. Duwez, J. Less Comm. Met. 6, 248 (1964).

105. C. Dasarathy and W. Hume-Rothery, Proc. Roy. Soc. (London) A286, 141 (1965).

106. H. L. Luo, Trans. TMS-AIME 239, 119 (1967).

107. H. L. Luo and P. Duwez, Appl. Phys. Lett. 2, 21 (1963).

108. J. H. Carpenter, J. Phys. Chem. 67, 2141 (1963).

109. S. Geller, Acta Cryst. 9, 885 (1956).

110. W. Klement, Jr., Can. J. Phys. 40, 1397 (1962).

111. R. C. Crewdson, California Institute of Technology Report, CALT-221-21 (1966).

112. K. C. Jain and S. Bhan, Trans. Indian Inst. Metals 19, 49 (1966).

113. B. C. Giessen, C. Borromee-Gautier, and N. J. Grant, unpublished results.

114. H. L. Luo, Ph.D. Thesis, California Institute of Technology, Pasadena, California (1964).

115. C. C. Tsuei, Phys. Rev. 170, 775 (1968).

116. R. H. Willens, J. Appl. Phys. 33, 3269 (1962).

117. R. C. Ruhl, B. C. Giessen, M. Cohen, and N. J. Grant, Mat. Sc. and Engin. 2, 314 (1968).

118. B. C. Giessen, R. H. Kane, and N. J. Grant, Nature 207, 854 (1965).

119. C. B. Jordan, J. Chem. Phys. 39, 1613 (1963).

120. T. Heumann and O. Alpaut, J. Less. Comm. Met. 6, 108 (1964).

121. P. N. Adler and H. Margolin, Atca Met. 14, 1645 (1966).

122. H. L. Luo and R. H. Willens, Phys. Rev. 154, 436 (1967).

123. N. I. Varich and B. N. Litvin, Fiz. Metal. Metalloved. 16, 526 (1963).

124. R. C. Ruhl, B. C. Giessen, M. Cohen, and N. J. Grant, J. Less Comm. Met. 13, 611 (1967).

125. E. Raub, J. Less Comm. Met. 1, 3 (1959).

126. I. I. Kornilov and E. N. Pylaeva, Russ. J. Inorg. Chem. 7, 300 (1962).

127. P. Duwez, R. H. Willens, and R. C. Crewdson, J. Appl. Phys. 36, 2267 (1965).

128. R. H. Willens, Abstr. Bull. IMD-AIME 2 (1), 22 (1967).

129. R. C. Crewdson, California Institute of Technology Report, CALT-221-20 (1966).

130. P. K. Srivastava, Ph.D. Thesis, Massachusetts Institute of Technology, Cambridge, Massachusetts (1966).

131. W. K. Goetz and J. H. Brophy, J. Less. Comm. Metals 6, 345 (1964).

132. P. K. Srivastava, B. C. Giessen, and N. J. Grant, to be published.

133. J. Vitek, B. C. Giessen, and N. J. Grant, unpublished results.

134. P. Esslinger, Z. Met. 57, 109 (1966).

135. P. T. Sarjeant and R. Roy, J. Appl. Phys. 38, 4540 (1967).

136. J. S. Benjamin, B. C. Giessen, and N. J. Grant, Trans. TMS-AIME 236, 224 (1966).

137. C. B. Shoemaker and D. P. Shoemaker, Acta Cryst. 23, 231 (1967).

138. R. C. Ruhl and M. Cohen, Trans. TMS-AIME 245, 253 (1968).

139. P. Duwez and S. C. H. Lin, J. Appl. Phys. 38, 4096 (1967).

140. C. C. Tsuei, G. Longworth, and S. C. H. Lin, Phys. Rev. 170, 603 (1968).

141. S. C. H. Lin, California Institute of Technology Report, CALT-221-53 (1968).

142. S. C. H. Lin, California Institute of Technology Report,
      CALT-221-58 (1968).

143. R. H. Willens and E. Buehler, J. Appl. Phys. 38, 405 (1967).

144. R. H. Willens, E. Buehler, and B. T. Matthias, Phys. Rev.
      159, 327 (1967).

145. E. A. Nesbitt, R. H. Willens, H. J. Williams, and
      R. C. Sherwood, J. Appl. Phys. 38, 1003 (1967).

146. R. H. Willens, E. Buehler, and E. A. Nesbitt, Rev. Sci. Instr.
      39, 194 (1968).

147. C. C. Tsuei and P. Duwez, J. Appl. Phys. 37, 435 (1966).

148. M. E. Weiner, Abstr. Bull. IMD-AIME 3 (2), 99A (1968).

149. P. Duwez, R. H. Willens, and W. Klement, Jr., J. Appl. Phys.
      31, 1500 (1960).

150. M. Heimendahl, Acta Met. 15, 1441 (1967).

SUBJECT INDEX

A

A1 structure type, 1, 19, 20, 31

A1 phases in metastable alloys, 233, 250, 256

A2 structure type, 1, 19, 20, 31

A3 structure type, 1, 19, 20, 31

A15 (β-W) structure, 110, 256

α' phases in metastable alloys, 234

α_1 phases in metastable alloys, 245

Actinide compounds with interstitial elements, 36-7

Actinides, 183 ff

AlB_2 structure, 13, 14, 210

Al_3Ho structure, 18

Alkali metals, 19,20
  in NaTl structure, 41-48

Aluminum compounds, 26-30

Amorphous phases, 228 ff, 258, 267, 268, 270

Area-of-stability plots for thoride compounds, 200

Atomic radii, 2, 127

Atomic sizes, 103, 148
  in η phases, 160
  in thoride compounds, 185 ff

Atomic volume in thoride compounds, 187, 223

$AuBe_5$ structure type, 222

Au_7Bi_8, metastable, 238

$AuCu_3$ structure, 10, 218
  filled, 141-151

B

β phases, Hume-Rothery, 3 ff

β phases, metastable, 265

$BaPb_3$ structure, 18